FUKUSHIMA ACCIDENT: RADIOACTIVITY IMPACT ON THE ENVIRONMENT

FUKUSHIMA ACCIDENT: RADIOACTIVITY IMPACT ON THE ENVIRONMENT

PAVEL P. POVINEC
Comenius University, Bratislava, Slovakia

KATSUMI HIROSE
Sophia University, Tokyo, Japan

MICHIO AOYAMA
Meteorological Research Institute, Tsukuba, Japan

ELSEVIER

AMSTERDAM • BOSTON • HEIDELBERG • LONDON
NEW YORK • OXFORD • PARIS • SAN DIEGO
SAN FRANCISCO • SYDNEY • TOKYO

Elsevier
225 Wyman Street, Waltham, MA 02451, USA
The Boulevard, Langford Lane, Kidlington, Oxford OX5 1GB, UK
Radarweg 29, PO Box 211, 1000 AE Amsterdam, The Netherlands

Library of Congress Cataloging-in-Publication Data
LCCN 2013943823

British Library Cataloguing-in-Publication Data
A catalogue record for this book is available from the British Library.

ISBN: 978-0-12-408132-1

For information on all Elsevier publications
visit our web site at store.elsevier.com

Printed and bound by CPI Group (UK) Ltd, Croydon, CR0 4YY

To our wives,
Elena, Marico and Keiko, for their kind understanding and support of spending so much time with our science

CONTENTS

FOREWORD

The strong Tohoku earthquake of magnitude nine on 11 March 2011, and unexpectedly high tsunami has damaged the Fukushima Dai-ichi Nuclear Power Plant (NPP) in Japan. This was the fourth largest earthquake registered on the Earth from 1900 with about 800 aftershocks following the earthquake within 24 h, and about 10,000 aftershocks observed in the succeeding month. Three nuclear reactors, which were in operation during the earthquake were shut down as expected, and diesel generators, triggered by the earthquake supplied enough electricity for cooling all reactors and cooling ponds at the Fukushima Dai-ichi NPP. However, the subsequent large tsunami damaged the diesel generators and cooling systems of the NPP. Following pressure venting of the reactor containments, hydrogen explosions and fires in the NPP, large atmospheric radionuclide releases occurred between 12 March and 5 April 2011. The total atmospheric releases of ^{131}I and ^{137}Cs were estimated to be between 150–160 PBq and 10–15 PBq, respectively.

The discharged radioactive material from the Fukushima NPP consisted of several radionuclides; except the mentioned ^{131}I and ^{137}Cs, it also included ^{134}Cs, ^{132}Te, ^{132}I, ^{136}Cs, etc., as well as radioactive noble gases (^{133}Xe and ^{135}Xe). Large radionuclide amounts were also directly released to the sea in the form of highly contaminated fresh and seawater; the total activities of ^{137}Cs and ^{90}Sr were estimated to be from 4 to 27 PBq, and from 0.1 to 1 PBq, respectively.

The radioactive plume formed after the Fukushima accident was due to the prevailing eastward wind direction transported over the Pacific Ocean, to be registered first by the Comprehensive Test-Ben Treaty Organization (CTBTO) radionuclide monitoring stations in Hawaii and at the west coast of USA. The elevated radionuclide levels were first detected in Europe by the CTBTO station at Iceland, and later by many national laboratories. It is expected that about 80% of the released activity was deposited over the Pacific, about 20% over Japan, and about 0.5% came to Europe.

Radioecologically, the most important radionuclides detected in Europe after the Fukushima accident were ^{137}Cs ($T_{1/2} = 30.17$ years), ^{134}Cs ($T_{1/2} = 2.06$ years) and ^{131}I ($T_{1/2} = 8.02$ days). They all represent the anthropogenic radionuclides of nuclear fission origin without natural contribution, which were released in large quantities after atmospheric nuclear weapons tests, as well from the Chernobyl nuclear reactor accident, and after authorized releases from nuclear reprocessing facilities. Under normal conditions, only ^{137}Cs due to its long half-life is still present in the environment. When comparing the Fukushima atmospheric radionuclide releases with the Chernobyl ones, we see that they were lower for ^{131}I and ^{137}Cs by a factor of about 12 and 7, respectively.

The Fukushima accident was classified on the International Nuclear and Radiological Event Scale at the maximum level of seven, similarly as the Chernobyl accident. The largest radioactivity (around 950 PBq of ^{137}Cs) released to the atmosphere till now was, however, from atmospheric tests of nuclear weapons. Although the radionuclide release rates were only by about a factor of 10 lower when compared with the Chernobyl accident, the environmental impact of the Fukushima accident outside of Japan has been hardly measurable.

The aim of this book has been to review the radioactivity impact of the Fukushima accident on the terrestrial and the marine environment. A wide range of topics have been covered: evaluation of the past and present sources of environmental radioactivity, focusing mainly on global fallout and the Chernobyl accident, and their impact on the Pacific Ocean, which has been the main depository of Fukushima-derived radionuclides; evaluation of radionuclide source terms in the damaged nuclear reactors and their fractions released to the environment; distribution of Fukushima-derived radionuclides in the global atmosphere, and comparison of observed radionuclide levels with those predicted by Lagrangian atmospheric models; distribution of radionuclides in the marine environment, their transport from coastal waters off Fukushima to the eastern Pacific Ocean, and comparison of measured radionuclide concentrations in surface waters and in the water column with the predicted levels obtained by Global Ocean Circulation Models; and estimation of radiation doses from Fukushima radionuclides delivered to the public and the marine biota, and their comparison with natural sources of radioactivity, as well as with other radiation doses to the public (e.g. medical applications).

The authors are indebted to numerous Japanese colleagues for sampling and analyzing atmospheric, soil, seawater and biota samples after the Fukushima accident. Many our colleagues, with whom we have been working hard during the last 2 years, investigating experimentally Fukushima radioactivity worldwide and simulating the distribution of Fukushima-derived radionuclides in the atmosphere and the ocean are highly acknowledged for fruitful collaboration. First, we would like to thank our Japanese colleagues, namely: H. Amano, M. Chino, T. Doi, M. Ebihara, S. Endo, Y. Hamajima, Y. Igarashi, Y. Kanai, C.-K. Kim, Y. Kumamoto, N. Momoshima, Y. Muramatsu, M. Nakano, Y. Onda, A. Sakaguchi, K. Sueki, K. Tagami, D. Tsumune, M. Uematsu, M. Yamamoto, C. Yonezawa, and J. Zheng. Then we thank J. Bartok and M. Gažák (MICRO-STEP, Bratislava), K.O. Buesseler (WHOI, Woods Hole), H. Florou (NCRS Demokritos, Athens), R. Golser and P. Steier (University of Vienna), X. Hou (Technical University of Denmark, Roskilde), A. Ioannidou (University of Thessaloniki), A.J.T. Jull (Univerzity of Arizona, Tucson), S.-H. Lee (KRISS, Daejeong), G. Lujaniené (SRI Center for Physical Sciences and Technology, Vilnius), O. Masson (IRSN, Cadarache), H. Nies and M.K. Pham (IAEA-EL, Monaco), L. Palcsu (ATOMKI, Debrecen), W. Plastino (University Roma Tre), S.C. Sheppard and S. Hisamatsu (Editors of JENVRAD), I. Štekl (CVUT, Prague), G. Wotawa (CIMG, Vienna) and R. Breier, K. Holý, M. Gera, M. Ješkovský, J. Kaizer, A. Kováčik, I. Sýkora, (Comenius University of Bratislava). We also thank G.H. Hong (KORDI, Seoul), G. Lujaniené and C. Papastefanou (University of Thessaloniki) for useful comments regarding the content of this book. The Ministry of Education, Culture, Sports, Science and Technology (MEXT) and the Tokyo Electric Power Company

(TEPCO) are acknowledged for provision of results on analysis of radionuclides in the terrestrial and marine environment. PPP acknowledges support provided by the EU Research and Development Operational Program funded by the ERDF (projects no. 26240120012, 26240120026 and 26240220004). KH acknowledges his colleagues of the former Nuclear Safety Commission to provide constructive information. MA acknowledges research fund support provided by MEXT, and his staff members, A Mori, Y. Yoshimura, T. Kudo and S. Shimada for preparation of databases, figures and tables used in this book. He also acknowledges M. Uesugi, a medical doctor, who operated him three times in 2012, and without his help MA could not be able to complete this book.

In publishing this book, we hope to stimulate further work on environmental radioactivity studies and applications of radionuclides as tracers of environmental processes, important for radioecological assessment of impacts of nuclear industry on humans and the terrestrial and marine environment.

Pavel P. Povinec
Comenius University, Bratislava, Slovakia

Katsumi Hirose
Sophia University, Tokyo, Japan

Michio Aoyama
Meteorological Research Institute, Tsukuba, Japan

PROLOGUE

It happened very often in the past that our civilization, especially in high technology developments, has been very vulnerable to human errors or environmental disasters, which had negative impact on its future. In this respect, for example in chemical industry, in weapons developments or their disarmaments, in river dams, or in other large-scale projects, heavy accidents occurred from time to time with hundreds of deaths of workers and civilians.

A specific case is represented by accidents in nuclear industry. As we cannot smelt radioactivity present in the air, or we cannot see the secondary effects from the accident, but also due to radioactivity phobia, which is very common in our present world, people are very sensitive to any nuclear accident with regional or global impact on the total environment (humans, fauna and flora). There were many low-scale accidents in the nuclear industry in the past, which had a very local character, usually with no radionuclide releases to the air or the aquatic environment (or with only very small amounts), and no casualties on workers and civilians occurred. This was nothing special, as usually with a new technology applied on industrial scale, small accidents occur during tuning of the technology, and training of the staff.

However, some of the accidents, which occurred in the past, such as accidents in the Mayak nuclear processing facility (1957, Russia), Windscale reprocessing facility (presently called Sellafield, 1957; United Kingdom), or the accident at the Three Miles Island nuclear power plant (1979, USA) triggered an attention of people asking for better regulations, and more safe operations of nuclear facilities. New and more strict regulations on the performance of nuclear industry installations have been applied, on both national and international levels.

All these activities, however, did not prevent the nuclear industry to fall even to nuclear accidents with global impacts, as was the accident at the Chernobyl nuclear power plant in April 1986 in the former Soviet Union (presently Ukraine), and more recently, the accident at the Fukushima Dai-ichi nuclear power plant. The nuclear community (International Atomic Energy Agency) developed an International Nuclear and Radiological Event Scale, which run up to the level of seven, which represents a nuclear accident with serious local and regional impacts, observable also on a global scale. Both the Chernobyl and the Fukushima accidents were classified at the maximum level of seven.

While in the Chernobyl accident human errors played a significant role (at least in the wrong planning and carrying out of the performance tests of the RBMK-1000 nuclear reactor, which had, however, some constructional weaknesses as well), the Fukushima accident was fully triggered by the environmental disaster caused by the large-scale Tohoku earthquake (magnitude nine), and subsequent unexpectedly high tsunami, which completely damaged the emergency electricity system. As the Fukushima nuclear power plant was designed for the maximum tsunami height of 7 m (a 100 years maximum), the observed tsunami height on 11 March 2011 of around 15 m did not give any chance to survive the diesel electrical generators, and to supply necessary electricity for cooling of three nuclear reactors and the nuclear fuel cooling ponds.

We need to learn therefore from both the Chernobyl and Fukushima accidents, and to improve not only the nuclear reactor technologies but also the system of nuclear regulations, as well as the management of possible nuclear accidents in the future. As it was always in the past with nuclear weapons tests and the nuclear facility accidents, we also need to better understand behavior of radionuclides released to the environment for minimizing their impact on humans, fauna and flora, as well as for using these radionuclides as traces for better understanding of environmental processes. And this has also been our attempt when working on this book to bring to the readers new information what we have learned when evaluating Fukushima-derived radionuclides in the environment.

1

INTRODUCTION

CHAPTER OUTLINE

1.1 Fukushima Accident

The Fukushima accident happened due to the failure of the cooling system of the Fukushima Dai-ichi nuclear power plant (NPP) after the Tohoku earthquake and the subsequent unexpectedly high tsunami waves on 11 March 2011 (IAEA, 2011; NSCJ, 2011; NISA, 2012; NERH, 2012; TEPCO, 2012). Due to the damage of the electrical network, as well as the emergency diesel generators, it was not possible to provide electricity to cool nuclear reactors and the fuel storage pools, which resulted in numerous explosions and total damage of the Fukushima Dai-ichi NPP (Fig. 1.1).

The atmospheric radionuclide releases during the Fukushima accident were estimated to be the highest for ^{131}I (153–160 PBq) and ^{137}Cs (13–15 PBq) (Chino et al., 2012). Stohl et al. (2012) estimated even higher atmospheric releases for ^{137}Cs (23–50 PBq). The discharged radioactive material, in addition to ^{131}I and ^{137}Cs, also included ^{134}Cs, ^{132}Te, ^{132}I, ^{136}Cs, and other radionuclides, as well as radioactive noble gases (^{133}Xe, ^{135}Xe) (Bowyer et al., 2011). The contribution of ^{134}Cs was similar to that of ^{137}Cs, as the ^{134}Cs/^{137}Cs activity ratio was close to 1 (Masson et al., 2011).

Fukushima Accident. http://dx.doi.org/10.1016/B978-0-12-408132-1.00001-2

The Fukushima accident was classified by the Government of Japan on the international nuclear and radiological event scale at the maximum level of 7, similar to the Chernobyl accident, which

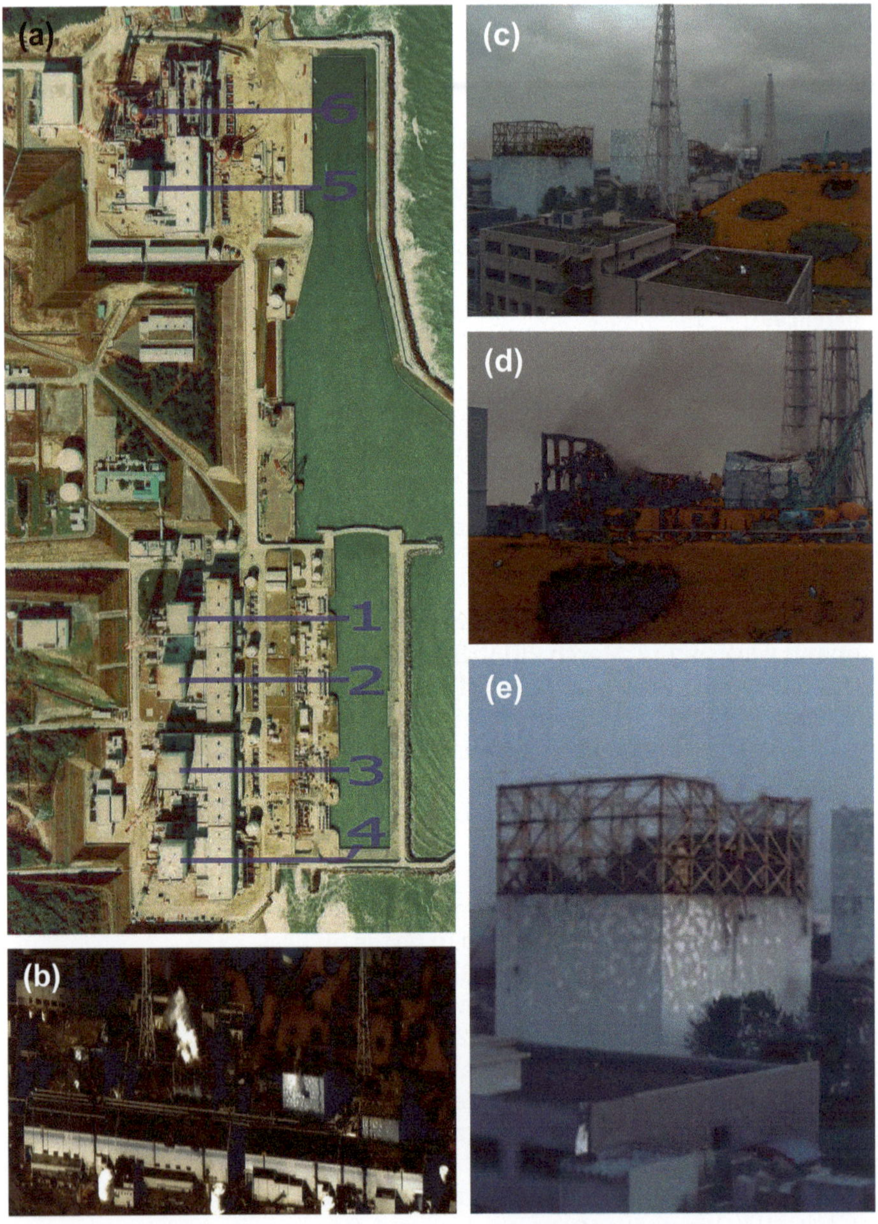

Figure 1.1 (a–j) Views of the damaged Fukushima Dai-ichi nuclear power plant. (Source: (a,b) - Wikipedia; (c–j) - Reprinted with permission from TEPCO).

Figure 1.1 Continued

happened in 1986 in the former Soviet Union (presently Ukraine) (IAEA, 2003; WHO, 2005).

Apart from the contamination of the Japanese territory (Hirose, 2012; Kanai, 2012; Tanaka et al., 2012), the Japan Sea (Inoue et al., 2012a), and the Korean Peninsula (Hernández-Ceballos et al., 2012; Lee et al., 2012), due to prevailing western winds, the radionuclides emitted to the atmosphere were mainly transported from Fukushima over the Pacific Ocean (Kameník et al., 2013), then to North America (Bowyer et al., 2011; Biegalski et al., 2012; Diaz-Leon et al., 2011; Sinclair et al., 2011), the Atlantic Ocean,

Europe (Baeza et al., 2012; Barsanti et al., 2012; Beresford et al., 2012; Bikit et al., 2012; Bossew et al., 2012; Carvalho et al., 2012; Clemenza et al., 2012; Cosma et al., 2012; Evrard et al., 2012; Ioannidou et al., 2012; Kritidis et al., 2012; Kirchner et al., 2012; Loaiza et al., 2012; Lujanienė et al., 2012a,b, 2013; Manolopoulou et al., 2011; Perrot et al., 2012; Pham et al., 2012; Piñero García and Ferro García, 2012; Pittauerova et al., 2011; Povinec et al., 2012a,c, Povinec et al., 2013a,b; Tositti and Brattich, 2012), Arctic (Paatero et al., 2012), and back into Asia. In the beginning of April, the global atmosphere had been labeled with Fukushima-derived radionuclides (Hernández-Ceballos et al., 2012; Masson et al., 2011; Povinec et al., 2013a). The released radionuclides were mostly deposited over the North Pacific Ocean (about 80%), about 20% was deposited over Japan, and less than about 1% was deposited over the Atlantic and Europe (Morino et al., 2011; Stohl et al., 2012; Yoshida and Kanda, 2012).

Except atmospheric radionuclide releases which occurred mostly due to hydrogen explosions at the Fukushima NPP, large amounts of liquid radioactive wastes were directly discharged from the Fukushima Dai-ichi NPP into the ocean. Large volume of contaminated water was produced during emergency cooling of reactors using fresh water, and later also by seawater. Some of this water was unintentionally discharged directly to the sea, which widely contaminated coastal waters off the Fukushima NPP, as reported by the Tokyo Electric Power Company and the Ministry of Education, Culture, Sports, Science and Technology, and other investigators (Aoyama et al., 2012, 2013; Buesseler et al., 2011, 2012; Honda et al., 2012; MEXT, 2012; Povinec et al., 2012a,c, 2013b; TEPCO, 2012; Tsumune et al., 2012). The total amounts of ^{137}Cs directly released into the sea have been estimated to range from 1 to 4 PBq (Dietze and Kriest, 2012; JG, 2011; Kawamura et al., 2011; Tsumune et al., 2012) to 16.2 ± 1.6 PBq (Rypina et al., 2013), and to 27 ± 15 PBq (Bailly du Bois et al., 2012). As the cooling water directly interacted with ruptured nuclear fuel rods, it has been estimated that $0.1 \sim 1$ PBq of ^{90}Sr has also been released to the ocean (Povinec et al., 2012b).

The direct discharge of contaminated water to the sea has significantly elevated radionuclide concentrations in coastal seawater, as well as in the northwestern (NW) Pacific Ocean. The peak ^{137}Cs values were observed at the discharge point of the Fukushima NPP to the sea on 30 March (47 kBq/l) and on 6 April (68 kBq/l) (TEPCO, 2012). Several papers have already discussed ^{134}Cs and ^{137}Cs concentrations in surface waters of the NW Pacific Ocean. In the open ocean, the ^{137}Cs activity concentrations ranged from a few millibecquerels per liter to a few becquerels per liter (Aoyama et al., 2012; Buesseler et al., 2011, 2012; Honda et al., 2012; Inoue et al., 2012b; Povinec et al., 2013b).

1.2 Sources of Radionuclides in the Environment

There are five main sources of radionuclides that could be found in the environment prior to the Fukushima accident:

1. Natural: cosmogenic radionuclides—results of interactions of cosmic rays with atoms in the atmosphere and their subsequent deposition on the Earth and the ocean surface (e.g. ^{3}H, ^{7}Be, ^{10}Be, ^{14}C, ^{26}Al, ^{53}Mn, etc.).
2. Natural: primordial radionuclides (e.g. ^{40}K, ^{238}U, ^{232}Th) and their decay products (e.g. ^{226}Ra, ^{230}Th, etc.) found in the Earth's crust; due to radon emanation, its decay products are also found in the atmosphere, and then after deposition, in the terrestrial and marine environments (e.g. ^{210}Po, ^{210}Pb, etc.).
3. Anthropogenic: global fallout radionuclides—produced during atmospheric tests of nuclear weapons (e.g. ^{3}H, ^{14}C, ^{90}Sr, ^{137}Cs, Pu isotopes, ^{241}Am, etc.).
4. Anthropogenic: radionuclides released from nuclear installations—mostly from reprocessing nuclear facilities (e.g. ^{3}H, ^{14}C, ^{90}Sr, ^{99}Tc, ^{129}I, ^{137}Cs, etc.).
5. Anthropogenic: radionuclides released during the Chernobyl accident which occurred in 1986 (e.g. ^{137}Cs, Pu isotopes, etc.)

The largest amount of radionuclides ($\sim$950 PBq of ^{137}Cs) released to the atmosphere up to now, representing the main source of anthropogenic radionuclides in the world ocean, has been, however, global fallout resulting from atmospheric tests of nuclear weapons carried out mainly in the 1950s and the early the 1960s (Livingston and Povinec, 2002; UNSCEAR, 1993, 2008). Although global fallout is the dominant source of anthropogenic radionuclides in the environment, large quantities of radioactive materials released to the atmosphere and coastal waters following a nuclear accident at the Fukushima Dai-ichi NPP increased considerably the ^{137}Cs concentrations in coastal seawater off Fukushima up to eight orders of magnitude above the global fallout background (TEPCO, 2012; MEXT, 2012).

We shall focus in this book only on a few anthropogenic radionuclides, specifically on those, that have been frequently studied after the Fukushima accident both in the atmosphere and in the marine environment (Table 1.1). Although ^{131}I was released after the Fukushima accident in largest amounts, it does not represent a radionuclide frequently studied in the terrestrial and marine environment due to its short half-life ($T_{1/2} = 8.02$ days). The most important radionuclide in the Fukushima case is ^{137}Cs, as it was released in large quantities, and it has a relatively long half-life ($T_{1/2} = 30.17$ years).

The relatively short-lived isotope of cesium (^{134}Cs, $T_{1/2} = 2.06$ years) because of its shorter half-life has also not been frequently studied in the environment; however, one advantage of its shorter half-life is that it can clearly identify cesium of Fukushima origin, as there is no remaining contribution from global fallout and the 1986 Chernobyl accident.

Anthropogenic tritium and ^{129}I have been recognized as ideal short-term (^{3}H half-life $T_{1/2} =$ 12.32 years) and long-term (^{129}I half-life $T_{1/2} = 15.7$ Million years) atmospheric and oceanographic tracers, important for investigation of circulation processes in the atmosphere and the ocean, as well as for the study of atmosphere–ocean exchange processes (Hou et al., 2000; Povinec et al., 2010, 2011; Raisbeck and Yiou, 1999; Schlosser et al., 1999).

^{137}Cs has been recognized radiologically as one of the most important long-lived radionuclides of anthropogenic origin, which has accumulated in the

Table 1.1. Anthropogenic Radionuclides of Interest in the Environment After the Fukushima accident

Radionuclide	Half-Life	Radiation Used for the Detection	Energy (keV)	Detection Method
^{3}H	12.32 years	Beta	18.6	Beta-spectrometry, ^{3}He ingrowth mass spectrometry
^{89}Sr	50.6 days	Beta	1492	Beta-spectrometry
^{90}Sr	28.15 years	Beta	546 (or ingrowth ^{90}Y: 2284)	Beta-spectrometry
^{99}Tc	$2.14 \bullet 10^{5}$ years	Beta	294	ICPMS
^{129}I	$1.57 \bullet 10^{7}$ years	Beta	151.2	AMS
^{131}I	8.02 days	Gamma	364; 606.3	Gamma-spectrometry
^{133}Xe	8.24 days	Beta	346.3	Beta-gamma-spectrometry
		Gamma	80.99	
^{134}Cs	2.06 years	Gamma	604.69; 795.84	Gamma-spectrometry
^{137}Cs	30.17 years	Gamma	661.66 (^{137m}Ba)	Gamma-spectrometry
^{238}Pu	87.74 years	Alpha	5499	Alpha-spectrometry
^{239}Pu	$2.411 \bullet 10^{4}$ years	Alpha	5156	Alpha-spectrometry[1], AMS
^{240}Pu	6563 years	Alpha	5168	Alpha-spectrometry[1], AMS
^{241}Pu	14.4 years	Beta	20.8	Beta-spectrometry

AMS, accelerator mass spectrometry; IPCMS, inductively coupled plasma mass spectrometry.
[1]Simultaneous $^{239,240}Pu$ analysis.

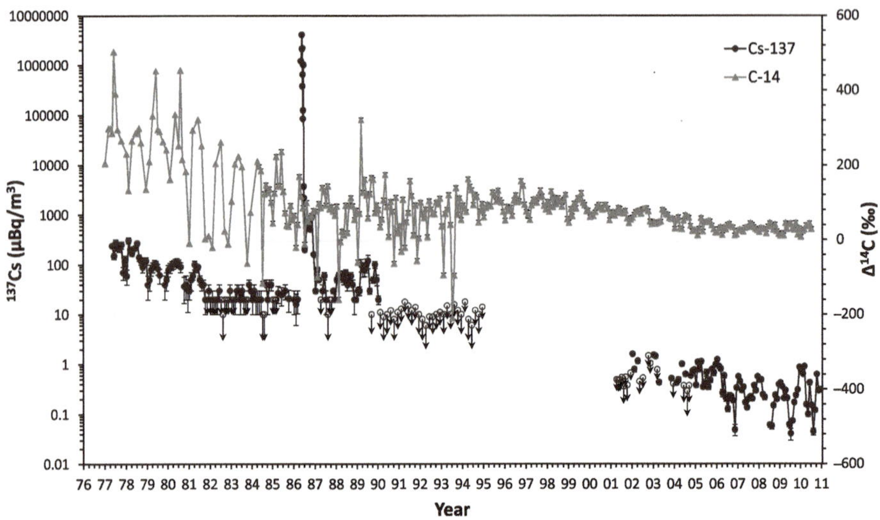

Figure 1.2 Record of ^{137}Cs in atmospheric aerosols and ^{14}C in atmospheric carbon dioxide in central Europe (Bratislava, Slovakia) during the pre-Fukushima time. (Adapted from Povinec et al., 2012a,c).

terrestrial and marine environment, and it is still measurable in the atmosphere as well (Livingston and Povinec, 2000, 2002). ^{137}Cs together with ^{14}C has been the most frequently studied radionuclide in the atmosphere. Figure 1.2 shows as an example the typical ^{137}Cs levels observed in atmospheric aerosols (and ^{14}C levels in atmospheric carbon dioxide) in Central Europe (Bratislava, Slovakia) during the pre-Fukushima time (Povinec et al., 2012b).

^{137}Cs has been considered to be the most important for the long-term radiological impact because of large releases, relatively long half-life, and its relative high bioavailability. Because of its accumulation in tissues, it has been important for delivering radiation doses to the public from the consumption of seafood (Aarkrog et al., 1997).

The most dominant source of ^{137}Cs in the atmosphere, the biosphere and the ocean has been global fallout originating from the atmospheric nuclear weapons testing (Table 1.2). Its main input into the ocean occurred in 1963 and afterward due to the wet and dry deposition of ^{137}Cs, after large-scale atmospheric nuclear weapons tests, which were carried out during 1961–1962 by the former Soviet Union at

Table 1.2. Pre-Fukushima Anthropogenic Radionuclide Inventories and Releases in the Atmosphere and Ocean (in PBq)

			Global Fallout Inventory in the Ocean[2]		Discharges from Reprocessing Facilities[2]		Chernobyl	
Nuclide	Natural Inventory[1]	Global Fallout Atmosphere[1]	Total inventory in the ocean	Inventory in 2010 in the ocean	Total inventory	Inventory in 2010	Atmosphere[3]	Ocean[2]
^{3}H	2200		113,000	8000	410	45		
^{90}Sr		600	380	105	7	4	1	
^{129}I	0.6×10^{-3}		0.3×10^{-3}	0.3×10^{-3}	0.04	0.04	0.013×10^{-3}	
^{131}I	—		—	—	39	—	1760	
^{137}Cs	—	950	600	170	40	26	85	16

[1]UNSCEAR, 2008.
[2]IAEA, 2005 (only discharges into the sea from the Sellafield (UK) and La Hague (France) nuclear reprocessing facilities).
[3]IAEA, 2003.

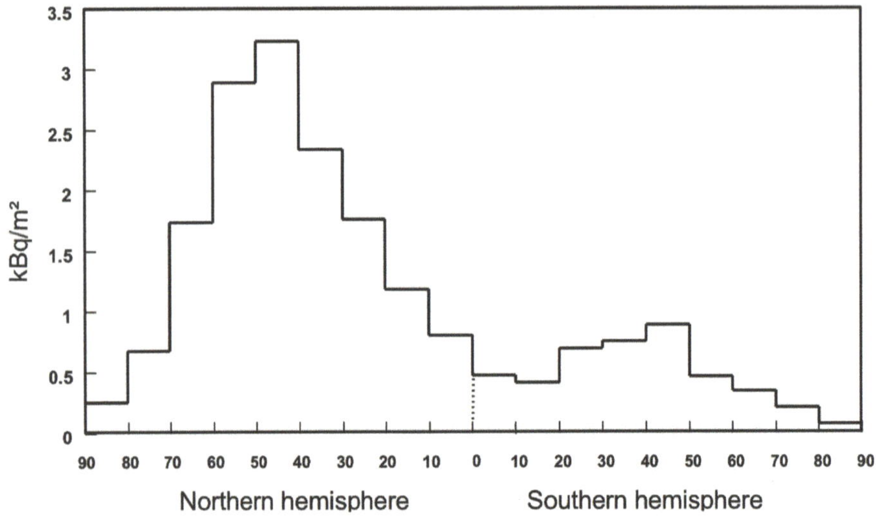

Figure 1.3 Integrated deposition density of ^{90}Sr from global fallout. (Adapted from IAEA, 2005).

Novaya Zemlya in the Kara Sea (Livingston and Povinec, 2002).

The major deposition of ^{137}Cs occurred in the midlatitudes of the northern hemisphere, similar to the case of ^{90}Sr (Fig. 1.3, IAEA, 2005).

The largest depositions due to specific meteorological conditions were observed, however, in the NW Pacific Ocean (Aoyama et al., 2006). The NW Pacific Ocean has therefore been well known as the area with the highest deposition of global fallout radionuclides into the ocean (Povinec et al., 2005; Inomata et al., 2009). Because of its huge size, the Pacific Ocean captured about 52% of ^{137}Cs. The Atlantic Ocean got 33%, the Indian Ocean got 14%, and the Arctic Ocean got only 1% (and similarly, e.g. for ^{90}Sr).

Large amounts of ^{137}Cs were also released from nuclear reprocessing facilities in Sellafield (situated on the western coast of England) and in La Hague (situated in the English Channel) (Table 1.2), which have mainly impacted the European seas, as shown in Fig. 1.4 (Povinec et al., 2003b).

^{137}Cs has been the most abundant anthropogenic radionuclide in the marine environment, as documented in Fig. 1.5. While the Irish Sea and the North Sea have been mostly influenced by the discharges from the Sellafield and the La Hague reprocessing facilities, ^{137}Cs in the Baltic Sea and the Black Sea has

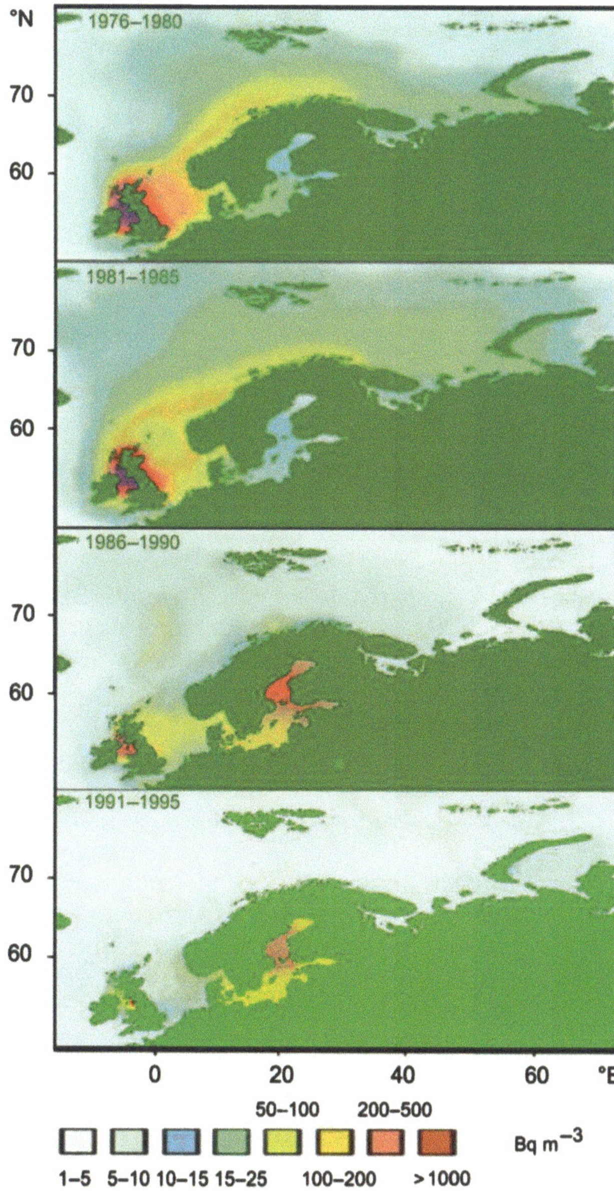

Figure 1.4 ^{137}Cs in surface waters of the northern European seas showing the main impact of authorized discharges from the Sellafield and La Hague reprocessing nuclear facilities. (Modified from Povinec et al., 2003b)

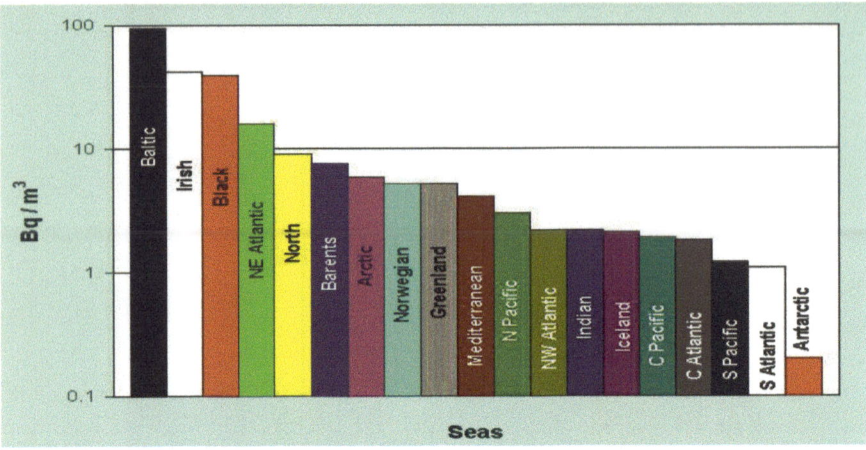

Figure 1.5 ^{137}Cs in surface waters of the oceans and seas. (Modified from IAEA, 2005)

been mostly due to the impact of the Chernobyl accident (IAEA, 2005).

After its release into the atmosphere, the ^{137}Cs is rapidly associated with aerosols, which represent a major reservoir of pollutants in the atmosphere (Lujanienė et al., 2009). The ^{137}Cs, present in the atmosphere mainly from global fallout and the Chernobyl accident, has primarily been deposited on the Earth's surface including the ocean by wet and dry deposition (Pham et al., 2011, 2012, 2013).

As ^{137}Cs is a conservative oceanographic tracer, it is mostly dissolved in seawater, and only <1% is attached to marine particles. The sediment distribution coefficient (K_d) for ^{137}Cs in coastal and open ocean waters is 4000 and 2000, respectively (IAEA, 2004). Its removal from the water column is mainly due to its radioactive decay and diffusion. It has been used frequently in water transport/mixing studies and in investigations of water column processes (Livingston and Povinec, 2002; Lujanienė, 2006; Ito et al., 2003).

1.3 Pacific Ocean: Hydrography Background

We shall focus in this book on the Pacific Ocean, specifically on its NW part, because the main impact of the Fukushima accident can be found there. There are specific circulation patterns in the NW Pacific

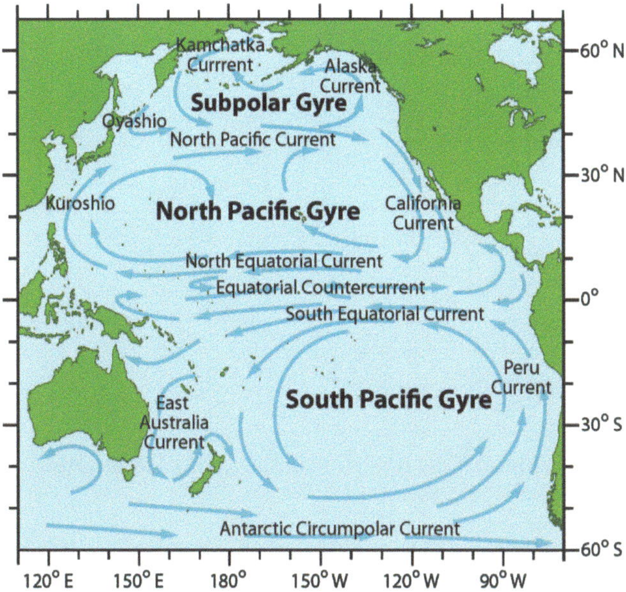

Figure 1.6 The main surface current system in the Pacific Ocean. (Source: National Oceanography Centre, www.noc.uk)

Ocean that make this area interesting for oceanographic investigations. The Pacific Ocean is the largest ocean, spreading from the northern boundary of the Bering Sea near 50°N to the boundary of the Southern Ocean at 60°S (Fig. 1.6). A typical global thermohaline circulation transports cold Circumpolar Bottom Waters from the South Pacific to the North Pacific and upwells deep water to shallower water in the North Pacific Ocean. Warm surface waters in the western North Pacific Ocean move to the Indian Ocean via the Indonesian seas. The current system in the western North Pacific Ocean comprises the Kuroshio Current, which is a typical western boundary current consisting of a part of the North Pacific subtropical circulation, and the Oyashio Current, which is included in the North Pacific Subpolar Gyre system. In the eastern North Pacific, the major currents are the Alaska Current included in the Alaskan Gyre, and the California Current, which is a part of the North Pacific Subtropical Gyre. The equatorial current system in the Pacific is complicated as it consists of the North Equatorial Current, the North Equatorial Countercurrent, the Northern

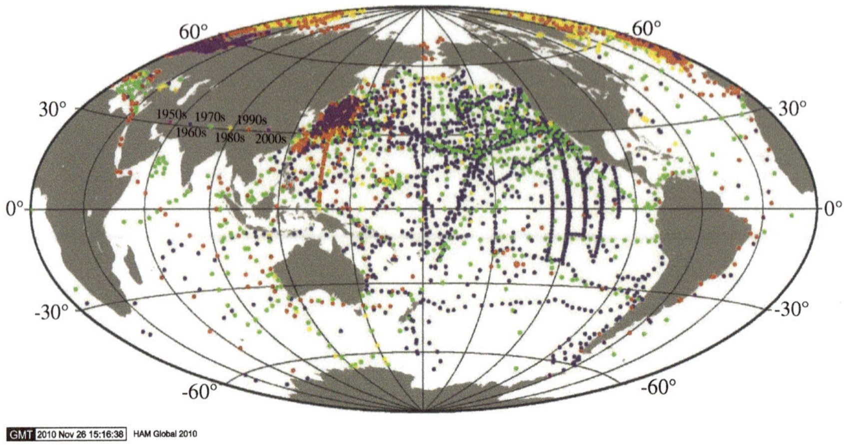

Figure 1.7 ^{137}Cs surface water data covered in the HAM marine radioactivity database. (Source: Aoyama M., Meteorological Research Institute, Tsukuba, Japan)

Subsurface Countercurrent, the Equatorial Undercurrent, the Southern Subsurface Countercurrent and the South Equatorial Current. Nevertheless, west-east ward flows are prevailing in the equatorial Pacific Ocean. In the South Pacific Ocean, the South Pacific Subtropical Gyre primarily governs the South Pacific Current system, including the East Australian Current and the Peru Current.

1.4 Pre-Fukushima Radionuclide Data for the Pacific Ocean

Fortunately, the pre-Fukushima ^{137}Cs concentrations in NW Pacific waters were well established (Aoyama et al., 2006; Povinec et al., 2005). Two marine radioactivity databases were developed: one at the International Atomic Energy Agency's (IAEA's) Environment Laboratories in Monaco (the GLOMARD/MARIS database; Povinec et al., 2004, 2012b) and the second one at the Meteorological Research Institute in Tsukuba (the HAM database; Aoyama and Hirose, 2004). The ^{137}Cs data can be downloaded from the IAEA website (www.iaea.org/maris). Figure 1.7 shows the coverage of ^{137}Cs concentrations in surface waters of the world ocean as obtained from the HAM database (Aoyama and Hirose, 2004). Summary of the ^{137}Cs input to the world ocean from global fallout is listed in

Table 1.3. Global Fallout ^{137}Cs Input to the World Ocean (in PBq)

Latitude Belt	Arctic Ocean	Atlantic Ocean	Indian Ocean	Pacific Ocean	Total
90–60°N	7.4	16.3	0.0	2.1	25.8
60–30°N	0.0	91.7	0.0	114.6	206.2
30–0°N	0.0	49.3	21.3	105.3	175.8
0–30°S	0.0	16.5	23.4	42.6	82.4
30–60°S	0.0	24.6	36	41.3	101.9
60–90°S	0.0	2.7	3.4	4.8	10.9
Total	7.4	201.1	84	310.6	603

Table 1.3. It can be seen that the north ocean in the latitude belt of 30–60°N received about one-third of the total of about 600 PBq of ^{137}Cs (IAEA, 2005).

Long-term changes in surface ^{137}Cs concentrations in two latitudinal belts at 25°–40°N (divided by the line at 180°E to the western and the eastern belt) of the North Pacific Ocean (representing upstream and downstream of the Kuroshio extension, where the water masses of the Kuroshio and Oyashio currents mix) are shown in Fig. 1.8. The surface ^{137}Cs concentrations decreased exponentially during the period 1970–2000 (except a small peak observed in 1986 due to the Chernobyl accident), with effective half-lives of 16.5 ± 0.9 and 10.6 ± 0.5 in the two regions. Higher ^{137}Cs concentrations in surface water occurred in the 1980s, probably due to liquid radioactive discharges from the nuclear fuel reprocessing plant at Tokai, which were, however, by several orders of magnitude lower than from the European reprocessing plants (Mizutani et al., 2009). By combining all ^{137}Cs data, the mean effective half-life of ^{137}Cs in North Pacific surface waters was determined to be 13 ± 1 years (Povinec et al., 2005).

We shall compare several results of radionuclide concentrations in the water column obtained during the IAEA'97 expedition carried out in the NW Pacific Ocean in November 1997 (at 30–35°N and 150–170°E). The cruise track is shown in Fig. 1.9.

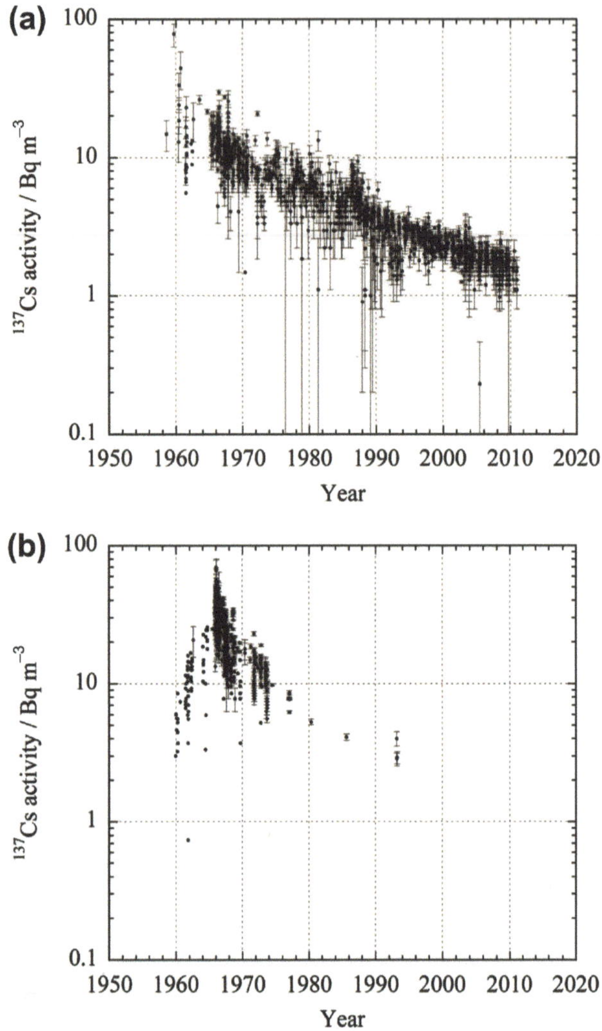

Figure 1.8 Long-term changes in surface ^{137}Cs concentrations in two latitudinal belts at 25°–40°N, divided by the line at 180°E to the western (a) and the eastern (b) belt of the north Pacific Ocean.

Some of the ^{137}Cs water profiles measured in the NW Pacific Ocean during the IAEA'97 expedition are presented in Fig. 1.10. Surface ^{137}Cs concentrations varied around 2.5 mBq/l, subsurface ^{137}Cs maxima around 3 mBq/l, a decrease down to 2 mBq/l was observed at 500 m, and levels <0.1 mBq/l were found at water depths below 1000 m (Povinec et al., 2003a). The corresponding decay-corrected concentrations

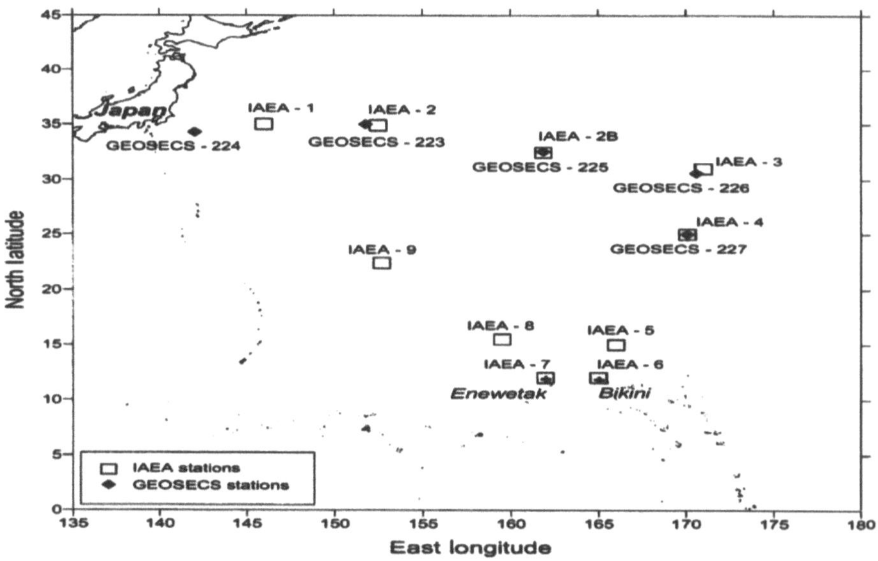

Figure 1.9 Seawater sampling stations during November 1997 in the north-western Pacific Ocean (IAEA'97 cruise. GEOSECS (Geochemical Ocean Sections Program) sampling stations (1973–1974) are also shown. (Modified from Povinec et al. (2010))

were 1.2, 1.4, 0.95, and <0.05 mBq/l in 2011, respectively, based on an effective half-life of 13 years of ^{137}Cs in the Pacific Ocean.

The distribution of ^{137}Cs in surface waters of the NW Pacific Ocean using data extracted from the MARIS database for the pre-Fukushima time is shown in Fig. 1.11. On the basis of the data stored in the MARIS database (www.iaea.org/MARIS), the average ^{137}Cs concentrations in NW and northeastern Pacific surface waters for the year 2010 are estimated to be 1.0 ± 0.1 mBq/l and 0.9 ± 0.2 mBq/l, respectively.

Iodine-129 has been introduced into the global ocean mainly from nuclear reprocessing facilities (Table 1.2). Because of its long half-life, it represents an alternative tracer to global fallout radionuclides (Hou et al., 2000, 2007; Povinec et al., 2010, 2011; Raisbeck and Yiou, 1999). As it is a soft beta-emitter with maximum beta-energy of 154 keV, and very long half-life, it is less important in the view of radiation protection of humans. However, it is an excellent analog to reconstruct levels and distribution of short-lived ^{131}I in the environment, which is usually released at largest quantities during nuclear accidents (Hou et al., 2013). ^{129}I is a useful environmental tracer for

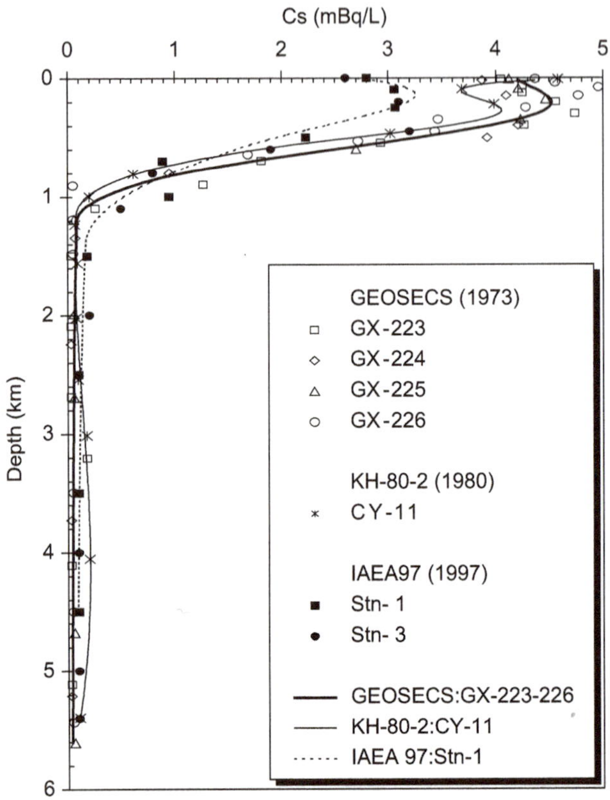

Figure 1.10 The ^{137}Cs water profiles measured in the north-western Pacific Ocean during the IAEA'97 expedition. (Modified from Povinec et al., 2003a)

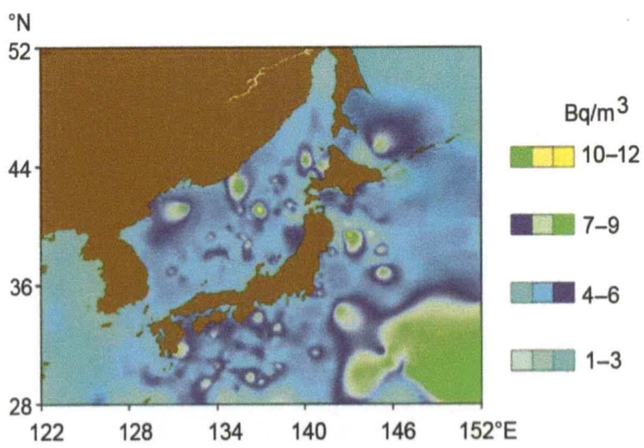

Figure 1.11 Distribution of ^{137}Cs in surface waters of the north-western Pacific Ocean during the pre-Fukushima time. (Data extracted from the MARIS database)

investigation of interaction of atmosphere and seawater, as well as biogeochemical cycles of stable iodine through chemical speciation analysis of ^{129}I and ^{127}I (Hou et al., 2001, 2009a, 2013). Unfortunately, there are only a few data available on the distribution of ^{129}I in NW Pacific waters. Povinec et al. (2000) reported results on ^{129}I concentrations in a water profile collected in 1995 during the Japan-South Korea-Russia-IAEA expedition on investigation of radioactive waste dumping sites in the NW Pacific Ocean. The samples collected offshore Kamchatka (52°30′N, 159°10′E) showed ^{129}I levels up to two orders of magnitude higher than the open ocean values, which were probably caused by leakages from the dumping site (similar to that observed in the Kara Sea (Arctic Ocean, Povinec et al., 2000). The IAEA'97 results in the NW Pacific Ocean showed surface ^{129}I values around 1.5×10^7 atoms/l (Fig. 1.12), subsurface maxima at around 2.1×10^7 atoms/l, a decrease down to 1.5×10^7 at 500 m, and a fast decrease $\leq 0.8 \times 10^7$ atoms/l at depths below 1000 m (Povinec et al., 2010). Higher ^{129}I levels (up to 3×10^7 atoms/l at 200 m water depth) were observed close to the Enewetak and Bikini Atolls (about 11°N, 162°E). Stations at both atolls showed concentration minima at around 500 m and secondary maxima (1.8×10^7 atoms/l) at about 1000 m. It has been concluded that the NW Pacific

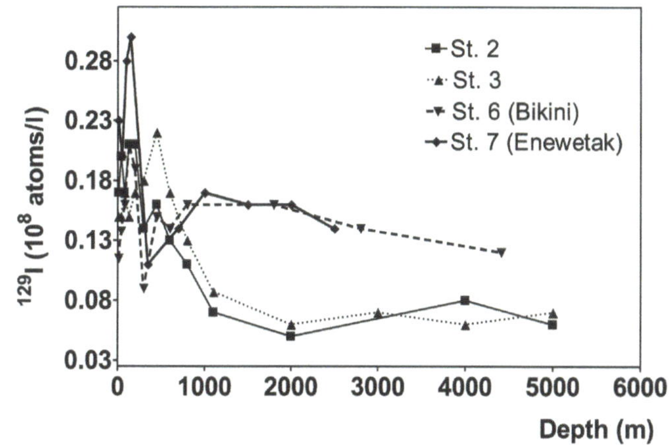

Figure 1.12 The ^{129}I concentrations measured in the water column during the IAEA'97 expedition in the north-western Pacific Ocean. (Modified from Povinec et al., 2010)

stations could be influenced by a local impact of nuclear weapons testing carried out on the Enewetak and Bikini Atolls during the 1950s. Suzuki et al. (2010) reported ^{129}I concentrations in the NW Pacific waters to be decreasing from the surface (2×10^7 atoms/l) to 100 m water depth (0.7×10^7 atoms/l), and then to $<0.2 \times 10^7$ atoms/l at water depths below 100 m. Corresponding $^{129}I/^{127}I$ ratios were 0.7×10^{-10}, 0.25×10^{-10}, and $<0.07 \times 10^{-10}$, respectively. The previous investigations thus showed that the expected pre-Fukushima ^{129}I concentration in surface NW Pacific waters should be $(2 \pm 0.2) \times 10^7$ atoms/l. High solubility and long residence time of iodine in the ocean makes ^{129}I an ideal oceanographic tracer for the investigation of water circulation on the global scale. Clearly, we need to understand how much its concentrations have changed in the Pacific Ocean after the Fukushima accident.

Tritium is an ideal tracer used extensively in oceanographic studies, as it is directly incorporated into the water molecule, usually as HTO (or T_2O), and has a suitable half-life (12.32 years). It is produced naturally by interactions of cosmic rays with nitrogen and oxygen atoms in the upper troposphere and lower stratosphere, but it has also been produced in large amounts in atmospheric nuclear weapons tests. Its peak concentration in the atmospheric moisture in 1963 was 1000 times higher than its natural cosmogenic background. It has also been released in large quantities from nuclear reprocessing facilities (Table 1.2). As it is a soft beta-emitter with maximum beta-energy of 18.6 keV only, it is less important in the view of radiation protection of humans (when compared with ^{137}Cs). The penetration of bomb tritium from surface waters into deeper layers of the ocean was used to study pathways and timescales of deep and bottom water formation (e.g. Schlosser et al., 1999). Tritium distribution in the NW Pacific Ocean was investigated recently in the framework of the WOMARS (Worldwide Marine Radioactivity Studies) project, carried out by the IAEA during 1995–2005 (IAEA, 2005; Povinec et al., 2004, 2005, 2010, 2011). The IAEA'97 results in the NW Pacific Ocean (Fig. 1.13) showed surface ^3H values around 1 TU, subsurface maxima (at 200–300 m) around 1.3

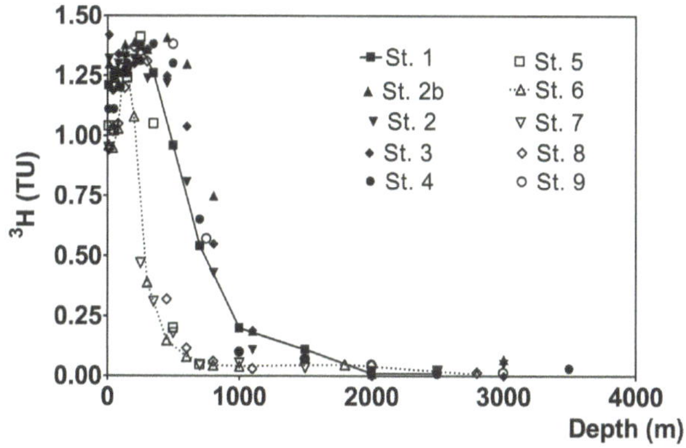

Figure 1.13 The ^3H concentrations measured in the water column during the IAEA'97 expedition in the north-western Pacific Ocean. (Modified from Povinec et al., 2010)

TU, a decrease down to 0.7 TU at 500 m, and a fast decrease down to 0.1 TU at depths below 1000 m (Povinec et al., 2010).

The most comprehensive tritium study was, however, the WOCE (World Ocean Circulation Experiment) program conducted in the 1980s and 1990s, which represents the most extensive coverage of ^3H in the World Ocean (www.eWOCE.org), Fig. 1.14. Tritium data are usually expressed in tritium units: 1 TU is $^3H/^1H$ atomic ratio of 1×10^{-18},

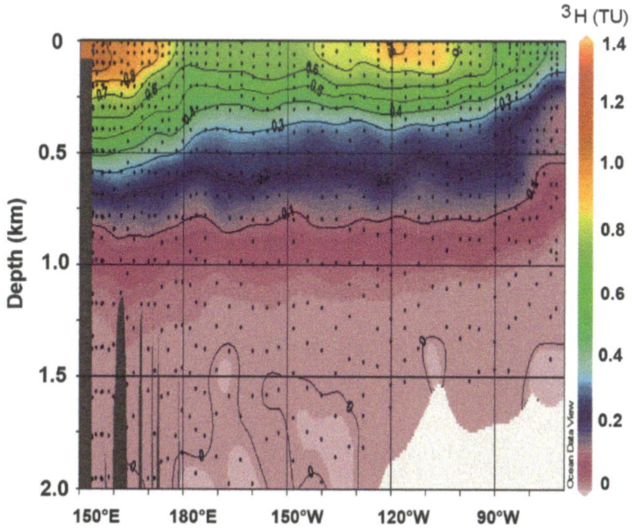

Figure 1.14 Distribution of tritium in the water column of the Pacific Ocean. (Constructed using tritium data from www.WOCE.org)

corresponding to an activity concentration of 0.118 Bq/l of water.

On the basis of the data stored in the MARIS database, the pre-Fukushima ^{3}H concentration in surface NW Pacific waters can be estimated to be (0.6 ± 0.1) TU. Although tritium is not an important radionuclide from the radiological point of view, as it has been used as an oceanographic tracer (also in climate change studies), we need to know how its concentrations have changed in the NW Pacific Ocean after the Fukushima accident.

We shall discuss in more details radionuclide records in the atmosphere and the world ocean, specifically in North Pacific waters. We shall also evaluate the past radionuclide source terms and budgets, so that we shall be able to assess an impact of the Fukushima accident on radionuclide levels in the global atmosphere and the Pacific Ocean. We shall explain the observed radionuclide patterns, and compare them with global fallout data and post-Chernobyl measurements carried out in the atmosphere and the NW Pacific Ocean. Finally, we shall estimate radiation doses to the public and compare them with radiation doses from natural sources and from applications of radiation, e.g. in the medicine.

References

Aarkrog, A., Baxter, M.S., Bettencourt, A.O., Bojanowski, R., Bologa, A., Charmasson, S., Cunha, I., Delfanti, R., Duran, E., Holm, E., Jeffree, R., Livingston, H.D., Mahapanyawong, S., Nies, H., Osvath, I., Pingyu, L., Povinec, P.P., Sanchez, A., Smith, J.N., Swift, D., 1997. A comparison of doses from ^{137}Cs and ^{210}Po in marine food: a major international study. J. Environ. Radioact. 34, 69–90.

Aoyama, M., Hirose, K., 2004. Artificial radionuclides database in the Pacific Ocean: HAM database. The Scientific World Journal 4, 200–215.

Aoyama, M., Hirose, K., Igarashi, Y., 2006. Re-construction and up-dating our understanding on the global weapons tests ^{137}Cs fallout. J. Environ. Monit. 8, 431–438.

Aoyama, M., Tsumune, D., Hamajima, Y., 2012. Distribution of ^{137}Cs and ^{134}Cs in the North Pacific Ocean: impacts of the TEPCO Fukushima Dai-chi NPP accident. J. Radioanal. Nucl. Chem. http://dx.doi.org/10.1007/s10967-012-2033-2.

Aoyama, M., Uematsu, M., Tsumune, D., Hamajima, Y., 2013. Surface pathway of radioactive plume of TEPCO Fukushima

NPP1 released 134Cs and 137Cs. Biogeosciences Discuss., doi: 10.5194/bgd-10-265-2013.
Baeza, A., Corbacho, J.A., Rodríguez, A., Galván, J., García-Tenorio, R., Manjón, G., Mantero, J., Vioque, I., Arnold, D., Grossi, C., Serrano, I., Vallés, I., Vargas, A., 2012. Influence of the Fukushima Dai-ichi nuclear accident on Spanish environmental radioactivity levels. J. Environ. Radioact. 114, 138–145.
Bailly du Bois, P., Laguionie, P., Boust, D., Korsakissok, I., Didier, D., Fievet, B., 2012. Estimation of marine source-term following Fukushima Dai-ichi accident. J. Environ. Radioact. 114, 2–9.
Barsanti, M., Conte, F., Delbono, I., Iurlaro, G., Battisti, P., Bartoluzzi, S., Lorenzelli, R., Salvi, S., Zicari, S., Papucci, C., Delfanti, R., 2012. Environmental radioactivity analyses in Italy following the Fukushima Dai-ichi nuclear accident. J. Environ. Radioact. 114, 126–130.
Biegalski, S.R., Bowyer, T.W., Eslinger, P.W., Friese, J.A., Greenwood, L.R., Haas, D.A., Hayes, J.C., Hoffman, I., Keillor, M., Miley, H.S., Moring, M., 2012. Analysis of data from sensitive U.S. monitoring stations for the Fukushima Dai-ichi nuclear reactor accident. J. Environ. Radioact. 114, 15–21.
Beresford, N.A., Barnett, C.L., Howard, B.J., Howard, D.C., Wells, C., Tyler, A.N., Bradley, S., Copplestone, D., 2012. Observations of Fukushima fallout in Great Britain. J. Environ. Radioact. 114, 48–53.
Bikit, I., Mrda, D., Todorovic, N., Nikolov, J., Krmar, M., Veskovic, M., Slivka, J., Hansman, J., Forkapic, S., Jovancevic, N., 2012. Airborne radioiodine in northern Serbia from Fukushima. J. Environ. Radioact. 114, 89–93.
Bossew, P., Kirchner, G., De Cort, M., De Vries, G., Nishev, A., De felice, L., 2012. Radioactivity from Fukushima Dai-ichi in air over Europe: part 1: spatio-temporal analysis. J. Environ. Radioact. 114, 22–34.
Bowyer, T.W., Biegalski, S.R., Cooper, M., Eslinger, P.W., Haas, D., Hayes, J.C., Miley, H.S., Strom, D.J., Woods, V., 2011. Elevated radioxenon detected remotely following the Fukushima nuclear accident. J. Environ. Radioact. 102, 681–687.
Buesseler, K., Aoyama, M., Fukasawa, M., 2011. Impacts of the Fukushima nuclear power plants on marine radioactivity. Environ. Sci. Technol. 45, 9931–9935.
Buesseler, K.O., Jayne, S.R., Fisher, N.S., Rypina, I.I., Baumann, H., Baumann, Z., Breier, C.F., Douglass, E.M., George, J., Macdonald, A.M., Miyamoto, H., Nishikawa, J., Pike, S.M., Yoshida, S., 2012. Fukushima-derived radionuclides in the ocean and biota off Japan. Proc. Natl. Acad. Sci. USA 109, 5984–5988.
Carvalho, F.P., Reis, M.C., Oliveira, J.M., Malta, M., Silva, L., 2012. Radioactivity from Fukushima nuclear accident detected in Lisbon. Portugal. J. Environ. Radioact. 114, 152–156.
Chino, M., Nakayama, H., Nagai, H., Terada, H., Katata, G., Yamazawa, H., 2011. Preliminary estimation of release amounts of ^{131}I and ^{137}Cs accidentally discharged from the

Fukushima Daiichi nuclear power plant into the atmosphere. J. Nucl. Sci. Technol. 48, 1129–1134.

Clemenza, M., Fiorini, E., Previtali, E., Sala, E., 2012. Measurement of airborne ^{131}I, ^{134}Cs and ^{137}Cs due to the Fukushima reactor incident in Milan (Italy). J. Environ. Radioact. 114, 113–118.

Cosma, C., Iurian, A.R., Nita, D.C., Begy, R., Cindea, C., 2012. Indicators of the Fukushima radioactive release in NW Romania. J. Environ. Radioact. 114, 94–99.

Diaz Leon, J., Jaffe, D.A., Kaspar, J., Knecht, A., Miller, M.L., Robertson, R.G.H., Schubert, A.G., 2011. Arrival time and magnitude of airborne fission products from the Fukushima, Japan, reactor incident as measured in Seattle, WA, USA. J. Environ. Radioact. http://dx.doi.org/10.1016/j.jenvrad.2011.06.005.

Dietze, H., Kriest, I., 2012. ^{137}Cs off Fukushima Dai-ichi, Japan: model based estimates of dilution and fate. Ocean Sci. 8, 319–332.

Evrard, O., Van Beek, P., Gateuille, D., Pont, V., Lefévre, I., Lansard, B., Bonté, P., 2012. Evidence of the radioactive fallout in France due to the Fukushima nuclear accident. J. Environ. Radioact. 114, 54–60.

Hernández-Ceballos, M.A., Hong, G.H., Lozano, R.L., Kim, Y.I., Lee, H.M., Kim, S.H., Yeh, S.W., Bolívar, J.P., Baskaran, M., 2012. Tracking the complete revolution of surface westerlies over Northern Hemisphere using radionuclides emitted from Fukushima. Sci. Total Environ. 438, 80–85.

Hirose, K., 2012. Fukushima Dai-ichi nuclear power plant accident: summary of regional radioactive deposition monitoring results. J. Environ. Radioact. 111, 13–17.

Honda, M., Aono, T., Aoyama, M., Hamajima, Y., Kawakami, H., Kitamura, M., Masumoto, Y., Miyazawa, Y., Takigawa, M., Saino, T., 2012. Dispersion of artificial caesium-134 and -137 in the Western North Pacific one month after the Fukushima accident. Geochem. J. 46, 1–9.

Hou, X.L., Dahlgaard, H., Nielsen, S.P., 2000. Iodine-129 time series in Danish, Norwegian and Northwest Greenland coast and the Baltic Sea by seaweed, Estuarine. Coastal Shelf Sci. 51, 571–584.

Hou, X.L., Dahlgaard, H., Nielsen, S.P., 2001. Chemical speciation analysis of ^{129}I in seawater and a preliminary investigation to use it as a tracer for geochemical cycle study of stable iodine. Marine Chem. 74, 145–155.

Hou, X.L., Aldahan, A., Nislen, S., Possnert, G., Nies, H., Hedfords, J., 2007. Speciation of ^{129}I and ^{127}I in seawater and implications for sources and transport pathways in North Sea. Environ. Sci. Technol. 41, 5993–5999.

Hou, X.L., Aldahan, A., Nielsen, S.P., Possnert, G., 2009a. Time series of ^{129}I and ^{127}I speciation in precipitation from Denmark. Environ. Sci. Technol. 43, 6522–6528.

Hou, X.L., Povinec, P.P.L.Y., Zhang, L.Y., Biddulph, D., Chang, C.-C., Fan, Y.K., Golser, R., Ješkovský, M., Jull, A.J.T., Liu, Q., Shi, K.L., Steier, P., Zhou, W.J., 2013. Iodine-129 in seawater

offshore Fukushima: distribution, speciation, sources, and budget. Environ. Sci. Technol. 47, 3091–3098.
IAEA, International Atomic Energy Agency, 2003. Chernobyl's Legacy: Health, Environmental and Socio-economic Impacts. IAEA, Vienna.
IAEA, International Atomic Energy Agency, 2004. Sediment Distribution Coefficients and Concentration Factors for Biota in the Marine Environment. Technical Reports Series no. 422. IAEA, Vienna.
IAEA, International Atomic Energy Agency, 2005. Worldwide Marine Radioactivity Studies (WOMARS). Radionuclide levels in oceans and sea, IAEA-TECDOC-1429. IAEA, Vienna.
IAEA, International Atomic Energy Agency, 2011. Briefings on Fukushima Nuclear Accident, www.iaea.org/.
Ioannidou, A., Manenti, S., Gini, L., Groppi, F., 2012. Fukushima fallout at Milano, Italy. J. Environ. Radioact. 114, 119–125.
Inomata, Y., Aoyama, M., Hirose, K., 2009. Analysis of 50-y record of surface ^{137}Cs concentrations in the global ocean using the HAM-global database. J. Environ. Monit. 11, 116–125.
Inoue, M., Kofuji, H., Nagao, S., Yamamoto, M., Hamajima, Y., Yoshida, K., Fujimoto, K., Takada, T., Isoda, Y., 2012a. Lateral variation of ^{134}Cs and ^{137}Cs concentrations in surface seawater in and around the Japan Sea after the Fukushima Dai-ichi nuclear power plant accident. J. Environ. Radioact. 109, 45–51.
Inoue, M., Kofuji, H., Hamajima, Y., Nagao, S., Yoshida, K., Yamamoto, M., 2012b. ^{134}Cs and ^{137}Cs activities in coastal seawater along Northern Sanriku and Tsugaru Strait, northeastern Japan, after Fukushima Dai-ichi nuclear power plant accident. J. Environ. Radioact. 111, 116–119.
Ito, T., Aramaki, T., Kitamura, T., Otosaka, S., Suzuki, T., Togawa, O., Kobayashi, T., Senjyu, T., Chaykovskaya, E.L., Karasev, E.V., Lishavskaya, T.S., Novichkov, V.P., Tkalin, A.V., Shcherbinin, A.F., Volkov, Y.N., 2003. Anthropogenic radionuclides in the Japan Sea: their distributions and transport processes. J. Environ. Radioact. 68, 249–267.
JG, Japanese Government Report, 2011. http://www.kantei.go.jp/jp/Topics/2011/iaea_houkokusho.html (accessed 10.07.12.).
Kameník, J., Dulaiova, H., Buesseler, K.O., Pike, S.M., Šťastná, K. Cesium-134 and 137 activities in the central North Pacific Ocean after the Fukushima Dai-ichi nuclear power plant accident. Biogeosciences (in press).
Kanai, Y., 2012. Monitoring of aerosol in Tsukuba after Fukushima nuclear power plant incident in 2011. J. Environ. Radioact. 111, 33–37.
Kawamura, H., Kobayashi, T., Furuno, A., In, T., Ishikawa, Y., Nakayama, T., Shima, S., Awaji, T., 2011. Preliminary numerical experiments on oceanic dispersion ^{131}I and ^{137}Cs discharged into the ocean because of the Fukushima Dai-ichi nuclear power plant disaster. J. Nucl. Sci. Technol. 48, 1349–1356.
Kirchner, G., Bossew, P., De Cort, M., 2012. Radioactivity from Fukushima Dai-ichi in air over Europe: part 2: what can it tell us about the accident? J. Environ. Radioact. 114, 35–40.

Kritidis, P., Florou, H., Eleftheriadis, K., Evangeliou, N., Gini, M., SotiroPoulou, M., Diapouli, E., Vratolis, 2012. Radioactive pollution of Athens, Greece due to the Fukushima nuclear accident. J. Environ. Radioact. 114, 100–104.

Lee, S.-H., Heo, D.-H., Kang, H.-B., Oh, P.-J., Lee, J.-M., Park, T.-S., Lee, K.B., Oh, J.S., Suh, J.-K., 2012. Distribution of ^{131}I, ^{134}Cs, ^{137}Cs and $^{239,240}Pu$ concentrations in Korean rainwater after the Fukushima nuclear power plant accident. J. Radioanal. Nucl. Chem. http://dx.doi.org/10.1007/s10967-012-2030-5.

Livingston, H.D., Povinec, P.P., 2000. Anthropogenic marine radioactivity. Ocean Coastal Manage. 43, 689–712.

Livingston, H.D., Povinec, P.P., 2002. A millennium perspective on the contribution of global fallout radionuclides to ocean science. Health Phys. 82, 656–668.

Loaiza, P., Brudanin, V., Piquemal, F., Reyss, J.-L., Stekl, I., Warot, G., Zampaolo, M., 2012. Air radioactivity levels following the Fukushima reactor accident measured at the Laboratoire Souterrain de Modane, France. J. Environ. Radioact. 114, 66–70.

Lujanienė, G., Jokšas, K., Šilobritienė, B., Morkūnienė, R., 2006. Physical and chemical characteristics of ^{137}Cs in the Baltic Sea. Radioactivity Environ. 8, 165–179.

Lujanienė, G., Aninkevicius, V., Lujanas, V., 2009. Artificial radionuclides in the atmosphere over Lithuania. J. Environ. Radioact. 100, 108–119.

Lujanienė, G., Byčenkienė, S., Povinec, P.P., Gera, M., 2012a. Radionuclides from the Fukushima accident in the air over Lithuania: measurement and modeling approaches. J. Environ. Radioact. 114, 71–80.

Lujanienė, G., Valiulis, D., Byčenkienė, S., Šakalys, J., Povinec, P.P., 2012b. Plutonium isotopes and ^{241}Am in the atmosphere of Lithuania: a comparison of different source terms. Atm. Environ. 61, 419–427.

Lujanienė, G., Valiulis, D., Byčenkienė, S., Šakalys, J., Povinec, P.P., 2013. Appl. Radiat. Isotopes, http://dx.doi.org/10.1016/j.apradiso.2013.03.072.

Manolopoulou, M., Vagena, E., Stoulos, A., Ioannidou, A., Papastefanou, C., 2011. Radioidine and radiocesium in Thessaloniki, Northern Greece due to the Fukushima nuclear accident. J. Environ. Radioact. 102, 796–797.

Masson, O., Baeza, A., Bieringer, J., Brudecki, K., Bucci, S., Cappai, M., Carvalho, F.P., Connan, O., Cosma, C., Dalheimer, A., Didier, D., Depuydt, G., De Geer, L.E., De Vismes, A., Gini, L., Groppi, F., Gudnason, K., Gurriaran, R., Hainz, D., Halldorsson, O., Hammond, D., Hanley, O., Holý, K., Homoki, Zs., Ioannidou, A., Isajenko, K., Jankovick, M., Katzlberger, C., Kettunen, M., Kierepko, R., Kontro, R., Kwakman, P.J.M., Lecomte, M., Leon Vintro, L., Leppänen, A.-P., Lind, B., Lujaniene, G., McGinnity, P., Mc Mahon, C., Mala, H., Manenti, S., Manolopoulou, M., Mattila, A., Mauring, A., Mietelski, J.W., Møller, B.S., Nielsen, P., Nikolick, J., Overwater, R.M.W., Palsson, S.E., Papastefanou, C., Penev, I., Pham, M.K., Povinec, P.P.,

Ramebäck, H., Reis, M.C., Ringer, W., Rodriguez, A., Rulík, P., Saey, P.R.J., Samsonov, V., Schlosser, C., Sgorbati, G., Silobritiene, B.V., Söderström, C., Sogni, R., Solier, L., Sonck, M., Steinhauser, G., Steinkopff, T., Steinmann, P., Stoulos, S., Sýkora, I., Todorovic, D., Tooloutalaie, N., Tositti, L., Tschiersch, J., Ugron, A., Vagena, E., Vargas, A., Wershofen, H., Zhukova, O., 2011. Tracking of airborne radionuclides from the damaged Fukushima Dai-ichi nuclear reactors by European networks. Environ. Sci. Technol. 45, 7670–7677.

MEXT, Ministry of Education, Culture, Sports, Science and Technology, 2012. Monitoring information of environmental radioactivity level. http://radioactivity.mext.go.jp/en/ (accessed 10.10.12.).

Mizutani, T., Koarashi, J., Takeishi, M., 2009. Monitoring of low-level radioactive liquid effluent in Tokai reprocessing plant. J. Nucl. Sci. Technol. 46, 665–672.

Morino, Y., Ohara, T., Nishizawa, M., 2011. Atmospheric behavior, deposition, and budget of radioactive materials from the Fukushima Dai-ichi nuclear power plant in March 2011. Geophys. Res. Lett. 38, L00G11. http://dx.doi.org/10.1029/2011GL048689.

NISA, Nuclear and Industrial Safety Agency, 2012. Regarding the Evaluation of the Conditions on Reactor Cores of Unit 1, 2 and 3 Related to the Accident at Fukushima Daiichi Nuclear Power Station. Tokyo Electric Power Co. Inc. http://www.nisa.meti.go.jp/english/press/2011/06/en20110615-5.pdf (accessed 10.11.12.).

NERH, Nuclear Emergency Response Headquarters, Government of Japan, 2012. Report of the Japanese Government to the IAEA Ministerial Conference on Nuclear Safety – The Accident at TEPCO's Fukushima Nuclear Power Stations, 7 June, Available at: http://www.iaea.org/newscenter/focus/fukushima/japan-report/ (accessed 10.10.12.).

NSCJ, Nuclear Safety Commission of Japan, 2011. Trial Estimation of Emission of Radioactive Materials (I-131, Cs-137) into the Atmosphere from Fukushima Dai-ichi Nuclear Power Station, Tokyo. Available at http://www.nsc.go.jp/NSCenglish/geje/2011%200412%20press.pdf.

Paatero, J., Vira, J., Siitari-Kauppi, M., Hatakka, J., Holmén, K., Viisanen, Y., 2012. Airborne fission products in the high Arctic after the Fukushima nuclear accident. J. Environ. Radioact. 114, 41–47.

Perrot, F., Hubert, Ph., Marquet, Ch., Pravikoff, M.S., Bourquin, P., Chiron, H., Guernion, P.-Y., Nachab, A., 2012. Evidence of ^{131}I and ^{134,137}Cs activities in Bordeaux, France due to the Fukushima nuclear accident. J. Environ. Radioact. 114, 61–65.

Pham, M.K., Betti, M., Nies, H., Povinec, P.P., 2011. Temporal changes of ^{7}Be, ^{137}Cs and ^{210}Pb activity concentrations in surface air at Monaco and their correlation with meteorological parameters. J. Environ. Radioact. 102, 1045–1054.

Pham, M.K., Eriksson, M., Levi, I., Nies, H., Osvath, I., Betti, M., 2012. Detection of Fukushima Dai-ichi nuclear power plant accident radioactive traces in Monaco. J. Environ. Radioact. 114, 131–137.

Pham, M.K., Povinec, P.P., Nies, H., Betti, M., 2013. Dry and wet deposition of ^{7}Be, ^{210}Pb and ^{137}Cs in Monaco air during 1997–2010: seasonal variations of deposition fluxes. J. Environ. Radioact. 120, 45–57.

Pittauerová, D., Hettwig, B., Fischer, H.W., 2011. Fukushima fallout in Northwest German environmental media. J. Environ. Radioact. 102, 877–880.

Piñero García, F., Ferro García, M.A., 2012. Traces of fission products in southeast Spain after the Fukushima nuclear accident. J. Environ. Radioact. 114, 146–151.

Povinec, P.P., Oregioni, B., Jull, A.J.T., Kieser, W.E., Zhao, X.-L., 2000. AMS measurements of ^{14}C and ^{129}I in seawater around radioactive waste dump sites. Nucl. Instrum. Methods Phys. Res. B 172, 672–678.

Povinec, P.P., Livingston, H.D., Shima, S., Aoyama, M., Gastaud, J., Goroncy, I., Hirose, K., Huyhn-Ngoc, L., Ikeuchi, Y., Ito, T., La Rosa, J., Liong Wee Kwong, L., Lee, S.-H., Moriya, H., Mulsow, S., Oregioni, B., Pettersson, H., Togawa, O., 2003a. IAEA'97 expedition to the NW Pacific Ocean – results of oceanographic and radionuclide investigations of the water column. Deep-Sea Res. 50 (II), 2607–2638.

Povinec, P.P., Bailly Du Bois, P., Kershaw, P.J., Nies, H., Scotto, P., 2003b. Temporal and spatial trends in the distribution of ^{137}Cs in surface waters of Northern European Seas – a record of 40 years of investigations. Deep-Sea Res. Part II Top. Stud. Oceanogr. 50, 2785–2801.

Povinec, P.P., Hirose, K., Honda, T., Ito, T., Scott, E.M., Togawa, O., 2004. Spatial distribution of ^{3}H, ^{90}Sr, ^{137}Cs and ^{239,240}Pu in surface waters of the Pacific and Indian Oceans – GLOMARD database. J. Environ. Radioact. 76, 113–137.

Povinec, P.P., Aarkrog, A., Buesseler, K.O., Delfanti, R., Hirose, K., Hong, G.H., Ito, T., Livingston, H.D., Nies, H., Noshkin, V.E., Shima, S., Togawa, O., 2005. ^{90}Sr, ^{137}Cs and ^{239,240}Pu concentration surface water time series in the Pacific and Indian Oceans – WOMARS results. J. Environ. Radioact. 81, 63–87.

Povinec, P.P., Lee, S.H., Liong Wee Kwong, L., Oregioni, B., Jull, A.J.T., Kieser, W.E., Morgenstern, U., Top, Z., 2010. Tritium, radiocarbon, ^{90}Sr and ^{129}I in the Pacific and Indian Oceans. Nucl. Instrum. Methods Phys. Res. B 268, 1214–1218.

Povinec, P.P., Breier, R., Coppola, L., Groening, M., Jeandel, C., Jull, A.J.T., Kieser, W.E., Top, Z., 2011. Tracing of water masses using a multi-isotope approach in the southern Indian Ocean. Earth Planetary Sci. Lett. 302, 14–26.

Povinec, P.P., Hirose, K., Aoyama, M., 2012a. Radiostronium in the western North Pacific: characteristics, behavior, and the Fukushima impact. Environ. Sci. Technol. 46, 10356–10363.

Povinec, P.P., Sýkora, I., Holý, K., Gera, M., Kováčik, A., Bres'áková, L., 2012b,c. Aerosol radioactivity record in Bratislava/Slovakia following the Fukushima accident –

a comparison with global fallout and the Chernobyl accident. J. Environ. Radioact. 114, 81–88.
Povinec, P.P., Holý, K., Chudý, M., Šivo, A., Sýkora, I., Ješkovský, M., Richtáriková, M., 2012b. Long-term variations of ^{14}C and ^{137}Cs in the Bratislava air – implications of different atmospheric transport processes. J. Environ. Radioact. 108, 33–40.
Povinec, P.P., Bartok, J., Gera, M., Hirose, K., Lujaniené, G., Nakano, M., Plastino, W. 2013a. Dispersion of Fukushima radionuclides in the global atmosphere and the ocean. Appl. Radiat. Isotopes, http://dx.doi.org/10.1016/j.apradiso.2013.03.058.
Povinec, P.P., Aoyama, M., Biddulph, D., Breier, R., Buesseler, K., Chang, C.C., Golser, R., Hou, X.L., Ješkovský, M., Jull, A.J.T., Kaizer, J., Nakano, M., Nies, H., Palcsu, L., Papp, L., Pham, M.K., Steier, P., Zhang, L.Y. Cesium, 2013b. iodine and tritium in NW Pacific waters – a comparison of the Fukushima impact with global fallout. Biogeosciences Discuss., http://dx.doi:10.5194/bgd-10-6377-2013
Raisbeck, G.M., Yiou, F., 1999. ^{129}I in the oceans: origins and applications. Sci. Total Environ. 237/238, 31–41.
Rypina, I.I., Jayne, S.R., Yoshida, S., Macdonald, A.M, Douglass, E., Buesseler, K., 2013. Short-term dispersal of Fukushima-derived radionuclides off Japan: Modeling efforts and model-data intercomparison. Biogeosciences Discuss., http://dx.doi.org/10.5194/bgd-10-1517-2013.
Schlosser, P., Bayer, R., Boenisch, G., Cooper, L.W., Ekwurzel, B., Jenkins, W.J., Khatiwala, S., Pfirman, S., Smethie, W.M., 1999. Pathways and residence times of dissolved pollutants in the ocean derived from transient tracers and stable isotopes. Sci. Total Environ. 237/238, 15–30.
Sinclair, L.E., Seywerd, H.C.J., Fortin, R., Carson, J.M., Saull, P.R.B., Coyle, M.J., Van Brabant, R.A., Buckle, J.L., Desjardins, S.M., Hall, R.M., 2011. Aerial measurement of radioxenon concentration off the west coast of Vancouver Island following the Fukushima reactor accident. J. Environ. Radioact. 102, 1018–1023.
Stohl, A., Seibert, P., Wotawa, G., Arnold, D., Burkhart, J.F., Eckhardt, S., Tapia, C., Vargas, A., Yasunari, T.J., 2012. Xenon-133 and caesium-137 releases into the atmosphere from the Fukushima Dai-ichi nuclear power plant: determination of the source term, atmospheric dispersion, and deposition. Atmos. Chem. Phys. 12, 2313–2343.
Suzuki, T., Minakawa, M., Amano, H., Togawa, O., 2010. The vertical profiles of iodine-129 in the Pacific Ocean and the Japan Sea before the routine operation of a new nuclear fuel reprocessing plant. Nucl. Instrum. Methods Phys. Res. B 268, 1229–1231.
Tanaka, K., Takahashi, Y., Sakaguchi, A., Umeo, M., Hayakawa, S., Tanida, H., Saito, T., Kana, Y., 2012. Vertical profiles of Iodine-131 and Cesium-137 in soils in Fukushima prefecture related to the Fukushima Daiichi nuclear power station accident. Geochem. J. 46, 73–76.
TEPCO, Tokyo Electric power Company, 2012. Fukushima Daiichi nuclear power station unit 2: countermeasures to stop the

outflow of contaminated water and the water amount flowed out into the sea. http://www.tepco.co.jp/en/press/corp-com/release/11042103-e.html (accessed 10.10.12.).

Tositti, L., Brattich, E., Cinelli, G., Previti, A., Mostacci, D., 2012. Comparison of radioactivity data measured in PM_{10} aerosol samples at two elevated stations in northern Italy during the Fukushima event. J. Environ. Radioact. 114, 105–112.

Tsumune, D., Tsubono, T., Aoyama, M., Hirose, K., 2012. Distribution of oceanic ^{137}Cs from the Fukushima Dai-ichi Nuclear Power Plant simulated numerically by a regional ocean model. J. Environ. Radioact. 111, 100–108.

UNSCEAR, 1993. United Nations Scientific Committee on the Effects of Atomic Radiation. Sources and Effects of Ionizing Radiation. Report to the General Assembly. United Nations, New York, USA.

UNSCEAR, United Nations Scientific Committee on the Effects of Atomic Radiation, 2008. Sources and Effects of Ionizing Radiation. Report to the General Assembly, United Nations, New York.

WHO, World Health Organization, 2005. Chernobyl Forum, Geneva, Switzerland.

Yoshida, N., Kanda, J., 2012. Tracking the Fukushima radionuclides. Science 336, 1115–1116.

FUKUSHIMA DAI-ICHI NUCLEAR POWER PLANT

2

CHAPTER OUTLINE

Fukushima Dai-ichi Nuclear Power Plant (hereinafter referred to as FNPP1) is located in Futaba, Fukushima Prefecture (Japan), facing the Pacific Ocean on the east side (as shown in Figs 2.1 and 2.2). The position is 37° 25.3′ N, 141° 01.9′ E. The site has an own port to load/unload heavy goods including containment structures. An aerial image obtained at Google web site is shown in Fig. 2.3. The site has a half oval shape with a long axis along the coastline. The site area is approximately 3.5 million m^2.

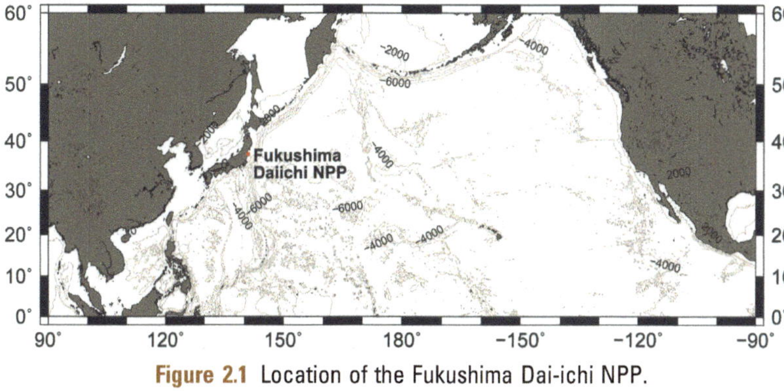

Figure 2.1 Location of the Fukushima Dai-ichi NPP.

Fukushima Accident. http://dx.doi.org/10.1016/B978-0-12-408132-1.00002-4

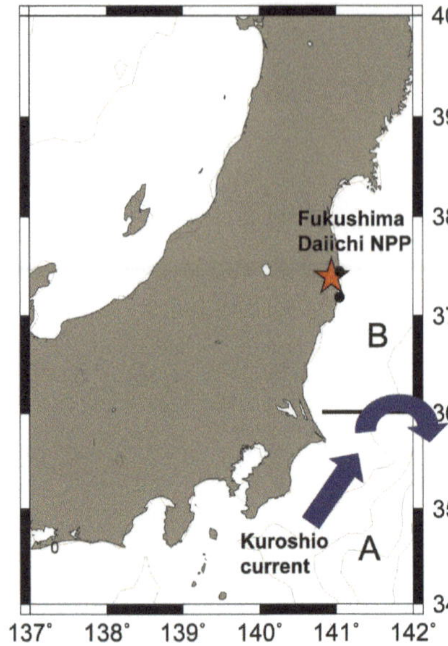

Figure 2.2 Location of the Fukushima Dai-ichi NPP (enlarged and with position of Kuroshio Current).

Figure 2.3 Photo of Fukushima Dai-ichi NPP. (Source: After GOOGLE)

2.1 Nuclear Reactors

This is the first nuclear power station constructed and operated by the Tokyo Electric Power Company, Incorporated (hereinafter referred to as TEPCO). Since the commissioning of Unit 1 in March 1971, additional reactors were constructed in sequence. There were six reactors at the site. A site layout for six reactors is shown in Fig. 2.4.

The total power generating capacity of the facilities was 4.696 million kW (Japanese Government, 7 June 2011). The Units 1–5 were built as Mark I type (light bulb torus) containment structures. The Unit 6 is a Mark II type (over/under) containment structure.

The reactor's emergency diesel generators and direct current batteries, in helping keep the reactors cool in the event of a power loss, were located in the basements of the reactor turbine buildings. TEPCO was strictly following the General Electric's design in the construction of the reactors because the Japanese government did not allow changing the General Electric's original design at that time. Key parameters for Units 1–6 are shown in Table 2.1.

Cross-section sketch of a boiling water reactor Mark I containment (U.S. Nuclear Regulatory Commission, 1975), as used in Units 1–5 at FNPP1, is shown in Fig. 2.5. The reactor core consists of fuel

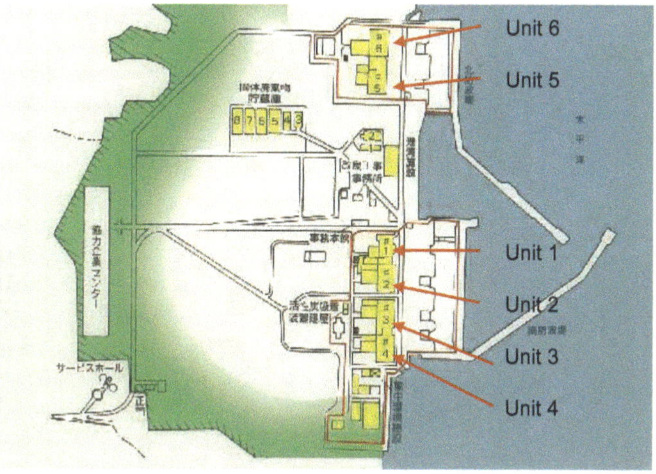

Figure 2.4 Site layouts of six units of the Fukushima Dai-ichi NPP. (Source: Japanese Government 2011)

Table 2.1. Parameters of Units 1–6 of FNPP1

	Unit 1	Unit 2	Unit 3	Unit 4	Unit 5	Unit 6
Electric output (10,000 kW)	46.0	78.4	78.4	78.4	78.4	110.0
Start of construction	Sep. 1967	May 1969	Oct. 1970	Sep. 1972	Dec. 1971	May 1973
Commissioning	Mar. 1971	Jul. 1974	Mar. 1976	Oct. 1978	Apr. 1978	Oct. 1979
Reactor type	BWR-3	BWR-4	BWR-4	BWR-4	BWR-4	BWR-5
Containment type	Mark I	Mark I	Mark I	Mark I	Mark I	Mark II
Number of fuel assemblies	400	548	548	548	548	764
Number of control rods	97	137	137	137	137	185

BWR, boiling water reactor.
Source: Japanese Government 2011.

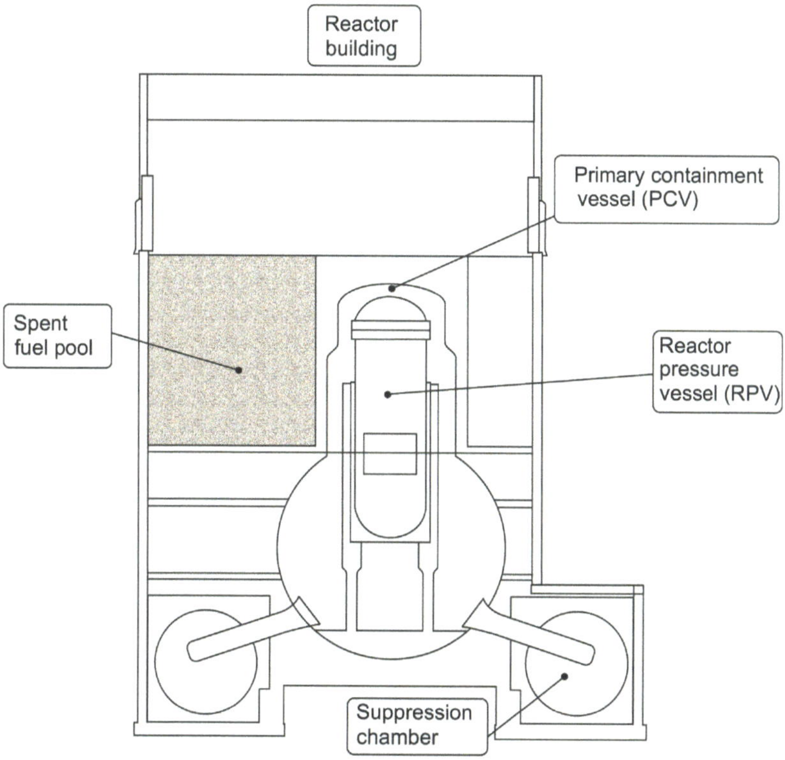

Figure 2.5 A cross-section sketch of a BWR Mark I containment, as used in Units 1–5 at the Fukushima Dai-ichi NPP. (Modified from the U.S. Nuclear Regulatory Commission, 1975)

rods and moderator rods, which were moved in and out by a device. Around the pressure vessel, there is an outer containment, which is closed by a concrete plug. When fuel rods are moved in or out, the cran will move this plug to the pool for facilities. Spent fuel pool and equipment pool are located at the top floor of the plant building.

2.2 Spent Fuel Pools

In the spent fuel pools of Units 1–6, as well as in a common pool, the used fuel rods were stored at the time of accident. A sketch of the spent fuel pool at Unit 4 is shown in Fig. 2.6. Numbers of used fuel rods in the spent fuel pools and estimated total decay heat for Units 1–6 and a common pool are shown in Table 2.2 (Japanese Government, 7 June 2011). Total

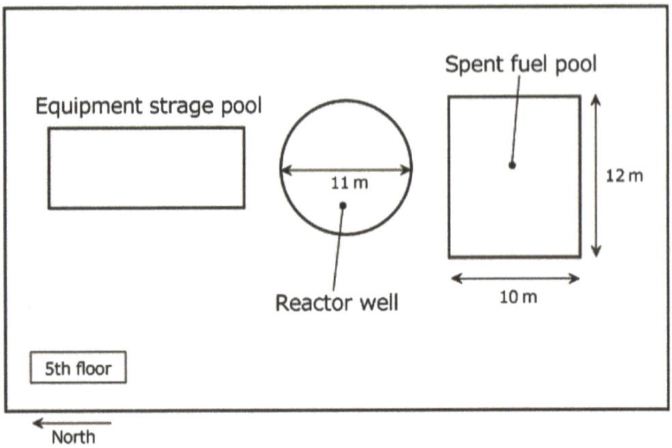

Figure 2.6 Layout of the spent fuel pool, core well and equipment storage pool at the fifth floor of Unit 4 of the Fukushima Dai-ichi NPP. (Source: After TEPCO)

number of used fuel rods at FNPP1 at the time of accident was 10,911 and the total amount of the estimated total decay heat was 6.6 MW on 11 March 2011.

2.3 Radionuclide Inventories

The amounts of radioactive nuclides present in the FNPP1 on 31 March 2011 were estimated using the ORIGEN2 code (Nishihara et al., 2012). The results are given for the irradiated uranium pellets, in the activated cladding tubes of zirconium alloys in the core, and in the spent fuel storage pools of the respective reactors. The evaluated values were weight, radioactivity, heat generation, photon generation and neutron generation rates. The weight and radioactivity for 16 fission products, 10 actinoids and 2 neutron activation products are summarized for cores of Units 1–3 and for the spent fuel pools of Units 1–4 in Tables 2.3–2.9. Activities were derived by weight given by Nishihara et al. (2012), and by mass number and half-life as shown in the tables. The radionuclide activity ratios to ^{137}Cs are also shown, as this will be used in discussions in the following chapters.

Inventories of major fission products of which mass numbers are from 85 to 144 in cores of Units

Table 2.2. Parameters of the Spent Fuel Pools of FNPP1 at the Time of the Accident

	Stored Fuel	Capacity	Decay Heat (MWt)	
Name of Pool	Number of Bundles	Number of Bundles	March 11	June 11
Unit 1	392 (100)	900	0.18	0.16
Unit 2	615 (28)	1240	0.62	0.52
Unit 3	556 (52)	1220	0.54	0.46
Unit 4	1535 (204)	1590	2.26	1.58
Unit 5	994 (48)	1590	1.00	0.76
Unit 6	940 (64)	1770	0.87	0.73
Common	6375 (0)	6840	1.13	1.12
Total	11 407 (496)	15 150	6.60	5.33

Numbers in parentheses indicate number of bundles of new fuel.
Source: Japanese Government 2011.

Table 2.3. Radionuclide Inventories of the Reactor Core of Unit 1 of FNPP1

Radionuclide	g	Half-life	Bq	Activity Ratio to ^{137}Cs
Fission Products				
^{85}Kr	1.59E + 03	10.756 years	2.31E + 16	0.11
^{89}Sr	1.27E + 03	50.53 days	1.36E + 18	6.74
^{90}Sr	2.92E + 04	28.74 years	1.50E + 17	0.74
^{99}Mo	1.44E + 02	65.94 h	2.57E + 18	12.67
^{99m}Tc	1.15E + 01	6.01 h	2.24E + 18	11.05
^{110m}Ag	2.30E + 01	249.8 days	4.04E + 15	0.02
^{125}Sb	2.83E + 02	2.758 years	1.09E + 16	0.05
^{129m}Te	3.94E + 01	33.6 days	4.40E + 16	0.22
^{131}I	2.93E + 02	8.02 days	1.35E + 18	6.65
^{132}Te	1.71E + 02	3.2 days	1.95E + 18	9.65
^{133}I	6.77E + 01	20.8 days	1.18E + 17	0.58
^{133}Xe	3.90E + 02	5.24 days	2.71E + 18	13.37
^{134}Cs	3.98E + 03	2.065 years	1.90E + 17	0.94
^{136}Cs	2.01E + 01	13.16 days	5.42E + 16	0.27
^{137}Cs	6.30E + 04	30.04 years	2.03E + 17	1.00
^{144}Ce	1.13E + 04	284.9 days	1.33E + 18	6.57

Actinides				
^{235}U	1.11E + 06	7.038 × 10^8 years		
^{236}U	2.42E + 05	2.342 × 10^7 years		
^{238}U	6.53E + 07	4.468 × 10^9 years		
^{238}Pu	7.30E + 03	87.74 years	4.63E + 15	
^{239}Pu	3.05E + 05	24,100 years	7.00E + 14	
^{240}Pu	1.05E + 05	6570 years	8.83E + 14	
^{241}Pu	5.86E + 04			
^{241}Am	4.42E + 03	433 years	5.61E + 14	
^{242}Cm	7.28E + 02	162.8 days	8.94E + 16	
^{244}Cm	9.04E + 02	18.11 years	2.71E + 15	
Activation Products				
^{54}Mn	2.25E − 01	312.12 days	6.47E + 13	0.00
^{60}Co	6.04E − 02	5.271 years	2.53E + 12	0.00

Table 2.4. Radionuclide Inventories of the Reactor Core of Unit 2 of FNPP1

Radionuclide	g	Half-life	Bq	Activity Ratio to ^{137}Cs
Fission Products				
^{85}Kr	2.15E + 03	10.756 years	3.11E + 16	0.12
^{89}Sr	2.06E + 03	50.53 days	2.21E + 18	8.64
^{90}Sr	3.73E + 04	28.74 years	1.91E + 17	0.75
^{99}Mo	2.49E + 02	65.94 h	4.43E + 18	17.33
^{99m}Tc	1.99E + 01	6.01 h	3.88E + 18	15.19
^{110m}Ag	3.64E + 01	249.8 days	6.41E + 15	0.03
^{125}Sb	4.30E + 02	2.758 years	1.65E + 16	0.06
^{129m}Te	6.33E + 01	33.6 days	7.06E + 16	0.28
^{131}I	5.09E + 02	8.02 days	2.34E + 18	9.15
^{132}Te	2.94E + 02	3.2 days	3.36E + 18	13.15
^{133}I	1.17E + 02	20.8 days	2.04E + 17	0.80
^{133}Xe	6.73E + 02	5.24 days	4.67E + 18	18.26
^{134}Cs	5.78E + 03	2.065 years	2.77E + 17	1.08
^{136}Cs	3.03E + 01	13.16 days	8.17E + 16	0.32
^{137}Cs	7.95E + 04	30.04 years	2.56E + 17	1.00
^{144}Ce	1.96E + 04	284.9 days	2.31E + 18	9.04

Actinides				
^{235}U	1.70E + 06	7.038 × 10^8 years		
^{236}U	2.99E + 05	2.342 × 10^7 years		
^{238}U	8.91E + 07	4.468 × 10^9 years		
^{238}Pu	7.22E + 03	87.74 years	4.58E + 15	
^{239}Pu	3.84E + 05	24,100 years	8.83E + 14	
^{240}Pu	1.23E + 05	6570 years	1.03E + 15	
^{241}Pu	7.38E + 04			
^{241}Am	3.42E + 03	433 years	4.34E + 14	
^{242}Cm	7.31E + 02	162.8 days	8.97E + 16	
^{244}Cm	1.07E + 03	18.11 years	3.22E + 15	
Activation Products				
^{54}Mn	3.84E − 01	312.12 days	1.10E + 14	0.00
^{60}Co	8.63E − 02	5.271 years	3.61E + 12	0.00

Table 2.5. Radionuclide Inventories of the Reactor Core of Unit 3 of FNPP1

Radionuclide	g	Half-life	Bq	Activity Ratio to ^{137}Cs
Fission products				
^{85}Kr	2.04E + 03	10.756 years	2.95E + 16	0.12
^{89}Sr	2.19E + 03	50.53 days	2.35E + 18	9.74
^{90}Sr	3.53E + 04	28.74 years	1.81E + 17	0.75
^{99}Mo	2.49E + 02	65.94 h	4.42E + 18	18.32
^{99m}Tc	1.98E + 01	6.01 h	3.86E + 18	15.98
^{110m}Ag	3.41E + 01	249.8 days	5.99E + 15	0.02
^{125}Sb	4.09E + 02	2.758 years	1.57E + 16	0.07
^{129m}Te	6.71E + 01	33.6 days	7.48E + 16	0.31
^{131}I	5.05E + 02	8.02 days	2.33E + 18	9.63
^{132}Te	2.95E + 02	3.2 days	3.37E + 18	13.96
^{133}I	1.17E + 02	20.8 days	2.04E + 17	0.85
^{133}Xe	6.73E + 02	5.24 days	4.67E + 18	19.34
^{134}Cs	5.26E + 03	2.065 years	2.52E + 17	1.04
^{136}Cs	3.03E + 01	13.16 days	8.18E + 16	0.34
^{137}Cs	7.50E + 04	30.04 years	2.41E + 17	1.00
^{144}Ce	1.93E + 04	284.9 days	2.28E + 18	9.42

Actinides				
^{235}U	1.69E + 06	7.038×10^{8} years		
^{236}U	2.85E + 05	2.342×10^{7} years		
^{238}U	8.91E + 07	4.468×10^{9} years		
^{238}Pu	8.73E + 03	87.74 years	5.54E + 15	
^{239}Pu	4.52E + 05	24,100 years	1.04E + 15	
^{240}Pu	1.61E + 05	6570 years	1.35E + 15	
^{241}Pu	8.26E + 04			
^{241}Am	4.40E + 03	433 years	5.58E + 14	
^{242}Cm	8.48E + 02	162.8 days	1.04E + 17	
^{244}Cm	9.05E + 02	18.11 years	2.71E + 15	
Activation Products				
^{54}Mn	3.77E − 01	312.12 days	1.08E + 14	0.00
^{60}Co	7.83E − 02	5.271 years	3.28E + 12	0.00

Table 2.6. Radionuclide Inventories of the Spent Fuel Pool of Unit 1 of FNPP1

Radionuclide	g	Half-life	Bq	Activity Ratio to ^{137}Cs
Fission Products				
^{85}Kr	9.66E + 02	10.756 years	1.40E + 16	0.09
^{89}Sr	1.46E + 00	50.53 days	1.57E + 15	0.01
^{90}Sr	2.22E + 04	28.74 years	1.14E + 17	0.70
^{99}Mo	0.00E + 00	65.94 h	0.00E + 00	0.00
^{99m}Tc	0.00E + 00	6.01 h	0.00E + 00	0.00
^{110m}Ag	4.65E + 00	249.8 days	8.18E + 14	0.00
^{125}Sb	1.08E + 02	2.758 years	4.15E + 15	0.03
^{129m}Te	5.21E − 03	33.6 days	5.81E + 12	0.00
^{131}I	3.64E − 12	8.02 days	1.68E + 04	0.00
^{132}Te	0.00E + 00	3.2 days	0.00E + 00	0.00
^{133}I	0.00E + 00	20.8 days	0.00E + 00	0.00
^{133}Xe	5.87E − 19	5.24 days	4.07E − 03	0.00
^{134}Cs	1.83E + 03	2.065 years	8.75E + 16	0.53
^{136}Cs	6.25E − 08	13.16 days	1.69E + 08	0.00
^{137}Cs	5.09E + 04	30.04 years	1.64E + 17	1.00
^{144}Ce	1.15E + 03	284.9 days	1.36E + 17	0.83

Actinides				
^{235}U	5.97E + 05	7.038 × 10^{8} years		
^{236}U	2.08E + 05	2.342 × 10^{7} years		
^{238}U	4.74E + 07	4.468 × 10^{9} years		
^{238}Pu	1.02E + 04	87.74 years	6.45E + 15	
^{239}Pu	2.44E + 05	24,100 years	5.62E + 14	
^{240}Pu	9.62E + 04	6570 years	8.08E + 14	
^{241}Pu	4.75E + 04			
^{241}Am	1.46E + 04	433 years	1.85E + 15	
^{242}Cm	8.51E + 01	162.8 days	1.04E + 16	
^{244}Cm	1.25E + 03	18.11 years	3.74E + 15	
Activation Products				
^{54}Mn	3.21E − 02	312.12 days	9.20E + 12	0.00
^{60}Co	4.00E − 02	5.271 years	1.67E + 12	0.00

Table 2.7. Radionuclide Inventories of the Spent Fuel Pool of Unit 2 of FNPP1

Radionuclide	g	Half-life	Bq	Activity Ratio to ^{137}Cs
Fission Products				
^{85}Kr	2.85E + 03	10.756 years	4.13E + 16	0.09
^{89}Sr	2.60E + 01	50.53 days	2.79E + 16	0.06
^{90}Sr	5.97E + 04	28.74 years	3.06E + 17	0.68
^{99}Mo	2.57E − 18	65.94 h	4.56E − 02	0.00
^{99m}Tc	2.24E − 19	6.01 h	4.37E − 02	0.00
^{110m}Ag	1.86E + 01	249.8 days	3.27E + 15	0.01
^{125}Sb	3.63E + 02	2.758 years	1.40E + 16	0.03
^{129m}Te	3.33E − 01	33.6 days	3.72E + 14	0.00
^{131}I	2.71E − 05	8.02 days	1.25E + 11	0.00
^{132}Te	1.72E − 15	3.2 days	1.97E + 01	0.00
^{133}I	0.00E + 00	20.8 days	0.00E + 00	0.00
^{133}Xe	1.35E − 08	5.24 days	9.36E + 07	0.00
^{134}Cs	6.06E + 03	2.065 years	2.90E + 17	0.65
^{136}Cs	1.26E − 03	13.16 days	3.41E + 12	0.00
^{137}Cs	1.40E + 05	30.04 years	4.49E + 17	1.00
^{144}Ce	4.16E + 03	284.9 days	4.90E + 17	1.09

Actinides				
^{235}U	7.14E + 05	7.038 × 10^8 years		
^{236}U	4.82E + 05	2.342 × 10^7 years		
^{238}U	9.40E + 07	4.468 × 10^9 years		
^{238}Pu	2.57E + 04	87.74 years	1.63E + 16	
^{239}Pu	5.25E + 05	24,100 years	1.21E + 15	
^{240}Pu	2.32E + 05	6570 years	1.95E + 15	
^{241}Pu	1.31E + 05			
^{241}Am	3.10E + 04	433 years	3.93E + 15	
^{242}Cm	2.83E + 02	162.8 days	3.48E + 16	
^{244}Cm	4.54E + 03	18.11 years	1.36E + 16	
Activation Products				
^{54}Mn	1.16E − 01	312.12 days	3.32E + 13	0.00
^{60}Co	1.37E − 01	5.271 years	5.75E + 12	0.00

Table 2.8. Radionuclide Inventories of the Spent Fuel Pool of Unit 3 of FNPP1

Radionuclide	g	Half-life	Bq	Activity ratio to ^{137}Cs
Fission Products				
^{85}Kr	2.61E + 03	10.756 years	3.78E + 16	0.10
^{89}Sr	1.42E + 01	50.53 days	1.53E + 16	0.04
^{90}Sr	5.25E + 04	28.74 years	2.69E + 17	0.69
^{99}Mo	0.00E + 00	65.94 h	0.00E + 00	0.00
^{99m}Tc	0.00E + 00	6.01 h	0.00E + 00	0.00
^{110m}Ag	1.95E + 01	249.8 days	3.43E + 15	0.01
^{125}Sb	3.73E + 02	2.758 years	1.43E + 16	0.04
^{129m}Te	9.10E − 02	33.6 days	1.01E + 14	0.00
^{131}I	1.89E − 08	8.02 days	8.68E + 07	0.00
^{132}Te	0.00E + 00	3.2 days	0.00E + 00	0.00
^{133}I	0.00E + 00	20.8 days	0.00E + 00	0.00
^{133}Xe	1.55E − 13	5.24 days	1.08E + 03	0.00
^{134}Cs	6.09E + 03	2.065 years	2.91E + 17	0.74
^{136}Cs	1.64E − 05	13.16 days	4.43E + 10	0.00
^{137}Cs	1.22E + 05	30.04 years	3.92E + 17	1.00
^{144}Ce	4.79E + 03	284.9 days	5.65E + 17	1.44

Actinides				
^{235}U	6.55E + 05	7.038 × 10^8 years		
^{236}U	4.20E + 05	2.342 × 10^7 years		
^{238}U	8.24E + 07	4.468 × 10^9 years		
^{238}Pu	1.98E + 04	87.74 years	1.25E + 16	
^{239}Pu	4.57E + 05	24,100 years	1.05E + 15	
^{240}Pu	2.16E + 05	6570 years	1.81E + 15	
^{241}Pu	1.07E + 05			
^{241}Am	2.35E + 04	433 years	2.98E + 15	
^{242}Cm	2.29E + 02	162.8 days	2.81E + 16	
^{244}Cm	3.65E + 03	18.11 years	1.09E + 16	
Activation Products				
^{54}Mn	1.29E − 01	312.12 days	3.70E + 13	0.00
^{60}Co	1.37E − 01	5.271 years	5.75E + 12	0.00

Table 2.9. Radionuclide Inventories of the Spent Fuel Pool of Unit 4 of FNPP1

Radionuclide	g	Half-life	Bq	Activity ratio to ^{137}Cs
Fission Products				
^{85}Kr	5.88E + 03	10.756 years	8.51E + 16	0.10
^{89}Sr	6.24E + 02	50.53 days	6.70E + 17	0.76
^{90}Sr	1.21E + 05	28.74 years	6.19E + 17	0.70
^{99}Mo	2.45E − 09	65.94 h	4.36E + 07	0.00
^{99m}Tc	2.14E − 10	6.01 h	4.17E + 07	0.00
^{110m}Ag	5.96E + 01	249.8 days	1.05E + 16	0.01
^{125}Sb	8.37E + 02	2.758 years	3.22E + 16	0.04
^{129m}Te	9.88E + 00	33.6 days	1.10E + 16	0.01
^{131}I	1.00E − 01	8.02 days	4.61E + 14	0.00
^{132}Te	1.11E − 07	3.2 days	1.27E + 09	0.00
^{133}I	0.00E + 00	20.8 days	0.00E + 00	0.00
^{133}Xe	1.54E − 03	5.24 days	1.07E + 13	0.00
^{134}Cs	1.26E + 04	2.065 years	6.05E + 17	0.68
^{136}Cs	2.39E − 01	13.16 days	6.46E + 14	0.00
^{137}Cs	2.75E + 05	30.04 years	8.85E + 17	1.00
^{144}Ce	2.29E + 04	284.9 days	2.70E + 18	3.05

Actinides				
^{235}U	2.25E + 06	7.038×10^{8} years		
^{236}U	1.01E + 06	2.342×10^{7} years		
^{238}U	2.14E + 08	4.468×10^{9} years		
^{238}Pu	4.35E + 04	87.74 years	2.76E + 16	
^{239}Pu	1.15E + 06	24,100 years	2.65E + 15	
^{240}Pu	4.79E + 05	6570 years	4.02E + 15	
^{241}Pu	2.46E + 05			
^{241}Am	6.34E + 04	433 years	8.05E + 15	
^{242}Cm	8.74E + 02	162.8 days	1.07E + 17	
^{244}Cm	7.13E + 03	18.11 years	2.14E + 16	
Activation Products				
^{54}Mn	5.00E − 01	312.12 days	1.43E + 14	0.00
^{60}Co	2.56E − 01	5.271 years	1.07E + 13	0.00

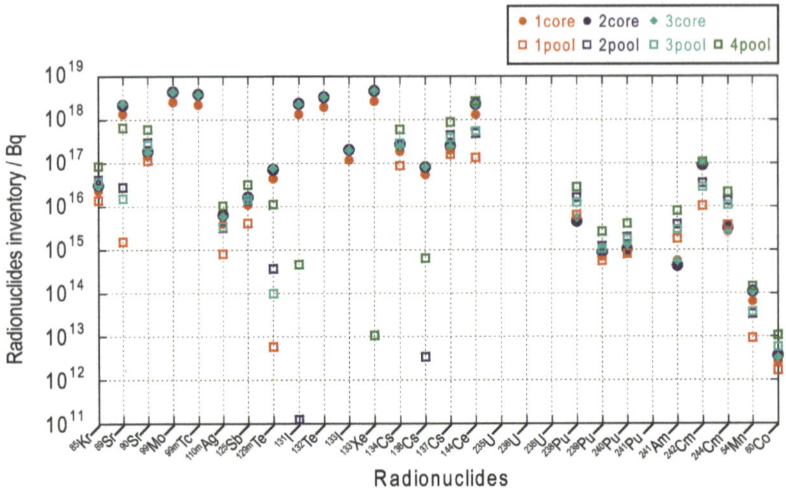

Figure 2.7 Radionuclide inventories in cores of Units 1–3, and spent fuel pools of Units 1–4 of the Fukushima Dai-ichi NPP.

1–3 ranged from 4.0 PBq for ^{110m}Ag in Unit 1 core to 4670 PBq for ^{133}Xe in Units 2 and 3 cores, as shown in Tables 2.3–2.9 and also in Fig. 2.7. Inventories of major fission products of which half-lives are longer than 1 year in the spent fuel pools of Units 1–4 ranged from 4.15 PBq for ^{125}Sb in Unit 1 pool to 885 PBq for ^{137}Cs in Unit 4 pool, as shown in Tables 2.3–2.9 and also in Fig. 2.7.

Inventories of two neutron activation products in cores of Units 1–3 ranged from 2.53 TBq for ^{60}Co in Unit 1 core to 110 TBq for ^{54}Mn in Unit 2 core, as shown in Tables 2.3–2.9 and also in Fig. 2.7. Inventories of two neutron activation products in the spent fuel pools of Units 1–4 ranged from 1.67 TBq for ^{60}Co in Unit 1 core to 143 TBq for ^{54}Mn in Unit 4 pool, as shown in Tables 2.3–2.9 and also in Fig. 2.7.

Radionuclide activity ratios to ^{137}Cs of inventories of 16 fission products and 2 neutron activation products in cores of Units 1–3 are shown in Fig. 2.8. Activity ratios of fission products in cores of Units 1–3 to ^{137}Cs ranged from ~0.01 to 20. For ^{134}Cs to ^{137}Cs activity ratios, they were 0.94 for Unit 1 core, 1.08 for Unit 2 core and 1.04 for Unit 3 core. For two neutron activation products, they were ~0.0001–0.0002 for ^{54}Mn and ~0.00001 for ^{60}Co.

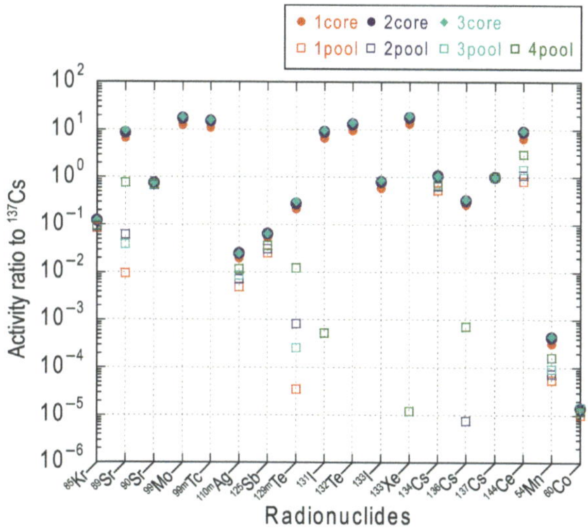

Figure 2.8 Radioactivity ratios of 16 fission products and 2 neutron activation products to ^{137}Cs in cores of Units 1–3 and spent fuel pools of Units 1–4 of the Fukushima Dai-ichi NPP.

Radionuclide activity ratios to ^{137}Cs of inventories of 16 fission products and 2 neutron activation products in spent fuel pools of Unit 1–4 are also shown in Fig. 2.8. Activity ratios of fission products of which half-lives are longer than 1 year in spent fuel pools of Units 1–4 to ^{137}Cs ranged from ~0.03 (^{125}Sb) to 0.74 (^{134}Cs). For ^{134}Cs to ^{137}Cs activity ratios, they were 0.53 for Unit 1 pool, 0.65 for Unit 2 pool, 0.74 for Unit 3 pool, and 0.68 for Unit 4 pool. For two neutron activation products, they were ~0.0001 for ^{54}Mn and ~0.00001 for ^{60}Co.

More information on radionuclide inventories, activity ratios and radionuclide release rate scenarios during the Fukushima accident can be found in Kirchner et al. (2012). International Atomic Energy Agency (IAEA, 2011) is running a Power Reactor Information System (www.iaea.org/programmes/a2/), where more information on radionuclide inventories in nuclear reactors of various designs can be found as well.

References

IAEA, International Atomic Energy Agency, 2011. International atomic energy agency power reactor information system PRIS. www.iaea.org/programmes/a2/ (accessed September 2011).

Japanese Government, 2011. Report of Japanese government to IAEA ministerial conference on nuclear safety – accident at TEPCO's Fukushima nuclear power stations. http://www.iaea.org/newscenter/focus/fukushima/japan-report/ (accessed 07.06.11.).

Kirchner, G., Bosew, P., De Cort, M., 2012. Radioactivity from Fukushima Dai-ichi in air over Europe; part 2: what can it tell us about the accident. J. Environ. Radioact. 114, 35–40.

Nishihara, K., Iwamoto, H., Suyama, K., 2012. Estimation of Fuel Compositions in Fukushima-Daiichi Nuclear Power Plant. Japan Atomic Energy Agency, Tokai, Japan.

U.S. Nuclear Regulatory Commission, 1975. Reactor safety study: an assessment of accident risks in U.S. commercial nuclear power plants, U.S. nuclear regulatory commission, Washington DC, NUREG-75/014 (WASH-1400).

3

FUKUSHIMA ACCIDENT

CHAPTER OUTLINE

On 11 March 2011, the Great East Japan Earthquake occurred which resulted in a gigantic tsunami in the Pacific-side coastal region of northeast Japan. The tsunami had devastating impact on the Tokyo Electric Power Company (TEPCO) Fukushima Dai-ichi nuclear power plant (FNPP1) situated at the coast. The combined effects of the earthquake and the ensuing tsunami at the FNPP1 represent the greatest external event challenges that any nuclear reactor had ever faced. Serious damages that occurred to the FNPP1 resulted in large radionuclide releases to the

Fukushima Accident. http://dx.doi.org/10.1016/B978-0-12-408132-1.00003-6

atmosphere. The Japanese government declared this accident of the level 7, the highest, on the international nuclear event scales (IAEA, 2009). The scale of the Fukushima accident was thus the same as that of the Chernobyl accident, which occurred in 1986 in the former Soviet Union (presently Ukraine). The Nuclear Accident Independent Investigation Commission (NAIIC, 2012), established by the Japanese government has concluded that this nuclear accident was a “man-made” disaster, although it was triggered by a natural disaster after a great earthquake and a gigantic tsunami. This was a result of reviewing the time course of the nuclear accident, as well as the history of the construction and maintenance of the TEPCO nuclear reactors and nuclear safety actions of the regulatory office. After the accident, the Japanese government conducted many protective actions to avoid radiological effects to the public.

3.1 Nature of the Accident

3.1.1 Earthquake and Tsunami

The Great East Japan Earthquake with a moment magnitude of 9.0 occurred at 2.46 pm (local time) on 11 March 2011. The epicenter of the earthquake with a hypocentral region about 500 km long and approximately 200 km wide (as shown in Fig. 3.1) was located 130 km offshore of the city of Sendai in Miyagi prefecture on the eastern coast of Honshu Island (the main part of Japan) at an underwater depth of approximately 24 km (METI, 2011; PMJC, 2011a). The earthquake, which was one of the five most powerful earthquakes in the world since modern record-keeping began in 1900, and the biggest earthquake which occurred in Japan, was a rare and complex double quake having duration of about 3 min. The earthquake moved Honshu Island 2.4 m east and shifted the Earth on its axis by estimates between 10 and 25 cm. The earthquake triggered gigantic tsunami waves that reached heights of up to 40.5 m in Miyako in Iwate prefecture, and which, in the Sendai area, traveled up to 10 km inland (METI, 2011).

As a result of the earthquake and the resulting tsunami, a Japanese National Police Agency report (21 November 2012) confirmed that 15,873 people

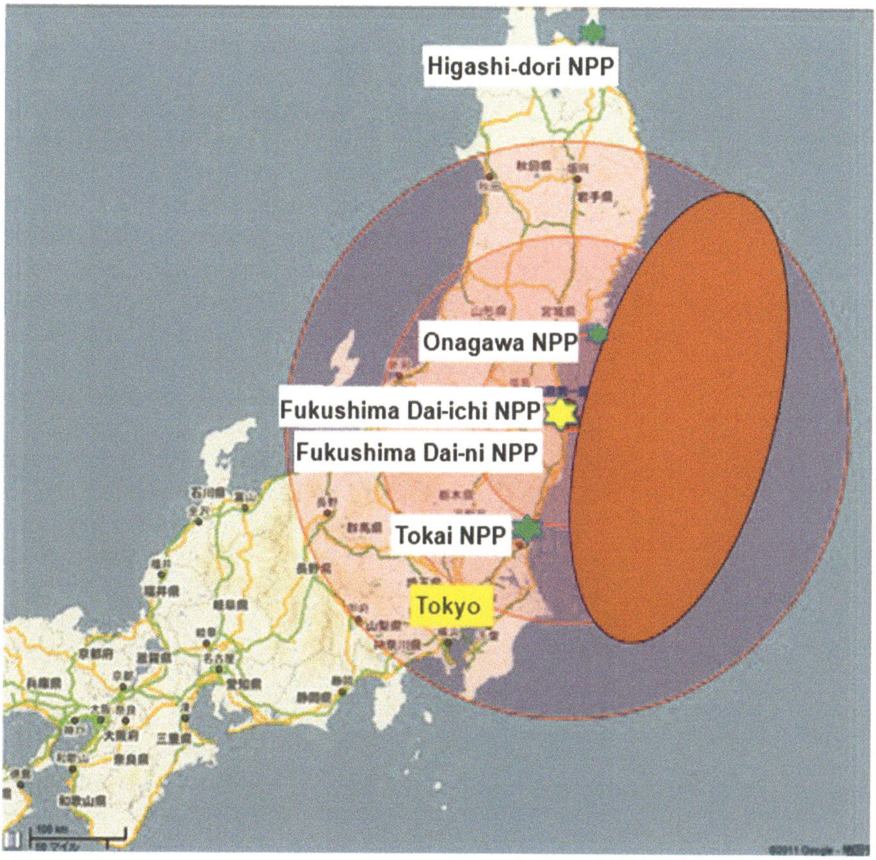

Figure 3.1 Epicenter of the Great East Japan earthquake (11 March 2011; hypocentral region: 500 km long, 200 km wide) and location of nuclear power plants along the Pacific Ocean coast. (Modified from TEPCO, 2011)

died, 6114 were injured, and 2744 were missing across 20 prefectures. Totally 129,627 buildings collapsed, with a further 266,404 buildings "half collapsed", and 728,583 buildings being partially damaged (NPAJ, 2012). The earthquake and tsunami also caused extensive and severe structural damage in northeastern Japan, including heavy damage to roads and railways, as well as caused fires in many areas and collapse of a dam.

3.1.2 Impacts of the Earthquake and Tsunami on Nuclear Power Plants

A total of 11 reactors at four nuclear power plants (NPPs) in the eastern coast of Honshu Island were in operation during the earthquake. All NPPs shutdown automatically when the quake hit them. The operating reactors which shutdown were the TEPCO's Fukushima Dai-ichi (Dai-ichi in Japanese means the first) NPP (Units 1, 2 and 3), the Fukushima Dai-ni (Dai-ni means the second) NPP (Units 1, 2, 3 and 4), the Tohoku's Onagawa NPP (Units 1, 2 and 3), and the Japan Atomic Energy Company's (JAPCO) Tokai NPP (Fig. 3.1), corresponding to a total of 9377 MW net. Although Onagawa Unit 1 briefly suffered a fire in the turbine building, the major problems initially centered on the Fukushima Dai-ichi reactors Units 1, 2 and 3. The Fukushima Dai-ichi reactors were of the boiling water reactor type of an early (1960s) design supplied by GE, Toshiba and Hitachi, known as the Mark I containment.

The detail story of the FNPP1 accident has been documented in government reports (METI, 2011; PMJC, 2011a). Furthermore, the Fukushima Nuclear Accident Independent Investigation Committee (NAIIC, 2012), which was the first independent commission created in the history of Japan's constitutional government, has published a document covering the FNPP1 accident: "When the earthquake occurred, Unit 1 of the FNPP1 was in normal operation at the related electricity output according to its specifications; Units 2 and 3 were in operation within the related heat parameters of their specifications; and Units 4, 5 and 6 were undergoing periodical inspections. The emergency shut-down feature, or SCRAM, went into operation at Units 1, 2 and 3 immediately after the commencement of the seismic activity (NAIIC, 2012)."

All nuclear reactors in the eastern coast of Honshu Island, including the Fukushima Dai-ichi Units 1, 2 and 3, almost proved to be seismically robust; however, the earthquake seriously damaged electricity transmission facilities. However, a possibility that the damage to Unit 1 was partly caused by the earthquake was indicated by the facts that

1. The largest tremor hit after the automatic shutdown (SCRAM: Safety Control Rod Axe Man).
2. Japan Nuclear Energy Safety Organization confirmed the possibility of a small-scale loss-of-coolant accident (LOCA).

3. Unit 1 operators were concerned about leakage of coolant from the valve.
4. The safety relief valve (SRV) was not operating (NAIIC, 2012).

To stabilize a nuclear reactor, it is necessary to safely achieve "cool shutdown" using the residual heat removal (RHR) system. Electric power from grid or backup generators were able to run the RHR system cooling pumps at eight reactors: Fukushima Dai-ni NPP (Units 1, 2, 3, and 4), Tohoku's Onagawa NPP (Units 1, 2, and 3) and JAPCO's Tokai NPP. All these rectors achieved "cool shutdown" within about 4 days, although some problems happened in each reactor.

Although initially cooling in the FNPP1 would have been maintained through the main stream circuit bypassing the turbine and going through the condensers, the three reactors of FNPP1 lost AC power at 3:42 pm (local time) due the first tsunami wave hitting as shown in Figs 3.2 and 3.3, almost an hour after the earthquake. Almost the entire site of FNPP1 lost the ability to maintain proper reactor cooling and water circulation functions due to being flooded by the tsunami with 15 m height (Fig. 3.4). This destroyed 12 of 13 backup generators on site, located in the basements of the turbine buildings, and also the heat exchangers for dumping the reactor heat to the sea. Electrical switchgear was also disabled. The reactor pressure and water levels in the reactor make it clear that a massive LOCA except in Unit 1 did not occur in the time between the earthquake and the tsunami.

The reactors at FNPP1 were vulnerable to the tsunami due to the actual tsunami height, which was more than two times higher than the provisional tsunami height for the FNPP1 (Fig. 3.4). It must be noted that the Onagawa NPP has been protected against hit of the tsunami wave, even though height of the tsunami wave was the same as that of the FNPP1 as shown in Fig. 3.2.

The seismic tremors damaged electricity transmission facilities between the TEPCO Shin-Fukushima Transformer Substations and the FNPP1, resulting in a total loss of off-site electricity. There was a backup 66-kV transmission line from the transmission network of Tohoku Electric Power Company,

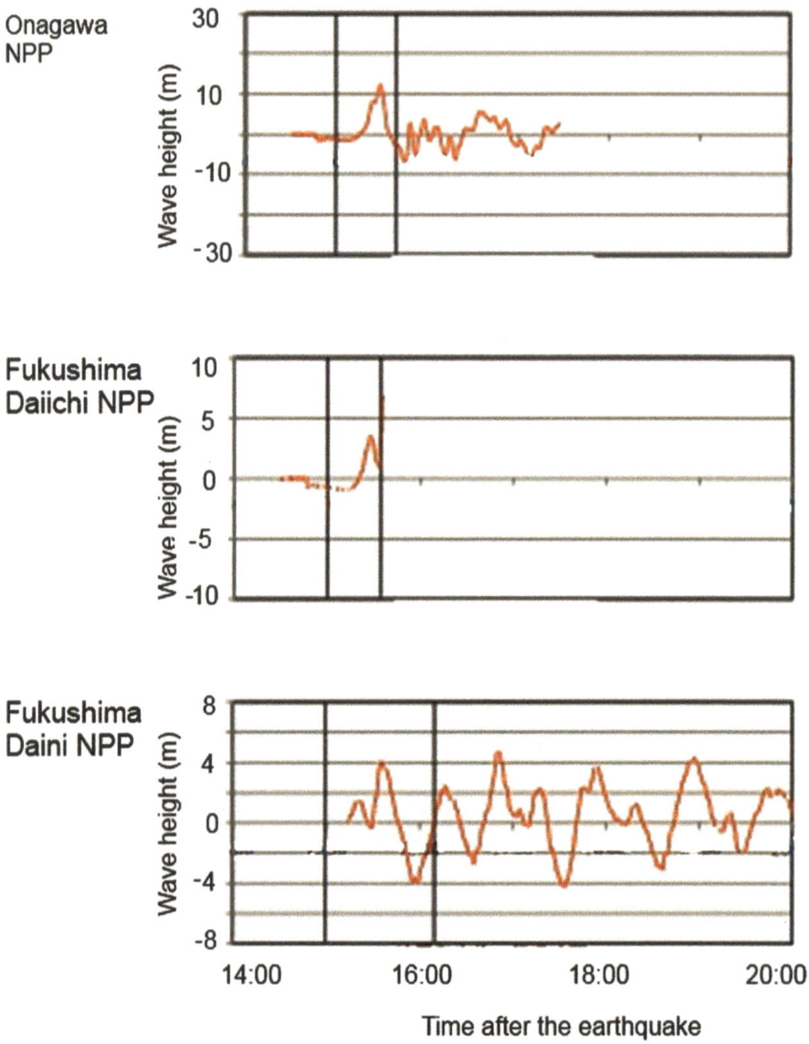

Figure 3.2 Heights of tsunami wave observed at Onagawa, Fukushima Dai-ichi, and Tokai Dai-ni NPPs. (Modified from PMJC, 2012)

but the backup line failed to feed Unit 1 via a metal-clad circuit of Unit 1 due to mismatched sockets.

The tsunami flooded and totally destroyed the emergency diesel generators, the seawater cooling pumps, the electric wiring system and the DC power supply for Units 1, 2 and 4, resulting in loss of all power, except for an external supply to Unit 6 from an air-cooled emergency diesel generator. In short,

(a)

(b)

Figure 3.3 Tsunami waves in front of the Fukushima Dai-ichi NPP on 11 March 2011 (a), and destroyed equipment of the seawater pump area (b). (Source: Reprinted with permission of TEPCO)

Units 1, 2 and 4 lost all power; Unit 3 lost all AC power, and later it lost DC before dawn on 13 March 2011. Unit 5 lost all AC power as well.

The tsunami also destroyed or washed away vehicles, heavy machinery, oil tanks, and gravel. It destroyed buildings, equipment, installations and other machinery. Seawater from the tsunami inundated the entire building areas and even reached the extremely high-pressure operating sections of Units 3

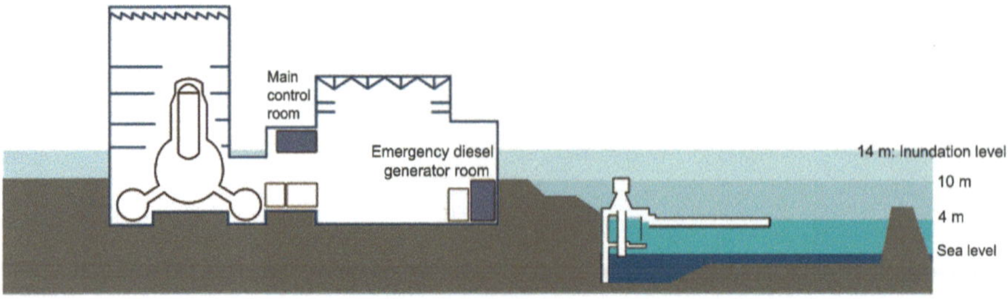

Figure 3.4 Schematic picture of the Fukushima Dai-ichi NPP and levels of tsunami wave, the projective tsunami height (5.7 m) and sea surface. (Modified from FNAIIC (2012))

and 4 and a supplemental operation common facility (Common Pool Building). After the water retreated, debris from the flooding was scattered all over the plant site, hindering movement. Manhole and ditch covers had disappeared, leaving gaping holes in the ground. In addition, the earthquake lifted, sank, and collapsed building interiors and pathways, so access to and within the plant site became extremely difficult. Recovery tasks were further interrupted as workers reacted to the intermittent and significant aftershocks and tsunami. The loss of electricity resulted in the sudden loss of monitoring equipment such as scales, monitors, meters and the control functions in the central control room. Lighting and communications were also heavily affected. The decisions and response to the accident had to be made on the spot by operational staff at the site, with no valid tools and manuals (NAIIC, 2012).

When the station blackout occurred at 3:42 pm on 11 March, the reactor cores were still producing from decay of fission products about 1.5% of their thermal power: about 22 MW in reactor No. 1 and 33 MW in reactor Units 2 and 3. The energy estimation was calculated using well-established models, although the exact amount depends on the average burn-up in each reactor core. Without heat removal by circulation to an outside heat exchanger, the cores produced a lot of steam in the reactor pressure vessels housing the cores, which was released into the dry containment (Primary Containment Vessel: PCV) through safety valves. As the pressure in the PCV increased, the steam was directed into the suppression chamber (S/C) under the reactor, within the containment, but the internal temperature and pressure rose quite rapidly. Water injection using the isolation condenser (IC) or reactor core isolation cooling (RCIC) and emergency core cooling system commenced. The IC system in reactor Unit 1 failed apparently on late 11 March, whereas the RCIC system in reactor Unit 2 ran for about 3 days. The RCIC system in reactor Unit 3 failed after 19 h, but the high-pressure coolant injection (HPCI) system then started automatically due to low water level, and it ran for 14 h.

After the built-in systems failed to cool the reactors, water injection to the Reactor Pressure Vessels (RPV) was done with fire pumps, but this required the internal pressures to be relieved by venting into the S/C/wetwell, connected into the RHR system.

Finally, meltdown of nuclear fuels occurred in FNPP1 Units 1, 2 and 3 due to failure of the cooling system of each reactor. On the other hand, Units 5 and 6, located in a separate building, also lost electric power on 11 March due to the tsunami. These reactors were in "cool shutdown" at that time, but still requiring cooling. One air-cooled diesel generator at FNPP1 Unit 6 was located at higher stage, so it survived the tsunami and after repairs on 19 March, it allowed a full restoration of the cooling for Units 5 and 6. While the power was stopped, the core temperature of Units 5 and 6 had risen to over 100 °C (128 °C in Unit 5) under pressure, so these units had to be cooled with fresh water injection, presumably with the RCIC system. Units 5 and 6 were restored to cool shutdown by the RHR system on 20 March, and the mains power was reinstated on 21–22 March.

As mentioned in the NAIIC (2012) report: "The loss of electricity made it very difficult to effectively cool down the reactors in a timely manner. Cooling the reactors and observing the results were heavily dependent on the electricity for high-pressure water injection, depressurizing the reactor, low-pressure water injection, the cooling and depressurizing of the reactor containers, and removal of decay heat at the final heat sink. The lack of access, as previously mentioned, obstructed the delivery of necessities such as alternative water injection using fire trucks, the recovery of electricity supply, the line configuration of the vent and its intermittent operation".

3.1.3 Accident Sequence of Each Reactor

3.1.3.1 Unit 1

There was normal operation of the Unit 1 reactor prior to the earthquake. After the earthquake, the reactor scrammed at 2:47 pm on 11 March due to the large acceleration of the earthquakes and automatically shut down as all control rods were inserted to bring the reactor into subcritical. The earthquake caused loss of external power supply due to damage of the receiving circuit breakers of the substation at the Okuma No. 1 and No. 2 power transmission lines. This led to automatic start of the two emergency diesel generators. For Unit 1, although cooling by the IC began at 2:52 pm on 11 March, just after the automatic shutdown of the nuclear rector, it was manually stopped at 3:03 pm by following the operation procedure documents because of rapid decrease of temperature of the RPV. After that, using only one system of the IC, start-up and shutdown was repeated manually. The loss of power supplies due to the following tsunami made it impossible to confirm the operating status of the IC. It is now provisionally understood that the water level dropped to the top of the fuel about 3 h after the SCRAM (at 6 pm), and to the bottom of the fuel at 7.30 pm. The Nuclear and Industrial Safety Agency (NISA) estimates put the timescales shorter, with core melt by 8 pm on 11 March. Sudden increase of the ambient dose rate occurred in the reactor and turbine buildings by 11 pm, whereas the ambient dose rate in the front gate of the FNPP1 was still at a background level (about 60 nSv/h), as shown in Fig. 3.5.

The RPV pressure in Unit 1 at 8 pm on 11 March was 6.9 MPa, which is the normal operating pressure. On 12 March at 2:46 am, the RPV pressure had come down to about 0.8 MPa, and was similar to the PCV pressure, which is twice higher than the PCV design level. The PCV temperature was 300 °C instead of the designed maximum temperature of 138 °C. In order to inject water into the RPV using the fire pumps, venting attempt was started at 1:30 am (under the authorization of the government). The ambient dose rate at the front gate suddenly increased from 70 to 130 nSv/h at 4:05 am, and exceeded 1 μSv/h at 4:50

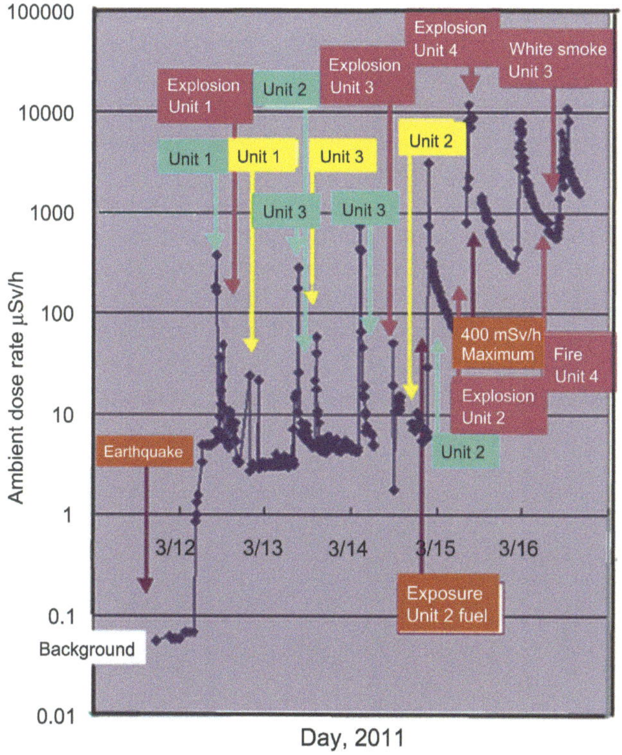

Figure 3.5 Temporal variation of ambient dose rate measured by a monitoring car at the frontal gate of the Fukushima Dai-ichi NPP. Blue and yellow arrows indicate ventilation of reactors and injection of seawater, respectively. (Modified from TEPCO, 2011)

am on 12 March, as shown in Fig. 3.5. Injection of fresh water into the RPV using the fire pumps started at 5:46 am on 12 March. About 80 m^3 of fresh water was injected into the reactor Unit 1 RPV over 9 h until the supply ran out. Seawater was then substituted, from about 7 pm. The venting was designed to be through an external stack, but in the absence of power, much of steam backflowed to the service floor at the top of the reactor building. The vented steam, noble gases and aerosols were accompanied by hydrogen, which was produced by the exothermic reaction of the fuel's zirconium cladding with steam and water at high temperatures. The ambient dose rate at the front gate of FNPP1 reached a peak of 390 µSv/h at 10:30 am. The high ambient dose rate of 1 mSv/h occurred near MP (Monitoring Post)-4 at

3:29 pm. At 3:36 pm on 12 March, a hydrogen explosion occurred on the service floor of the building above reactor Unit 1 containment, blowing off the roof and cladding on the top part of the building (Fig. 3.6), after the vented hydrogen mixed with air at critical point and ignited.

Assemblies (292) of spent fuel and 100 assemblies of fresh fuel were stored in the spent fuel pond (SFP) of Unit 1. Decay heat was calculated at 0.18 MW as of 11 March and at 0.16 MW as of 11 June. After the tsunami hit, all AC power and consequently seawater pump function were lost, and subsequently the cooling and makeup water functions of the SFP were lost. Water

Figure 3.6 Damaged Unit 1 of the Fukushima Dai-ichi NPP after the hydrogen explosion on 12 March 2011. (Source: Reprinted with permission of TEPCO)

supply had been conducted with concrete pump truck from 31 March to 22 May, by which the total amount of supplied fresh water was 240 t. However, it was uncertain whether the effective water injection had been made or not by the spraying of concrete pump truck. The measurement of radionuclide concentrations in the skimmer surge tank (SST) water was performed on 22 June 2011; ^{131}I and ^{137}Cs concentrations in the SST water were 68 and 1.4×10^4 MBq/m^3, respectively, which corresponded to an $^{131}I/^{137}Cs$ activity ratio of 33 at the time of the reactor cease. According to water balance and radionuclide analysis of the SST pond, Ministry of Economy, Trade and Industry (METI) concluded that uncovering of the spent nuclear fuels was highly unlikely (METI, 2011).

3.1.3.2 Unit 2

There was normal operation of Unit 2 reactor prior to the earthquake. After the earthquake, the reactor scrammed at 2:47 pm on 11 March due to the large acceleration of the earthquakes and automatically shut down as all control rods were inserted to bring the reactor into subcritical. The earthquake caused loss of the external power supply due to damage of the receiving circuit breakers of the substation at the Okuma No.1 and No.2 power transmission lines. This led to automatic activation of the two emergency diesel generators. Since the closing of the main stream isolation valve led to a rise in RPV pressure, in accordance with the operation manual, the RCIC was activated manually at 2:50 pm on 11 March. Then the reactor repeated automatic RCIC shutdown due to high reactor water level and manual activation. From 10:00 pm of 11 March to around 0:00 pm of 14 March, the reactor water level measure remained at a stable level (more than 3000 mm) sufficiently above the top of available fuel. The reactor pressure was controlled by the switch of the main steam SRV. Moreover, as operation of the SRV and RCIC led to a rise in the S/C temperature, the RHR pumps were moved in succession from 3:00 pm to 3:07 pm on 11 March in order to cool the S/C water. Although the S/C then showed an increasing tendency of temperature since past 3:30 pm, the RHR pumps successively shut down at around 3:36 pm on 11 March. This function failure is presumed to be caused by the tsunami. As a result of the tsunami, two emergency diesel generators stopped operating and all AC power supply was lost due to submersion of the cooling seawater pump, the power distribution panel, and the emergency bus bar by the flooding. Furthermore, information on parameters of the reactor operation could not be obtained due to the loss of direct electrical current function. In addition, damage of the RHR system pump function led to the loss of RHR function, and thus the decay heat was unable to be transferred to the sea, the ultimate heat sink. At 10:00 pm on 11 March, the reactor water level could be watched. Since the water level was measured to be stable, it could be presumed that the water injection by RCIC was successful. However, the reactor pressure was slightly lower than the rated pressure at 6 MPa gage. In Unit 2, the RPV pressure at 0:55 pm on 12 March was 6.1 MPa, which is the normal operating pressure. The RPV pressure increased to 9.0 MPa at 1:34 pm on 14 March and gradually decreased to 5.4 MPa at 6:06 pm. After that, it rapidly came down to 0.61 MPa. Water injection using the steam-driven RCIC system failed on 14 March, which occurred about 6 h before a fire pump started injecting seawater into the RPV. Before the fire pump could be used, the RPV pressure had to be relieved via the wetwell, which required power and nitrogen, hence the delay occurred. On 14 March afternoon, the reactor water level dropped rapidly after the RCIC cooling was lost, so that core damage started at about 8 pm. It is now provisionally understood that much of the fuel then melted and probably fell into the water at the bottom of the RPV. NISA similarly estimates that much of the fuel had melted about 100 h after the SCRAM. Pressure in the RPV was vented at 11 am on 13 March, and again at 0:02 am on 15 March, and meanwhile, the blowout panel near the top of the building was opened to avoid a repetition of situation which occurred in reactor No. 1 with hydrogen explosion. The pressure S/C under the actual reactor apparently ruptured at 6:10 on 15 March, possibly due to a hydrogen explosion there, releasing significant radioactivity, and dropping the drywell (D/W) containment pressure inside. The ambient dose rate at the front gate increased to 580 μSv/h at 6:50 am and reached a peak of 12 mSv/h at 9:00 am on 15 March, as shown in Fig. 3.5. Later analysis suggested that a leak of the PCV developed about midday of 12 March, about 21 h after the earthquake. In this connection, the containment pressure S/C should have failed until noon 12 March 2011 by only the seismic load due to the huge earthquake of 11 March or by the combination of seismic deterioration and dynamic load due to steam flowing in through the SRV (Tanabe, 2012a). Opening of the SRVs on 14 March at 9:18 pm should have resulted in emission of a large amount of radioactive materials through the S/C breach, which corresponded to ambient dose rate peak at 9:34 pm near the front gate of FNPP1. The containment D/W should have failed at 6:25 am on 15 March at the cable penetration seal due to the high temperature caused by heating up of the fuel materials on the floor of the D/W, which had flowed out from the reactor pressure vessel. Then large amounts of radioactive materials were emitted in the environment through the D/W breach.

Assemblies (587) of spent fuel and 28 assemblies of fresh fuel were stored in the SFP of Unit 2. Decay heat was calculated 0.62 MW as of 11 March and at 0.52 MW as of 11 June. After the tsunami hit, all AC power and consequently seawater pump function were lost, and subsequently the cooling and makeup water functions of the SFP were lost. Water injection was conducted using SFP piping on 20 March. The result of water balance suggested that exposure of spent nuclear fuels was highly unlikely (METI, 2011). For measurement of radionuclide concentrations in the SST water, sample water was collected on 16 April 2011; ^{131}I and ^{137}Cs concentrations in the SST water were 4.1×10^3 and 1.5×10^5 MBq/m^3, respectively, which corresponded to an $^{131}I/^{137}Cs$ activity ratio of 0.56 at the time of reactor cessation. The radionuclide analysis of the SST water suggests that some problems occurred in the spent fuel assemblies because the ^{137}Cs concentration in the SST water was greater than that in Unit 1 and the four SFPs, and the $^{131}I/^{137}Cs$ ratio was significantly lower than that in the atmospheric emissions.

3.1.3.3 Unit 3

There was normal operation of the Unit 3 reactor prior to the earthquake. After the earthquake, the reactor scrammed at 2:47 pm on 11 March due to the large acceleration of the earthquakes and automatically shut down as all control rods were inserted to bring the reactor into subcritical. In addition to the power line of Okuma Line 3, to which no power was supplied due to repair work that started before the earthquake, the breaker at Shin-Fukushima Substation tripped and the breaker for receiving electricity at the switchyard in the power station was damaged, disrupting the power supply from Okuma Line 4 and leading to loss of off-site power. As a result, two emergency diesel generators were activated automatically. In Unit 3, the RCIC system manually started at 3:07 pm on 11 March and was stopped due to the high water level in the reactor at 3:25 pm. At 3:38 pm, as a result of the tsunami, two emergency diesel generators stopped operating due to submersion of the cooling seawater pumps, the power distribution panel and the emergency bus at Unit 3 by flooding, resulting in station blackout. In addition, damage of the RHR system pump function due to tsunami led to the loss of the RHR function, which failed to transfer the decay heat in the PCV to the sea, the ultimate heat sink. However, the diesel generator bus of Unit 3 did not get flooded. Although power was not supplied through AC-DC transfer from the AC bus, the backup storage batteries supplied power to the loads (RCIC valves, recorders, etc.) (Fig. 3.7). Because of the decrease of water level resulting from the shutdown of the RCIC at 3:25 pm on 11 March, the RCIC restarted at 4:03 pm on 11 March. The RCIC system tripped at 11:36 am on 12 March. The HPCI system started at 0:35 pm on 12 March. On 13 March, at 5:08 am, water injection using the HPCI system failed, and water levels dropped dramatically. The RPV pressure was reduced by venting steam into the wetwell, allowing injection of seawater using a fire pump at 9:25 am on 13 March. At 8:35 am, venting of the S/C and the containment was successfully undertaken. It is now provisionally understood that the core damage started about 9 am, and much or all of the fuel melted on

Figure 3.7 Reactor control room on 22 March 2011, with batteries supplying a power to run instruments in the control room. (Source: Reprinted with permission of TEPCO)

the morning of 13 March. The fuel possibly fell into the water at the bottom of the RPV or was retained on the core support plate within the shroud. On 14 March at 5:20 am, PCV venting was repeated, although both RPV and D/W pressure remained at about 0.5 MPa. The venting evidently backflowed to the service floor of the building, and at 11:01 am, a very large hydrogen explosion occurred above the Unit 3 containment, which blew off much of the roof and walls and demolished the top part of the building (Fig. 3.8). The explosion created a lot of debris, and the ground near Unit 3 was extensively contaminated by radionuclides released from the reactor.

There is no description of the accident sequence after 17 March in the government report (METI, 2011; PMJC, 2011a). Tanabe (2012b) suggested that remelting of nuclear fuel that had already melted down in the containment vessel occurred during 21–24 March (Fig. 3.9) due to depression of supply of water, which was caused by increased pressure in the RPV. This event resulted in the emission of a large amount of radioactive materials in the environment, which corresponded to high radioactivity in the Kanto and Tohoku areas during 21–24 March 2011.

Assemblies (514) of spent fuel and 52 assemblies of fresh fuel were stored in the SFP of Unit 3. Decay

Figure 3.8 Units 1, 2, 3 and 4 of the Fukushima Dai-ichi NPP after a hydrogen explosion in Unit 3 on 15 March 2011. (Source: Reprinted with permission of TEPCO)

heat was calculated at 0.54 MW as of 11 March and at 0.46 MW as of 11 June. After the tsunami hit, all AC power and consequently seawater pump function were lost, and subsequently the cooling and makeup water functions of the SFP were lost. On 17 March, Self-Defense Forces started seawater spraying to the

Figure 3.9 Damaged Unit 3 of the Fukushima Dai-ichi NPP after a hydrogen explosion on 21 March 2011 with emission of smoke. (Source: Reprinted with permission of TEPCO)

upper side of the reactor buildings with helicopter. On 17 March, water spraying was started to the SFP by police's water spraying trucks. Since then till 25 March, water spraying had been conducted to the SFP by fire department's water spraying trucks with bending arms. After that, about 815 t of water was injected with pump trucks from 27 March to 22 April. The result of water balance suggested that an exposure of spent nuclear fuels was highly unlikely (METI, 2011). For measurement of radionuclide concentrations in the SFP water, sample water was collected on 8 May 2011; ^{131}I and ^{137}Cs concentrations in the SFP water were 1.1×10^4 and 1.5×10^5 MBq/m^3, respectively, which corresponded to an $^{131}I/^{137}Cs$ activity ratio of 10 at the time of the reactor cease. The radionuclide analysis of the SFP water suggests that some problems occurred in spent fuel assemblies because the ^{137}Cs concentration in the Unit 3 SFP was greater than that in Units 1 and 4 SFPs and the $^{131}I/^{137}Cs$ ratio was lower than that in the atmospheric emission. In April 2011, Unit 3 SFP was confirmed to be full of water (Fig. 3.10).

3.1.3.4 Unit 4

When the earthquake hit, Unit 4 was in periodic inspection and all nuclear fuel assemblies were transferred from the reactor to the SFP to carry out shroud replacing works. In Unit 4, the loss of all AC power, and the consequent loss of the possibility of using seawater pumping, resulted in cooling problems in this unit. This SFP had a particularly high heat load (3 MW) from the 1331 used fuel assemblies stored in the pond, so about 100 m^3/day water was additionally needed to replenish after circulation of cooling ceased. At about 6 am on 15 March, an explosion occurred in the top part of the building, near the SFP, which destroyed the top of the Unit 4 building (Fig. 3.11) and damaged Unit 3 superstructure further. This explosion was apparently derived from hydrogen released in Unit 3 and by backflow in shared ducts when venting was done in Unit 3.

Then, there was a fire, and soon the radiation level near the building reached 400 mSv/h, as shown in Fig. 3.5. The fire was extinguished in 3 h. However,

Figure 3.10 Views of the Unit 3 spent fuel pond in the Fukushima Dai-ichi NPP. (Source: Reprinted with permission of TEPCO)

after analyzing the results, including radionuclide measurements of the spent fuel pool water sample (^{131}I and ^{137}Cs concentrations in the SFP water collected on 12 April 2011 were 220 and 93 MBq/m^3, respectively, which corresponded to an ^{131}I/^{137}Cs activity ratio of 34 at time of the reactor cease), most of the fuel inside of the pool appears to be in sound conditions (METI, 2011). It is presumed that systematic mass damage to the fuel has not occurred. At 10 pm on 15 March, TEPCO was told to implement injection of water to the Unit 4 SFP. The spraying of

Figure 3.11 Damaged Unit 4 of the Fukushima Dai-ichi NPP after a hydrogen explosion on 16 March 2011. (Source: Reprinted with permission of TEPCO)

freshwater by Self-Defense Force's water spraying trucks began on 20 March and has been conducted periodically ever since (Fig. 3.12). Figure 3.13 shows the situation at Unit 4 during demolition works on 11 September 2011.

The NAIIC (2012) committee suggested that a major cause of safety storage of used fuels in the Unit 4 SFP was inflow of water from the reactor cavity to the used fuel pond via machine storage pit, which was filled with water during the planning cessation period of the reactor.

Figure 3.12 Damaged Unit 4 of the Fukushima Dai-ichi NPP after a hydrogen explosion on 22 March with water supply conducted by concrete pump truck. (Source: Reprinted with permission of TEPCO)

Figure 3.13 Unit 4 of the Fukushima Dai-ichi NPP on 11 September 2012. (Source: Reprinted with permission of TEPCO)

3.1.3.5 Unit 5

Unit 5 had been in outage for periodic inspection since 3 January 2011. On 11 March 2011, an RPV pressure leakage test had been conducted to the reactor loaded with nuclear fuel. At 3:40 pm on 11 March, as a result of the tsunami, two emergency diesel generators stopped their operation due to damage of the cooling seawater pumps and the power distribution panel by flooding, and all AC power was lost. Moreover, the damage of cooling seawater pump function led to the loss of the RHR function, resulting in a failure to transfer the decay heat to the sea, the ultimate heat sink. As for the reactor, the reactor pressure was increased to 7.2 MPa gage for the pressure leakage test. Then the pressure moderately rose because of the decay heat, and about 8 MPa gage of reactor pressure was maintained. On 13 March, water was successfully injected into the reactor using the condensate transfer pump of Unit 5, which was supplied power from the emergency diesel generator of Unit 6. Accordingly, after 5 am of 14 March, the reactor pressure and the water level were controlled by reducing pressure with the SRV along with repeated refilling of the reactor with water from the condensate storage tank (CST) through the condensate transfer pump. On 19 March, a temporary seawater pump was installed and started cooling, using the RHR system. Cold shutdown of Unit 5 was achieved at 2:30 pm on 20 March.

As of 11 March, 946 assemblies of spent fuel and 48 assemblies of fresh fuel were stored in SFP of Unit 5. Decay heat was calculated to be 1.01 MW on 11 March and 0.76 MW on 11 June. Due to the tsunami caused by the earthquake at 14:46 pm on 11 March, station blackout occurred and consequently seawater pump function was lost, and cooling function and makeup water function of SFP were lost. Water temperature of SFP tended to be increasing. However, increase of the water temperature was stopped at 68.8 °C since a temporary cooling facility started its operation in full scale at 5:00 am on 19 March. After that, SFP was maintained at a stable cooling status. Since transition to the reactor shutdown cooling (SHC) mode was carried out on 6 May and independent operation turned out to be possible on 25 June, more stable cooling condition of SFP was accomplished and finally water temperature in the SFP has been stabilized at around 30 °C.

3.1.3.6 Unit 6

Unit 6 had been in outage for periodic inspection since 14 August 2010. On 11 March 2011, the reactor, which was loaded with nuclear fuel, was in cold shutdown. At 3:40 pm, as a result of the tsunami, two emergency diesel generators (6A and 6H) stopped supplying power due to flooding of the cooling seawater pumps and the power distribution panel; however, another emergency diesel generator (6B), which was installed in the diesel generator building located at a relatively high site than the turbine building, remained operating. Therefore, there was AC power in Unit 6, although the function of cooling seawater pumps was lost. On 13 March, water was successfully injected into the reactor using the condensate transfer pump with power from the emergency diesel generator. Thus, after 14 March, the reactor pressure and water level were controlled by depressurization by SRV along with repeated refilling of the reactor with water from the CST through the condensate transfer pump. On 19 March, a temporary seawater pump was installed to activate the RHR system. The reactor could reach the cold shutdown at 7:27 pm on 20 March.

As of 11 March, 876 assemblies of spent fuel and 64 assemblies of fresh fuel were stored in the SFP of Unit 6. Decay heat was calculated to be 0.87 MW on 11 March and 0.73 MW on 11 June 2011. The tsunami caused loss of function of the seawater pump for cooling the SFP, although an emergency diesel generator (6B) still survived. As a result, cooling function and makeup water function of SFP were lost. The water temperature of SFP was increasing. However, increase of the water temperature was stopped at 67.5 °C of the highest value since a temporary cooling facility started its operation in full scale at 10:00 pm on 19 March. After that SFP was maintained at a stable cooling status. Transition to the SHC mode was carried out on 6 May, and because of the effect of an increase in air temperature, the water temperature has been stabilized between 30 °C and 50 °C.

The NAIIC (2012) committee evaluated operational problems related to the FNPP1 accident as follows: "There were many problems with on-site operation during the accident. Events make it clear that if there are no response measures for a severe accident in place, the steps that can be taken on-site in the event of a station blackout are very limited. Recovery work, such as confirming the operation of the isolation condenser (IC) in Unit 1, should have been conducted swiftly because of the loss of DC power, but was not. TEPCO did not plan measures for the IC operation, and had no manual or training regimes, so these were clearly organizational problems. Regarding the vent line composition, conducting line configuration work in a situation with no power and soaring radiation levels must have been extremely difficult and time consuming. On top of this, sections in the diagrams of the severe accident instruction manual were missing. Workers not only had to work using this flawed manual, but they were pressed for time, and working in the dark with flashlights as their only light source. The Kantei's (Prime Minister's Office) distrust of TEPCO management was exacerbated by the slow response, but the actual work being done was extremely difficult. Many layers of the security system were breached simultaneously, and the power to four reactors was lost at the same time. Had there not been coincidental events—such as the RCIC in Unit 2 operating for so many hours, the blowout panel failing out and releasing pressure, and the speed with which subcontractors cleaned up wreckage—Units 2 and 3 would have been in an even more precarious situation. We have concluded that—given the deficiencies in training and preparation—once the total station blackout occurred, including the loss of a direct power source, it was impossible to change the course of events".

3.2 Management of the Aftermath of the Accident

3.2.1 Protective Actions

Just after the loss of all AC power in Unit 1, the TEPCO informed the responsible regulators of the FNPP1 incident at 3:42 pm on 11 March, according to the Article 10 notification, corresponding to "site emergency action level" (IAEA, 2007), and the NISA acted according to the Act of Special Measures Concerning Nuclear Emergency Preparedness. At 4:36 am, about 1 h later, the TEPCO informed the responsible regulators of the FNPP1 incident, according to the Article 15 notification, corresponding to the "general emergency" situation (IAEA, 2007). The Japanese Government announced Declaration of a Nuclear Emergency Situation at 7:03 pm, which was a necessary step in launching the emergency response. The Precautionary Action Zone (PAZ) (IAEA, 2007), where urgent protective actions were taken in the event of a nuclear or radiological emergency to reduce the risk of severe deterministic effects off the site, was not legally established in Japan, although the introduction of the PAZ has been described as an example in the guidelines of the Nuclear Safety Committee (NSC) (NSC, 2010). The Nuclear Emergency Response Headquarters (NERH) of the Fukushima Local Government decided to evacuate the residents within 2-km zone from the FNPP1 at 8:50 pm. The Prime Minister (the Director General of the NERH) gave instructions to mayors of cities and villages for evacuation of the residents within the 3-km zone of FNPP1 and sheltering within 10-km zone from the FNPP1 at 9:23 pm. The Director General of the NERH extended the evacuation area from 3 km to 10 km from the FNPP1 site at 5:44 am on 12 March to avoid increasing radiological effects for the residents. After the explosion of Unit 1, the Director General of the NERH extended the evacuation area from 10 km to 20 km of the FNPP1 site at 6:25 pm, with restricted access to the evacuated area. At 11 am on 15 March, the Director General of the NERH ordered the evacuation of members of the emergency center near the FNPP1 and shelter in place for the residents in a zone between 20 km and 30 km from the FNPP1 site. Since the order continued for several weeks, the residents in the 20- to 30-km zone suffered greatly from lack of communications and necessities. As a result, the shelter-in-place order was then revised to voluntary evacuation. A total of 146,520 residents were evacuated as a result of the government evacuation orders.

After the major deposition of radioactivity, IAEA (2007) recommended implementation of urgent preventive actions according to operational intervention level (OIL) based on monitoring data. On 22 April, the government established the area where the annual cumulative dose might reach 20 mSv since the start of the accident, as the Deliberate Evacuation Area, as shown in Fig. 3.14. The residents in this area have almost completed evacuation to date. The Evacuation-Prepared Area in Case of Emergency, in which shelter in place and/or evacuation was required in case of emergency, was established on 22 April. The commission (NAIIC, 2012) evaluated evacuation issues as follows: "The central Government was not only slow in informing municipal governments about the nuclear power plant accident, but also failed to convey the severity of the accident. Similarly, the speed of information in the evacuation area varied significantly depending on the distance from the plant. Specifically, only 20 percent of the residents of the town hosting the plant knew about the accident when evacuation from the 3 km zone was ordered at 9:23 pm on the evening of 11 March. Most residents within the 10 km zone learned about the accident when the evacuation

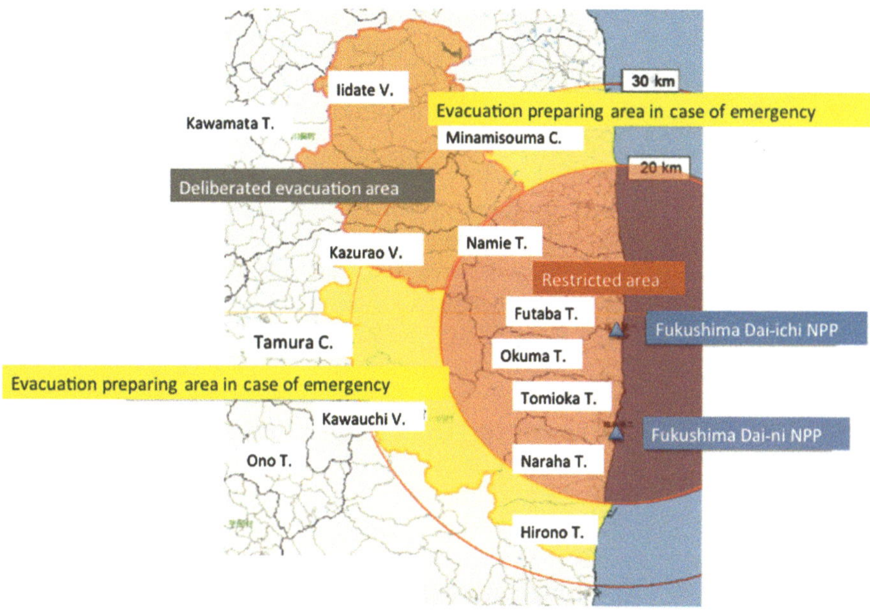

Figure 3.14 Scheme of the evacuation zone (restricted area: within 20 km of the FDNPP), deliberated evacuation area including Namie town and Iidate village, and the evacuation preparing area in case of emergency for the Fukushima Dai-ichi NPP accident. (Modified from PMJC, 2012)

order was issued at 5:44 am on 12 March, more than 12 hours after the Article 15 notification—but received no further explanation of the accident or evacuation directions. Many residents had to flee with only the barest necessities and were forced to move multiple times, sometimes to areas with elevated radiation levels. There was great confusion over the evacuation, caused by prolonged shelter-in-place orders and voluntary evacuation orders. Some residents were evacuated to high dosage areas because a radiation monitoring information was not provided. Some people were evacuated to areas with high levels of radiation, receiving no further evacuation orders until April. The commission had verified that there was a lag in upgrading unclear emergency preparedness and complex disaster countermeasures, and attributes this to regulators' negative attitudes toward revising and improving existing emergency plans" (NAIIC, 2012).

The Emergency Response Support System (ERSS) and the System for Prediction of Environmental Emergency Dose Information (SPEEDI) (Chino et al., 1984), whose data were automatically sent to the off-site emergency center near each NPP, have been constructed to provide urgent information and to implement more adequate protective actions to a nuclear accident. The SPEEDI had been operated by the Ministry of Education, Culture, Sports, Science and Technology (MEXT). In the case of the FNPP1 accident, however, information from the ERSS and SPEEDI was never used for protective actions such as evacuation of the residents because the regulatory office (NISA) could not get any data from the ERSS due to the station blackout. The SPEEDI calculations could not be performed using the ERSS data, although the visual display units of the SPEEDI are located in MEXT, NISA, NSC, the Fukushima Local Nuclear Emergency Response Headquarters (LNERH) and the Fukushima prefectural government office. On 23 March, the NSC released a map of cumulative radiation doses calculated using the SPEEDI system, and the data on total emissions of major radionuclides, estimated from the monitoring data and the dispersion model (Chino et al., 2011). High radiation levels exceeding 20 mSv/years spread in the northwest direction, more than 30 km from the FNPP1 site, as shown in Fig. 3.15.

The predictive mechanism based on the decision support system such as SPEEDI has been introduced in Japanese nuclear emergency protective system; however, IAEA (2007) has pointed out that "The decision support systems, including those using computer models, may not be able to predict the size and timing of a radioactive release (the source term), the movements of plumes, deposition levels or resulting doses sufficiently quickly or accurately during an emergency that they could provide the sole basis for deciding on initial urgent protective action. This is particularly true for emergencies, for which protective actions must be initiated before or shortly after a release, to be effective or for which a release is by an unmonitored pathway. Consequently, for such emergencies, immediate protective actions should be implemented out to predetermined distance from the

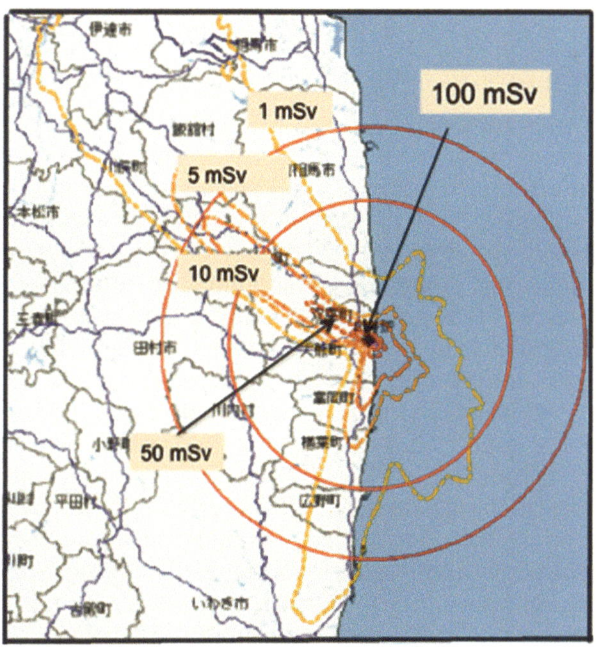

Figure 3.15 Cumulative effective dose during the period 12 March to 5 April calculated by the SPEEDI. (Modified from NSC, 2011)

facility in all directions when severe conditions are detected in the facility".

The NAIIC (2012) commission concluded on the evacuation issues: "...that residents' confusion over the evacuation stemmed from the regulations' negligence and failure over the year to implement adequate measures against a nuclear disaster, as well as a lack of action by previous governments and regulators focused on crisis management. The crisis management system that existed for the Kantei and the regulators should protect the health and safety of the public, but it failed in this function".

After the FNPP1 accident, about 850 patients were hospitalized in seven major hospitals, which have been built within the 20-km zone from the FNPP1. The NAIIC commission (2012) documented that at least 60 severely sick patients died during the evacuation course until the end of March 2011. This finding suggests that the evacuation itself has a considerable health risk due to the transport of seriously sick patients at worse living circumstances.

3.2.2 Emergency Radioactivity Monitoring

In emergency exposure situations, depending on the severity of an accident, all three types of radiation monitoring, source monitoring, environmental monitoring and individual monitoring, should be carried out. The overall strategy for the emergency monitoring should be preplanned to address the needs of assessors, decision makers and responders over time and geographical location (IAEA, 2005).

The emergency radioactivity monitoring, just after the accident, should be carried out by TEPCO for source monitoring, and by the Fukushima Nuclear Emergency Center (FNEC) for environmental and individual monitoring. The center of accident response (Off-Site Center) was built in Okuma town, located at about 5 km of the FNPP1, where the LNERH was initially installed. However, 23 monitoring posts of 24, which automatically measure the ambient dose rate, failed due to the earthquake and the resulting tsunami, so the FNEC lost the role of the

Off-Site Center due to the loss of electricity and communication. At 3:20 am on 12 March, the function of the Off-Site Center restarted, after its emergency power supply was recovered, and satellite communication system was established. However, the emergency response mechanisms, such as information on the plant conditions, ERSS, SPEEDI and provision of monitoring results were not available in the Off-Site Center. Due to the progress of nuclear emergency and lack of logistics (such as fuel, food and other necessities), it was hard to operate the protective actions such as the LNERH. On 15 March, the LNERH was moved to the Fukushima Prefectural Building situated in Fukushima City, about 60 km far from the FNPP1.

Source monitoring is an important issue in order to implement urgent protective actions for the nuclear emergency (IAEA, 2005). The monitoring posts installed in the FNPP1 are shown in Fig. 3.16. Although just after the earthquake no anomalous radionuclide concentrations were observed in the air stack monitors of each Unit of the FNPP1, the station blackout caused cease of the operation of the sampling system to measure radionuclides, monitoring systems to measure dose rates, and to carry out meteorological observations. Therefore, initially there was no information about amounts of radionuclides emitted from FNPP1, although meteorological data were obtained using a monitoring car. TEPCO was not able to measure radioactivity at the monitoring posts at FNPP1 due to the loss of external power supply on 11 March, except measurements of the ambient dose rates, which were carried out using a monitoring car, results of which are shown in Fig. 3.5. On 19 March, TEPCO started radionuclide measurements in surface air samples, which were collected at the west gate of FNPP1 (Fig. 3.16) using air sampler in the monitoring car. On 23 March, TEPCO restarted radiation measurements using three temporary monitoring posts (Fig. 3.16).

On 15 March, the MEXT, which is responsible for environmental radioactivity monitoring in Japan, started measurements of ambient dose rate (in cooperation with the Japan Atomic Energy Agency (JAEA)). The radioactivity monitoring was conducted

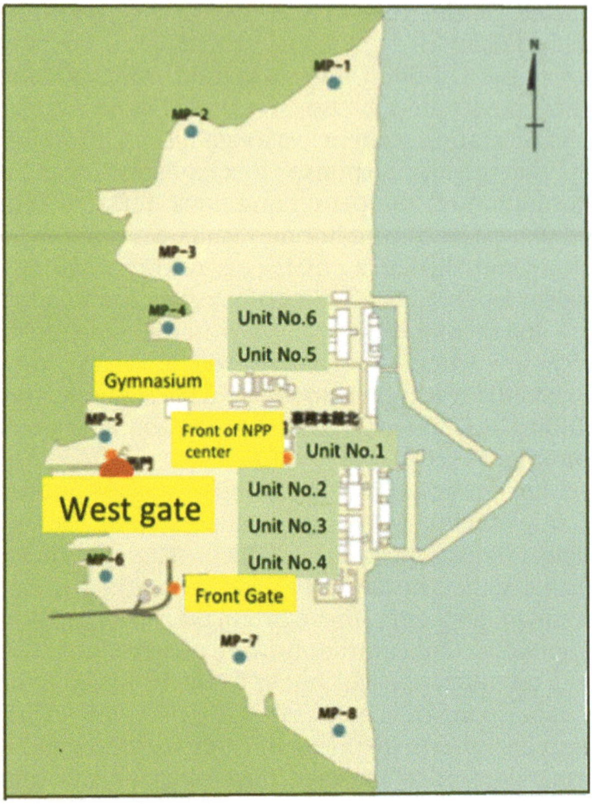

Figure 3.16 Monitoring posts (from MP-1 to MP-8) at the Fukushima Dai-ichi NPP. The temporary monitoring sites are shown by brown dots. (Modified from TEPCO, 2011)

using up to 15 monitoring cars in the area outside of the 20-km zone from the FNPP1. For the measurements of the ambient dose rate, a Geiger–Müller counter, an ionization chamber and a NaI scintillation counter were used. Since March 23, daily cumulative doses were monitored using portable-type dose counters at 15 sites. In the evacuation zone within the 20-km zone of FNPP1, there was no available environmental monitoring system, except the site monitoring carried out since 11 March. MEXT has measured the ambient dose rate and activities of radionuclides in soil samples collected in the evacuation zone in cooperation with electric power companies. This information on radioactive contamination in the evacuation zone was an important issue to assess temporary return of the evacuated residents. A map of cumulative ambient dose integrated throughout March 2011 was depicted from the monitoring results of the ambient dose rate as shown in Fig. 3.17. The map revealed that the area with the high cumulative ambient dose of more than 100 mSv spread northwest of FNPP1, whose distribution almost coincided with the SPEEDI simulation (Fig. 3.15). This map was used to establish the "Deliberated Evacuation Area" and to lift the "Evacuation-Prepared Area in Case of Emergency" on 9 August 2011 (Fig. 3.14).

Airborne monitoring (aerial monitoring system) is a powerful tool to depict highly radioactivity-contaminated area. Early campaign of airborne monitoring, which surveyed most of the area within

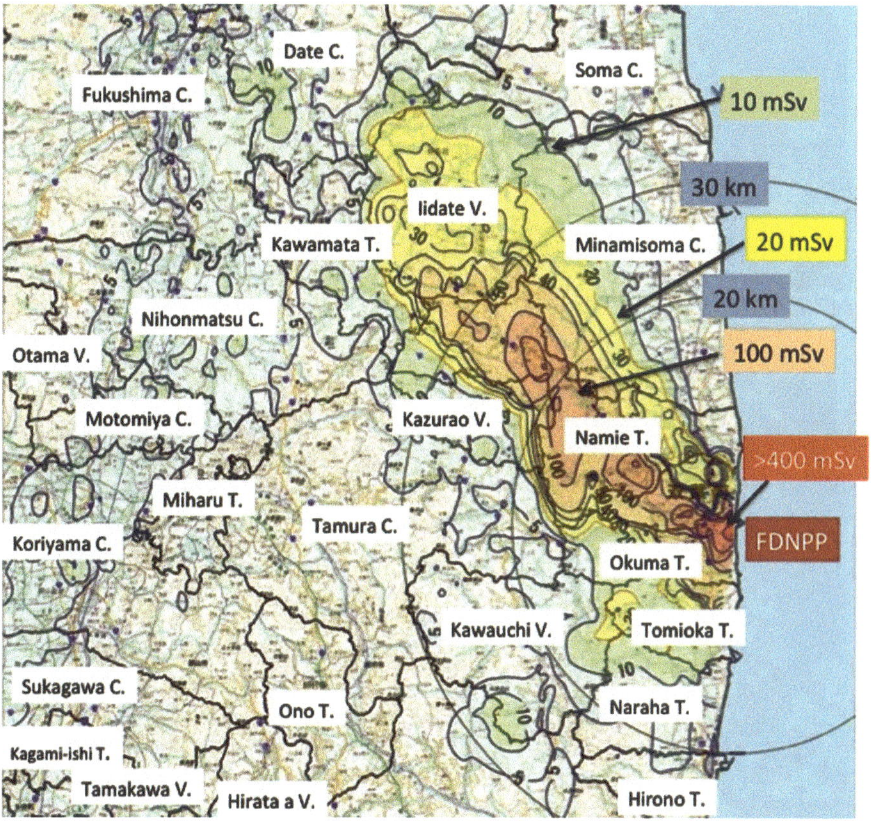

Figure 3.17 Mapping of the cumulative ambient dose, integrated throughout March 2011. (Modified from METI, 2011)

30 km of the FNPP1, was conducted by the US Department of Energy/National Nuclear Security Administration (US DOE/NNSA) from 17 to 19 March (Lyons and Colton, 2012). Preliminary results, published on the DOE web site on 22 March, showed highly contaminated areas northwest of FNPP1, which corresponds to Namie town, Iidate village and Minamisoma Country. MEXT and US DOE conducted the first airborne monitoring in the area within 80 km from the FNPP1 in the period from 6 to 29 April (MEXT, 2011a), results of which are shown in Fig. 3.18. For the airborne monitoring, the ambient dose rate was measured using NaI scintillation counters carried by aircrafts at heights between 150 and 300 m. The ambient dose rate at a height of 1 m, and the cumulative amounts of radionuclide depositions on land surface, were estimated from the airborne monitoring data. Mapping of the radioactivity-contaminated area was effective for establishing the Planning Evacuation Zone. In the period from April 2011 to March 2012, several campaigns of airborne monitoring were conducted by MEXT, from which radioactivity contamination maps were constructed of most of the Japanese mainland, including Honshu, Hokkaido, Kyushu and Shikoku Islands.

On 18 March, MEXT (2011b) started measurements of activity concentrations of gamma-emitting radionuclides in environmental samples, including dust, deposition, soil, grass, and pond waters in areas outside of 20 km from the FNPP1. Daily air dust and deposition samples were measured at 46 monitoring stations all over Japan. The dust sampling in surface air in Fukushima prefecture was performed with a portable dust sampler using monitoring cars, in which sampling times at each site were from 10 to 20 min. The dust sampling in the upper atmosphere was conducted by the Ministry of Defense since 24 March, to elucidate vertical distributions of the FNPP1-derived radionuclides. Measurements of activity concentrations of radionuclides in the environmental samples were done with gamma-spectrometers with Ge detectors. The counting time at the emergency stage was from 1000 to 3600 s.

On 23 March, MEXT started marine radioactivity monitoring, in which activity concentrations of

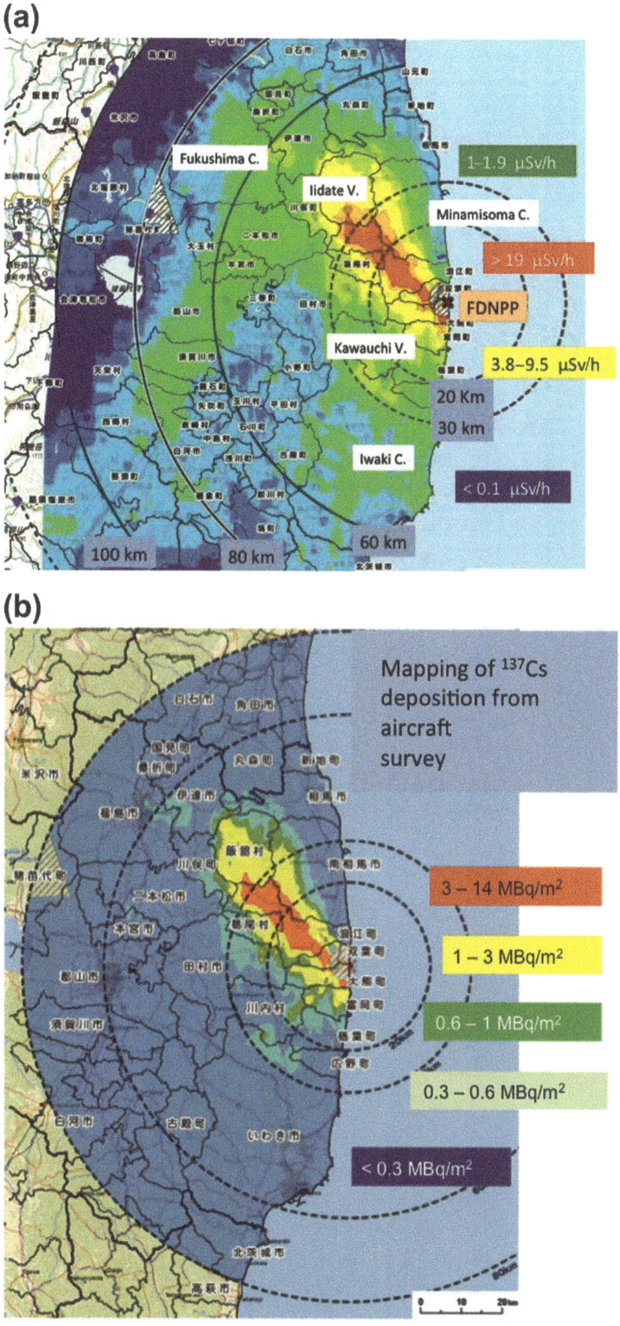

Figure 3.18 Mapping of ambient dose rate (1 m above the ground) (a) and cumulative ^{137}Cs deposition (b) obtained by the first airborne monitoring survey. The cumulative ^{137}Cs deposition was calculated from observed ambient dose rate by using conversion factors. (Modified from MEXT, 2011)

gamma-emitting radionuclides in seawater, bottom sediments and dust samples collected on shipboard were determined. Seawater and sediment samples were collected near the coast of the Fukushima and Ibaraki prefectures and in the sea area 30 km off the coast using the research vessel of the Japan Agency for Marine-Earth Science and Technology (JAMSTEC). Since 21 March, TEPCO measured the activity concentrations of gamma-emitting radionuclides in seawater and bottom sediment samples collected from the coastal area within 20 km from the Fukushima Dai-ichi and Dai-ni NPPs. After the discharge of stagnant water in the accident reactors in early April, MEXT announced to increase the monitoring sea area and sampling points on April 5. According to "the plan to step up environmental monitoring" constructed by the government NERH on 25 April, the MEXT showed enlarged sea area of radioactivity monitoring on 25 April. Seawater samples (0.5 l) were collected at three layers of each sampling point (surface, middle layer and bottom layer) using a CTD (Conductivity Temperature Depth profiler) Rosette sampler. Radioactivity measurements in dust and seawater samples collected on board were carried out by JAEA with gamma-spectrometer comprising a Ge detector.

3.2.3 Monitoring and Protective Actions on Foodstuffs and Drinking Water

3.2.3.1 Agricultural Foodstuff and Drinking Water

Radiological contamination of agricultural foodstuffs and drinking water was of concern after large atmospheric emissions of the FNPP1-derived radionuclides. Hamada and Ogino (2012) documented response of the Japan Government to food safety regulations after the nuclear accident. The NSC issued guidelines for nuclear disaster countermeasures in June 1980, where "index values" were provided as evaluation criteria to launch discussion on whether NERH needs to restrict food consumption or not (NSC, 1980). Since then, the index values have been revised three times until 2010. In 1980, the initial index value for ^{131}I was set to not exceed a total dose equivalent to the thyroid gland of 15 mSv when infants take contaminated milk, drinking water and leafy vegetables. The NSC assumption was based on provisional nuclear accidents such as the Three Mile Island accident in the United States, in which only ^{131}I and radioactive noble gases were the environmental emissions.

After the Chernobyl accident (1986), the index value for ^{131}I was changed in 1998. New index values (in which the protective action level of provisional effective dose was 5 mSv/years) were established for radiocesium, plutonium and other alpha emitters. Following the JCO (Japan Nuclear Fuel Conversion Co.) criticality accident (1999), the index value of uranium was added in 2005. After the FNPP1 accident, the Ministry of Health, Labor and Welfare (MHLW) notified the provisional "guideline values for food and drink intake restrictions" (Table 3.1) to each prefectural government on 17 March, which were based on the technical advice of the NSC. Any foodstuff and drinking water containing radioactivity exceeding regulation values were prohibited from consumption according to Item 2, Article 6 of the Food Sanitation Law.

Monitoring of radionuclides in agricultural foodstuffs and drinking water (tap water) was carried out by prefectural governments. MHLW gathered the monitoring data and released them to the public via the MEXT webpage. On 19 March 2011, MHLW notified the local government authorities and regional water suppliers that tap water contaminated above the provisional regulation value should not be taken, but that it can be drunk even by infants if its substitute is unavailable. This notice also indicated that use of contaminated water in daily life (e.g. for washing hand, bathing and laundry) is safely allowable because the estimated dose may be of the order of nanosievert. In late March 2011, high ^{131}I concentrations in tap water were measured in Fukushima, Ibaraki, Chiba, Tokyo, Tochigi and Saitama, where the maximum ^{131}I concentrations were 960, 300, 370, 210, 140 and 120 Bq/kg, respectively. Restrictions of tap water consumption for infants were conducted in each local government; the duration of restriction was several days except for the

Table 3.1. The Provisional Regulation Values for Protective Actions of Food Intake

Foodstuff Name	^{131}I (Bq/kg)	Radiocesium ($^{134}Cs + ^{137}Cs$) (Bq/kg)
Provisional Regulation Value (March 2011–March 2012)		
Drinking water	300 (100 for infants)	200
Milk	300 (100 for infants)	200
Vegetables	2000	500
Crops		500
Meat, egg, fish, others	2000	500
New Standard Limits (April 2012–)		
Drinking water		10
Milk		50
General foods		100
Infant foods		50

Fukushima prefecture (the maximum duration was 51 days). In Fukushima and Ibaraki prefectures, 1190 and 1700 Bq/kg of ^{131}I were detected in unprocessed raw milk collected on 16 and 19 March, respectively.

Radionuclide monitoring revealed that their levels in vegetables, typically for spinach, in Ibaraki and Fukushima prefectures exceeded the provisional regulation values on 18 and 21 March, respectively. On 21 March, since the foodstuffs (vegetables and raw milk) having radioactivity exceeding the provisional regulation values were considered to spread to wide areas, the Director General of the NERH provided instructions to restrict consumption of food, agricultural products and drinking water to relevant governors of the prefectures, according to advice of the NSC. The Ministry of Agriculture, Forestry and Fisheries issued guidelines on disposal of radioactivity-contaminated vegetables and raw milk, based on technical advice of the Emergency Technical Advisory Body of the NSC on 25 March, 26

April and 6 May. After setting the provisional regulation values, the NERH reviewed the monitoring plan and mechanism to lift restriction of consumption of food and drinking water. On 4 April, the NERH issued the following guidelines on consumption of food and its monitoring in restricted areas:

1. The production-restricted area should be prepared as a unit of local government.
2. The radioactivity monitoring of food in the production-restricted area should be done weekly, and the restrictions should be lifted in the case when the monitoring data continue to be below the provisional regulation values for three successive times.

On 4 April, MHLW provided more general safety measures about radioactivity contamination of tap water by developing the "future monitoring policy on radioactive materials in tap water" with the aim to enhance the monitoring of tap water in Fukushima prefecture. After 8 April, the restrictions on the consumption of radioactivity-contaminated foods including vegetables and raw milk were lifted.

The MHLW notified the standards for radioactive materials in feed to ensure that the value in produced meat and milk does not exceed the provisional regulation values stipulated in the Food Sanitation Act on 14 April. After that, the MHLW notified a guideline on the production and utilization of feed for avoiding radioactive contamination of cattle beef and milk. On 8–9 July 2011, however, concentrations of radiocesium (^{134}Cs and ^{137}Cs) exceeding the provisional regulation levels of the Food Sanitation Act were found in the beef cattle of 11 heads shipped from Minamisoma city, Fukushima prefecture (PMJC, 2011b). Investigation of radiocesium contamination rates in cattle indicated that the cattle were shipped from Asakawa town, Fukushima prefecture. They were fed with rice straw containing a high concentration of radiocesium on 14 July. In July 2011, radiocesium exceeding provisional regulation values was detected in beef shipped from Iwate, Miyagi and Tochigi prefectures in addition to the Fukushima prefecture. The NERH notified shipping restrictions to cattle fed rice straw within the Fukushima prefecture on 19 July, within Miyagi on 28

July, within Iwate on 1 August and within Tochigi on 2 August. On 19 August, the NERH temporally lifted restrictions on the transfer of cattle fed in the Miyagi prefecture to other prefectures and their shipment to slaughterhouses. On 25 August, the restrictions on the transfer of cattle fed in Fukushima, Iwate and Tochigi prefectures to other prefectures and their shipment to slaughterhouses were also lifted. Since beef with radiocesium exceeding the provisional regulation values was commercially distributed, and a dominant source of radiocesium-contaminated beef was contaminated rice straw feed, the following steps were taken under the Act on Special Measure Concerning Nuclear Energy Preparations: all beef from the prefectures of Fukushima, Miyagi, Iwate and Tochigi was restricted before entering commercial channels until radiocesium in beef animals and products were proven not to exceed the provisional regulation values.

Since the Fukushima accident occurred before planting of rice, which is a major food in Japan, the NERH issued guidelines about rice planting to radioactivity-contaminated fields according to the technical advice of the NRC on 8 April. On 22 April, the Director General of the NERH provided instructions about rice planting restrictions to relevant governors, in which a regulation value in surface soil of rice field was established not to exceed 5000 Bq/kg.

3.2.3.2 Marine Food

At the time of the Fukushima accident, there were no guidelines about provisional regulation levels on radioactive iodine contamination of fishery products. On 1 April, a high ^{131}I concentration of 4080 Bq/kg was measured in fish (Konago: juvenile sand lance, *Ammodytes personatus*) sampled offshore Ibaraki prefecture. After the detection of high concentrations of radioiodine in fishery products, the MHLW decided to use the same provisional regulation values of radioiodine for fishery products as applied for vegetables, referring to a technical advice from the NRC.

On April 2012, the MHLW (MHLW, 2012) lowered the provisional regulation value of radiocesium for internal exposure from food intake as shown in Table 3.1. According to this provisional regulation value, restriction of fish catchment and mushroom collection production has been continued in Fukushima and Ibaraki prefectures in 2012. Some public emotionally responded to this change; lower the concentration of radiocesium in food, more safe it is for health. Producers of some foods (fish, rice and vegetables) supplied foods with lower regulation values, but consumers required foods with zero concentrations of radiocesium.

3.2.4 Protective Actions on Radiation Exposure of Individuals

3.2.4.1 Protective Actions to Residents

Atmospheric emission of large amounts of radionuclides and explosions in the FNPP1 resulted in radiological concerns for workers, evacuated residents and members of the public. In the case of radioiodine (^{131}I, ^{132}I and others), which is the major radionuclide emitted during a nuclear reactor accident, a major concern is the internal radiation exposure following incorporation and uptake of radioiodine in the thyroid. This will occur through inhalation of contaminated air and ingestion of contaminated food and drink.

The IAEA (2007) suggested protective actions for the public for general emergency involving a nuclear reactor, including a provision of stable iodine for thyroid blocking of residents within the PAZ and UPZ (urgent protective action planning zone), and monitoring evacuated people and determining whether decontamination or medical treatment is needed.

In the case of the Japanese radiation protective system, stable iodine has been stored in the offices of cities and villages. The Fukushima Local Government distributed about 1.51 million of iodine pills and about 6.1 kg of iodine powder to cities and villages within the 50-km zone from the FNPP1. In the process of evacuation since 12 March, elevated radiation exposure was of concern to evacuated residents due to hydrogen explosion at Unit 3, followed by increase in the ambient dose rate at the

FNPP1 site. On 16 March, the Director General of the Local NERC provided instructions that the Governors of Fukushima Local Government should urgently deliver stable iodine to the evacuated residents, taking into account the technical advice of the NSC on 13 March, which recommended that stable iodine be preferentially administered to residents remaining in the evacuation zone. Intake of stable iodine should be implemented in the presence of medical experts because side effects such as gastrointestinal effects or hypersensitivity reactions cannot be ignored, although the risk of severe side effects from a single administration of stable iodine is very low, about 10^{-7} (WMO, 1999). Although intake of stable iodine was implemented to residents of Tomioka, Futaba, Okuma and Miharu towns, actually, most of the evacuated residents could not take stable iodine because the evacuation had already been completed at this time. Stable iodine prophylaxis with the aim to reduce the dose to the thyroid should be applied promptly because the intake of radioiodine by inhalation begins when the radioactive cloud arrives at a location (WMO, 1999).

In emergencies involving airborne releases, prompt assessment of the external contamination of individuals with radionuclides is required, which might be useful for a first screening to determine whether they are candidates for rigorous surveillance (e.g. by internal dosimetry or medical inspection). Individual measurements for evacuated residents started on 12 March. However, initially it was very difficult to measure several tens of thousands people at the screening level of 13,000 cpm. On 13 March, the Fukushima Local Government decided a tentative screening level of 100,000 cpm for decontamination of the whole body. A partial decontamination of individuals by wiping would be performed for screening levels between 13,000 and 100,000 cpm, based on suggestions of experts in radiation medicine dispatched by the MEXT, medical experts from the National Institute of Radiological Sciences (NIRS), and guidelines of the Fukushima Medical University. On 19 March, the NSC applied the screening level of 10,000 cpm for decontamination of contaminated residents. In this connection, the IAEA (2009) recommended the revised screening level of 1 μSv/h (dose rate at a distance of 10 cm) as OIL for decontamination of body surface for general residents.

In order to elucidate radioactivity contamination of the public, the Fukushima Local Government, in cooperation with the LNERH, implemented individual monitoring of residents in the prefecture involving people in the evacuation area within 20 km from the NPP site. Most of the 195,354 people monitored until 11 May were found to be below the screening level of 100,000 cpm. Decontamination was performed on 102 people, showing larger values than the adopted screening level (PMJC, 2011a).

Higher internal dose was of concern in more than 100 residents in the evacuated zone (52 residents) and Namie Town, Iitate Village, and part of Kawamata Town (122 residents). As a part of the preliminary inspection of the Health Management Survey for the residents in Fukushima prefecture, therefore, the internal exposure dose was measured by the whole body counter (WBC) and the bioassay method using urine at the NIRS from the end of June. Regarding the internal exposure dose to ^{134}Cs and ^{137}Cs, the depository effective dose had been assessed to be as low as less than 1 mSv in all the checked residents. Also, measurement of the internal contamination dose at JAEA was begun in July, and measurement in about 3200 residents in the target area was completed by the end of August.

3.2.4.2 Protective Actions to Children

In the period 26–30 March 2011, the LNERH with the Fukushima local government implemented individual monitoring of children on thyroid exposure in Iwaki and Kawamata cities and Iidate village in order to have better understanding of the current precious exposure dose and possible stochastic effects to children who have higher risk of thyroid cancer. In this case, individual monitoring to children, who have been staying in shelter-in-place or corresponding to higher equivalent doses of thyroid estimated by SPPEDI, was conducted. A technical advice on the method of individual monitoring was provided by the NSC. The individual monitoring of 1080 children aged

0–15 years was conducted for thyroid exposure. There was no child who exceeded the screening level of 0.2 μSv/h (equivalent to the thyroid equivalent dose of 100 mSv for a 1-year-old infant).

3.2.4.3 Protective Actions to Emergency Workers

In the case of normal authorized operation of the FNPP1, the dose limit for workers is legally regulated at an effective dose of 100 mSv within 5 years, and 50 mSv per year, based on the International Commission on Radiological Protection (ICRP) 1990 Recommendation (ICRP, 1991). In the emergency situation, the dose limit was set at 100 mSv. The Fukushima nuclear accident needed to revise the dose limit for radiation workers to prevent further worsening of the nuclear disaster. On 14 March, the effective dose limit was elevated to 250 mSv in the period from the date of “the Declaration of Nuclear Emergency” to “the Declaration of Cancellation” (PMJC, 2011a). In this connection, ICRP (1991) stipulated a dose limit of 500 mSv for persons voluntarily engaged in emergency rescue operations in order to avoid definite impact. The MHLW and METI consulted the Radiation Review Council in MEXT on justification of this revised regulation value, which stated that the revised regulation value is applicable for the FNPP1 accident.

Individual dosimeters such as alarm pocket dosimeter (APD), in which the external radiation dose of the APDs was read at the end of the daily work and automatically recorded and stored in an integrated file, were used for measurements of individual external gamma dose for emergency workers following a severe nuclear accident. Since the radiation exposure control system of the FNPP1 workers was damaged by the tsunami and the levels of radiation and radioactivity contamination at FNPP1 increased, it was stipulated that all these tasks were performed in an integrated manner at TEPCO’s response headquarters established in the “seismic isolation building”. The radiation control of the emergency workers, who operated at the FNPP1 site without passing through the “seismic isolation building”, has been performed by the radiation exposure control system in the “J Village” soccer practice facility located about 20 km south of FNPP1. The radiation dose records at J Village had been calculated manually after the FNPP1 accident, but in early June, TEPCO introduced a dose control system that adopted a personal authentication system using bar codes, following the introduction of this same system in the "seismic isolation building" in the middle of April. Since 29 July 2011, the management of individual radiation dose has been thoroughly strengthened by checking worker IDs and issuing permits to enter the site of the FNPP1 for workers who have stayed in J Village. In this case, photo identification system as personal identification, which has been introduced before the accident, has been used. On the other hand, as for some of the employees who were working in the site at the time of the FNPP1 accident, due to the fact that information on their exposure dose is unknown, an investigation team was established, and this team has been conducting the hearing of the employees who are presumed to have been engaged in the same works with those related employees, based on the name of companies such as primary contractors etc. as well as on the time of their entries. The investigation was envisioned to be completed by the end of September.

In order to assess internal exposure dose of the emergency workers, three in-vehicle WBCs (NaI scintillation detectors) were rented from JAEA, of which two were set up at Onahama Harbor in Fukushima prefecture and one was deployed for taking measurements around the Tokyo metropolitan area. In July 2011, one of the in-vehicle WBCs was moved from Onahama to the Hirono Town Soccer Ground (HTSG) adjacent to J Village, which is a major entrance point into the FNPP1 as well as the 20-km zone. A stationary WBC (NaI scintillation detector) was newly installed, and one stationary WBC (plastic scintillation detector) was moved from Fukushima Dai-ni NPS. Thus, three WBCs in total were in operation at the HTSG. In August 2011, three stationary WBCs (plastic scintillation detectors) were moved from FNPP1 to HTSG and, as a result, six WBCs were in operation.

About 7800 workers entered the FNPP1 site as emergency operators until 23 May, whose average dose was approximately 7.7 mSv (PMJC, 2011a).

About 167 workers were exposed to more than 100 mSv while dealing with the accident (NAIIC, 2012). Assessments of both external and internal exposure doses of approximately 9900 workers engaged in the emergency work at FNPP1 had been completed as of 10 August (METI, 2011). The maximum external exposure dose for all the emergency workers was less than 200 mSv. The external exposure doses of 39 workers in March 2011 were in the range of 100–200 mSv. On the other hand, the internal exposure dose (provisional value) of five workers exceeded 250 mSv in March 2011. The maximum internal exposure dose of individual, recorded in March, was 590 mSv. The monthly average values of the total exposure dose, which was given as the sum of the external and internal exposure doses for the emergency workers, were evaluated to be 22.4 mSv in March, 3.9 mSv in April, and 3.1 mSv in May, which showed a declining tendency. The maximum total exposure dose of individuals from March to May 2011 was 670.4 mSv, which occurred in March. The workers with doses exceeding 100 mSv were concentrated in March during the early period of the severe accident; the total number was 167 workers (146 TEPCO employees and 21 workers from the cooperative firms). The average cumulative effective dose of the emergency workers, which was given as an average value of the sum of external exposure from March to July and internal exposure from March to May, was evaluated to be 10.4 mSv; the maximum cumulative dose of individual was 672.27 mSv. As a result, it was confirmed that six people among emergency workers in March 2011 were exposed to radiation exceeding the dose limit of 250 mSv for emergency work.

The commission (NAIIC, 2012) evaluated public health and welfare issues related to the Fukushima accident as follows: "The Commission recognized that the residents in the affected area are still struggling from the effects of the accident. They continue to face grave concerns, including the health effects of radiation exposure, displacement, the dissolution of families, disruption of their lives and lifestyle and the radioactive contamination of vast area of the environment. There is no foreseeable end to the decontamination and restoration activities that are essential for rebuilding communities. The Commission concluded that the Government and the regulators are not fully committed to protecting public health and safety, and that they have not acted properly to protect the health of the residents and to restore their welfare".

References

Chino, M., Ishikawa, H., Kai, M., Honma, T., Hidaka, A., Imai, K., Iigima, T., Moriuchi, S., Asai, K., Nakamura, Y., Okuda, M., Horiuchi, K., 1984. SPEEDI: System for Prediction of Environmental Emergency Dose Information. Report JAERI-M 84-050. JAERI, Tokyo.

Chino, M., Nakayama, H., Nagai, H., Terada, H., Katata, G., Yamazawa, H., 2011. Preliminary estimation of released amounts of ^{131}I and ^{137}Cs accidentally discharged from the Fukushima Dai-ichi nuclear power plant into the atmosphere. J. Nucl. Sci. Technol. 48, 1129–1134.

Hamada, N., Ogino, H., 2012. Food safety regulations: what we learned from the Fukushima nuclear accident. J. Environ. Radioact. 111, 83–99.

IAEA, International Atomic Energy Agency, 2005. Environmental and Source Monitoring for Purposes of Radiation Protection. IAEA Safety Standards for Protecting People and the Environment. Safety Guide No. RS-G-1.8. IAEA, Vienna.

IAEA, International Atomic Energy Agency, 2007. Arrangements for Preparedness for a Nuclear or Radiological Emergency. IAEA Safety Standards for Protecting People and the Environment. Safety Guide No. GS-G-2.1. IAEA, Vienna.

IAEA, International Atomic Energy Agency, 2009. INES The International Nuclear and Radiological Event Scale User's Manual. IAEA, Vienna.

ICRP, International Commission on Radiological Protection, 1991. Recommendations of the International Commission on Radiological Protection. ICRP Publication 60. ICRP, Paris.

Lyons, C., Colton, D., 2012. Aerial measuring system in Japan. Health Phys. 102, 509–515.

METI, Ministry of Economy, Trade and Industry, 2011. Additional report of the Japanese government to the IAEA – the accident at TEPCO's Fukushima nuclear power stations – (second report). http://www.meti.go.jp/english/earthquake/nuclear/iaea/iaea_110911.html.

MEXT, Ministry of Education, Culture, Sports, Science and Technology, 2011a. Results of airborne monitoring in restricted areas and deliberate evacuation areas. http://radioactivity.mext.go.jp/ja/contents/4000/3708/24/1305819_0616_1.pdf; http://radioactivity.mext.go.jp/ja/contents/4000/3709/24/1305819_0616_2.pdf; http://radioactivity.mext.go.jp/en/list/203/list-1.html.

MEXT, Ministry of Education, Culture, Sports, Science and Technology, 2011b. Enlargement of national environmental radioactivity monitoring at emergency of the Fukushima Dai-ichi and Daini nuclear power plants. http://www.mext.go.jp/b_menu/houdou/23/03/1303843.htm.

MHLW, Ministry of Health, Labour and Welfare, 2012. New standard limits for radionuclides in foods. http://www.mhlw.go.jp/english/topics/2011eq/dl/new_standard.pdf.

NAIIC, Nuclear Accident Independent Investigation Commission, 2012. The official report of the Fukushima nuclear accident independent investigation commission. (Tokyo).

NPAJ, National Police Agency of Japan, 2012. Damage situation and police countermeasures associated with 2011 Tohoku distinct – off the Pacific Ocean earthquake. http://www.npa.go.jp/archive/keibi/biki/higaijyokyo_e.pdf.

NSC, Nuclear Safety Committee, 1980. Guideline for nuclear disaster measures. http://www.nsc.go.jp/shinsashishin/pdf/history/59-15.pdf.

NSC, Nuclear Safety Committee, 2010. Guideline for nuclear disaster measures, guideline of nuclear safety committee, Taisei-shuppan, Tokyo, pp. 1313–1544. (in Japanese).

NSC, Nuclear Safety Committee, 2011. http://www.nsc.go.jp.

PMJC, Prime Minister of Japan and his Cabinet, 2011a. Report of Japanese government to the IAEA ministerial conference on nuclear safety – the accident at TEPCO's Fukushima nuclear power stations. http://www.kantei.go.jp/jp/Topics/2011/iaea_houkokusho.html.

PMJC, Prime Minister of Japan and his Cabinet, 2011b. Government actions to ensure the safety of beef and other food. http://www.kantei.go.jp/foreign/kan/topics/201108/measures_beef.pdf.

Tanabe, F., 2012a. A scenario of large amount of radioactive materials discharge to the air from the Unit 2 reactor in the Fukushima Dai-ichi NPP accident. J. Nucl. Sci. Technol. 49, 360–365.

Tanabe, F., 2012b. Analysis of core melt and re-melt in the Fukushima Dai-ichi nuclear reactors. J. Nucl. Sci. Technol. 49, 18–36.

WMO, 1999. Guidelines for Iodine Prophylaxis Following Nuclear Accident Up-to-date 1999. Report WMO/SDE/PHE/99.6. WMO, New York.

4

RADIONUCLIDE RELEASES INTO THE ENVIRONMENT

CHAPTER OUTLINE

Radioactive materials were released to the environment from the Tokyo Electric Power Company (TEPCO) Fukushima Dai-ichi Nuclear Power Plant (hereafter FNPP1) as a result of reactor accidents caused by a total loss of electric power (black out) after the Tohoku earthquake and tsunami on 11 March 2011. Radioactive materials were emitted into the atmosphere and transferred to the land and ocean through wet and dry deposition (Stohl, 2011; Terada et al., 2012; Aoyama et al., 2012, 2013). In addition, highly contaminated fresh and seawater was directly released to the ocean. A schematic diagram of radioactive releases to the environment is shown in Fig. 4.1 for the site area, and in Fig. 4.2 for the Pacific Ocean.

Fukushima Accident. http://dx.doi.org/10.1016/B978-0-12-408132-1.00004-8

Figure 4.1 A schematic diagram of radionuclides releases into the atmosphere and the ocean from the Fukushima Dai-ichi NPP.

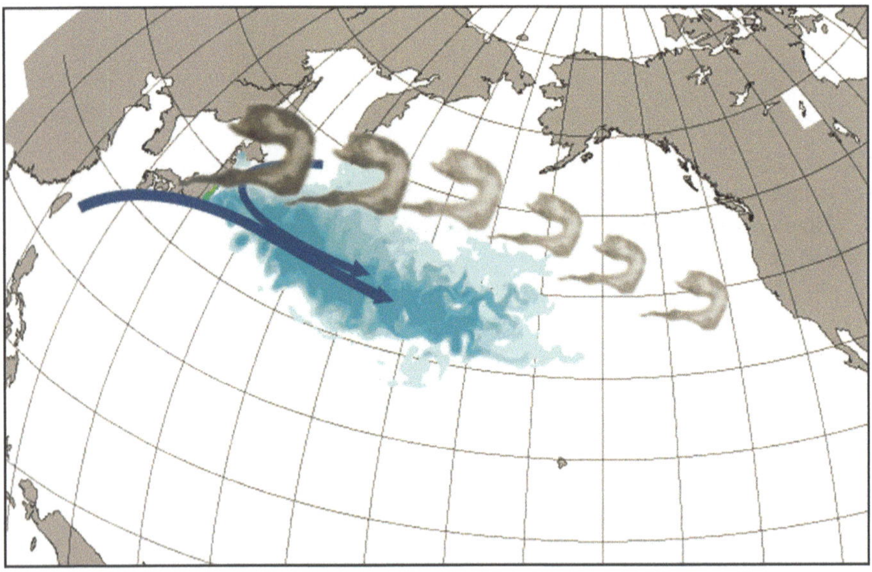

Figure 4.2 Transport of radionuclides released from the Fukushima Dai-ichi NPP by atmospheric deposition (in grey) and surface oceanic currents (in blue) in the Pacific Ocean.

Radionuclides released to the atmosphere were mainly transported to the east from the FNPP1 by low-pressure systems and strong westerly wind, typically occurring in the mid latitudes of the Northern Hemisphere. In the case of the ocean, the radionuclides were transported at mid latitudes to the east by ocean currents, mainly by the Kuroshio Current and the Kuroshio Extensions, as illustrated in Fig. 4.2. A detailed description of transport processes in the atmosphere and the ocean is presented in Chapter 5.

4.1 Atmospheric Releases of Radionuclides

It is important to estimate the source term of the radionuclides released during the accident for the assessment of radiological doses to the public, as well as for better understanding of atmospheric dispersion processes. Several researchers already published the radionuclide source terms by coupling atmospheric dispersion simulations and environmental monitoring data (Chino et al., 2011; Kawamura et al., 2011; Stohl, 2011; Katata et al., 2012a,b; Terada et al., 2012).

In this chapter, we shall adopt the latest results of the source terms of radioactive iodine and cesium from 12 March 2011 to 1 May 2011 (as discussed by Terada et al., 2012), as they represent at the moment the most reliable source terms. The release rates estimated for ^{131}I and ^{137}Cs by Terada et al. (2012) are shown in Table 4.1 and Figs 4.3 and 4.4.

Two major releases of radionuclides on the morning and afternoon of 15 March 2011, which contributed to the highest dose rate zones around FNPP1, were estimated by comparing the calculated air dose rates attributed to radionuclides deposited on the ground surface (ground shine) with the observed ones (Katata et al., 2012b). Katata et al. (2012a) also revised the release rates of Chino et al. (2011) from the morning of 12 March 2011 to late in the night of 14 March 2011 using the environmental monitoring data for air dose rate and air concentrations of radionuclides, which were additionally disclosed by TEPCO on 28 May 2011 (TEPCO, 2011), and the Ministry of Economy, Trade and Industry (METI) on 3 June 2011 (METI, 2011). After two major releases on 15 March 2011, another two major releases on 22 and 23 March 2011 and on 30 and 31 March 2011

Table 4.1. Radionuclide Release Rates Estimated or Refined for the Period between 5:00 (JST) on 12 March and 0:00 (JST) on 1 May 2011. The ^{137}Cs Release was Derived by the ^{131}I Release Rate and the $^{131}I/^{137}Cs$ Activity Ratio. (Data from Terada et al., 2012)

No.	Release Period Month/Day Time JST	Release Duration (h)	Release Rate of ^{131}I (Bq/h)	$^{131}I/^{137}Cs$	Release Rate of ^{137}Cs (Bq/h)	Release Height (m)	References
1	3/12 05:00–3/12 09:30	4.5	3.7E + 13	10	3.7E + 12	20	1
2	3/12 09:30–3/12 15:30	6	1.7E + 13	10	1.7E + 12	120	1
3	3/12 15:30–3/12 16:00	0.5	3.0E + 15	10	3.0E + 14	Volume source[a]	1
4	3/12 16:00–3/13 23:00	31	8.4E + 13	10	8.4E + 12	120	1
5	3/13 23:00–3/14 11:00	12	3.6E + 13	10	3.6E + 12	120	1
6	3/14 11:00–3/14 11:30	0.5	3.0E + 15	10	3.0E + 14[e]	Volume source[a]	1
7	3/14 11:30–3/14 21:30	10	2.3E + 13	10	2.3E + 12	20	2
8	3/14 21:30–3/15 00:00	2.5	1.3E + 15	10	1.3E + 14	120	2
9	3/15 00:00–3/15 07:00	7	3.5E + 14	8.8	4.0E + 13	120	2
10	3/15 07:00–3/15 10:00	3	3.0E + 15	10	3.0E + 14	20	2
11	3/15 10:00–3/15 13:00	3	8.0E + 13	10	8.0E + 12	20	2
12	3/15 13:00–3/15 17:00	4	4.0E + 15	10	4.0E + 14	20	2
13	3/15 17:00–3/17 06:00	37	2.1E + 14	70	3.0E + 12	20	2

14	3/17 06:00–3/19 15:00	57	4.1E + 14	41	1.0E + 13	20	2
15	3/19 15:00–3/21 03:00	36	3.8E + 14	11	3.5E + 13	20	2
16[b]	3/21 03:00–3/21 21:00	18	1.4E + 14	10	1.4E + 13	20	3
17	3/21 21:00–3/22 23:00	26	4.1E + 14	87	4.7E + 12	20	2
18	3/22 23:00–3/24 00:00	25	7.1E + 14	80	8.9E + 12	20	2
19	3/24 00:00–3/25 00:00	24	1.9E + 14	66	2.9E + 12	20	2
20	3/25 00:00–3/26 11:00	35	5.6E + 13	45	1.2E + 12	20	2
21	3/26 11:00–3/28 10:00	47	4.0E + 12	23	1.7E + 11	20	2
22	3/28 10:00–3/29 21:00	35	7.5E + 12	1.6	4.7E + 12	20	3
23[c]	3/29 21:00–3/30 11:00	14	1.5E + 13	1.7	8.8E + 12	20	3
24	3/30 11:00–3/31 00:00	13	1.8E + 14	1.3	1.4E + 14	20	3
25	3/31 00:00–3/31 22:00	22	2.4E + 13	5.3	4.5E + 12	20	2
26	3/31 22:00–4/2 09:00	35	1.8E + 12	1.1	1.6E + 12	20	2
27	4/2 09:00–4/4 09:00	48	1.8E + 12	3.1	5.8E + 11	20	2
28	4/4 09:00–4/7 17:00	80	7.0E + 11	4.9	1.4E + 11	20	3
29[d]	4/7 17:00–4/13 23:00	150	7.0E + 11	2.0	3.5E + 11	20	3
30[d]	4/13 23:00–5/1 00:00	409	7.0E + 11	4.0	1.8E + 11	20	3

[a]Volume sources with the sizes of (*x*, *y*, *z*). (100, 100, 100 m) and (100, 100, 300 m) were assumed for hydrogen explosions at Unit 1 (No. 3) and Unit 3 (No. 6), respectively.
[b]The $^{131}I/^{137}Cs$ radioactivity ratio was derived by Terada et al., 2012 on the basis of dust sampling data at FNPP2 (TEPCO, 2011), Tokai and Setagaya.
[c]The ^{131}I release rate was estimated by Terada et al., 2012 from the concentration of ^{131}I at FNPP2 by dust sampling from 9:27 to 9:35 on 30 March to be 1490 Bq/m^3 (TEPCO, 2011), and that by calculation assuming unit release to be 1.0×10^{-10} Bq/m^3. The release time of sampled air was estimated to be 8:00 on 30 March.
[d]The $^{131}I/^{137}Cs$ radioactivity ratios were determined by Terada et al., 2012 from the dust sampling data at Tokai. Release times of the plumes for Nos. 29 and 30 were estimated at 0:00 on 10 April and 22:00 on 17 April, respectively.
[e]Assumed to be the same as for No. 3.
[1]Data from Katata et al. (2012a).
[2]Data from Chino et al. (2011).
[3]Data from Terada et al. (2012).

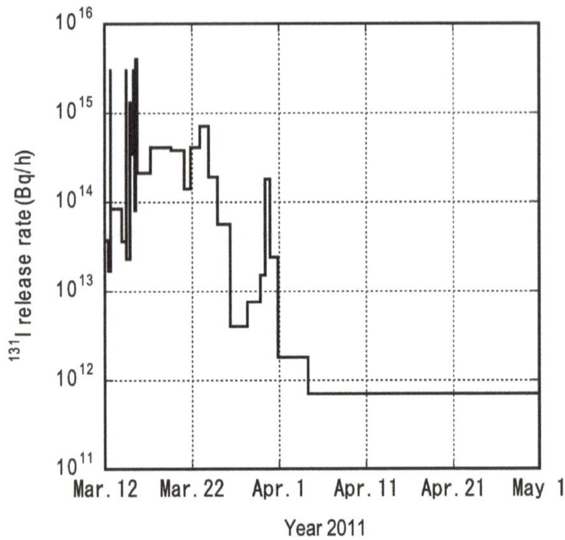

Figure 4.3 Estimated temporal variation of release rates of ^{131}I.

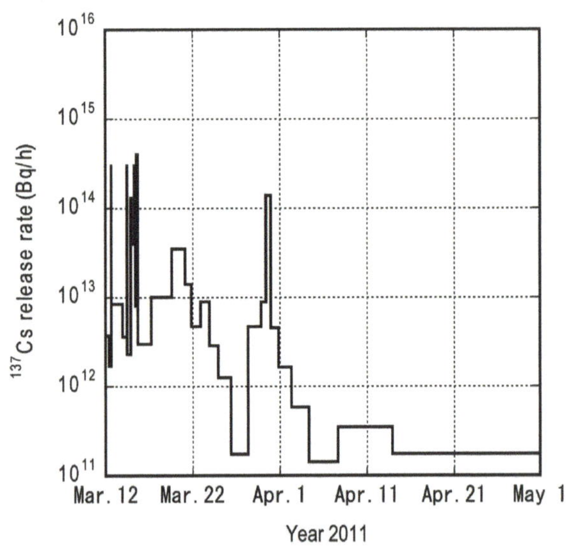

Figure 4.4 Estimated temporal variation of release rates of ^{137}Cs.

were also estimated, as shown in Table 4.1 and Figs 4.3 and 4.4.

Another possibility of estimation of source terms of released radionuclides is based on ^{133}Xe results from Comprehensive Nuclear-Test-Ban Treaty

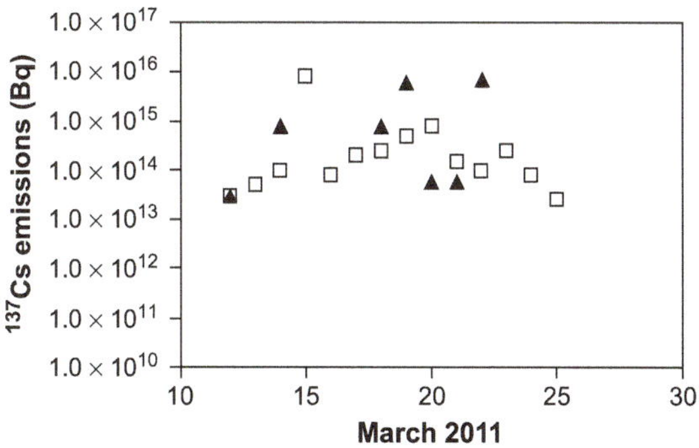

Figure 4.5 A comparison of ^{137}Cs release rates from the Fukushima NPP obtained from experimental data (squares, Kawamura et al., 2011) and reconstructed from ^{133}Xe data obtained at Japan and China CTBTO stations (triangles, Schöppner et al., 2013). Adapted from Povinec et al., 2013)

Organization (CTBTO) stations. The Fukushima ^{133}Xe source term has been reconstructed and compared with previously estimated ^{137}Cs and ^{131}I source terms (Schöppner et al., 2012, 2013). The reconstruction is accomplished by applying atmospheric transport modeling (ATM) and an adapted least square error method. The time-dependent relationship between a source that is emitting particles and a receptor in a remote location is described with a source–receptor sensitivity matrix. To determine the sensitivity between the FNPP1 and the affected International Monitoring System (IMS) stations, ATM has been utilized. A comparison of ^{137}Cs source term (Fig. 4.5), as obtained from Japanese data (Kawamura et al., 2011) and reconstructed ^{133}Xe data from the Japan and China CTBTO stations, shows a reasonable agreement between such very different approaches.

A model simulation using the source term as shown in Table 4.1 was conducted, and movement of radioactive plume was described in detail for the period from 12 March 2011 to 31 March 2011 as shown in the following nine time frames (Terada et al., 2012).

4.1.1 12–14 March 2011

The radioactive plume discharged from FNPP1 first flowed toward the Pacific Ocean over the early morning to the afternoon on 12 March 2011. The high-concentration plume around the hydrogen explosion, which occurred at Unit 1 of FNPP1 at 15:30 Japan Standard Time (JST) on 12 March 2011, flowed in the north–northwest direction from FNPP1 and caused a large amount of dry deposition. An increase in air dose rate was observed at the monitoring post of Minamisoma 4.5 h after the explosion (Fukushima Prefecture, 2011). The plume then flowed in the north–northeast direction and produced a large amount of dry deposition in the northeast coastal area of Miyagi Prefecture. Air dose rates observed at the nuclear power plant in Onagawa were elevated at 0:00 (JST) on 13 March when the plume passed through (Tohoku Electric Power, 2011). Afterward, the plume was transported mainly toward the ocean by the south-westerly wind until the night of 14 March 2011.

4.1.2 15 March 2011

The high-radionuclide concentration plume started to flow in the south–southwest direction from FNPP1 from late at night on 14 March 2011 and caused a large amount of dry deposition along the southeast coastal area of Fukushima Prefecture and the northeast area of Ibaraki Prefecture on the morning of 15 March 2011. The plume successively dispersed and caused dry radionuclide depositions in Tokyo, Saitama, and Kanagawa Prefectures, although the quantity was smaller than in the above-mentioned regions. From the afternoon of 15 March 2011, precipitation was observed broadly across eastern Japan, due to a low-pressure system passing through Honshu Island. The calculated plume encountered the rain band over the areas of Niigata, Gunma, Tochigi, and Fukushima Prefectures, and this caused a large amount of wet radionuclide deposition in the afternoon. Then, the plume changed its direction clockwise to the northwest of FNPP1, and passed through Miyagi, Yamagata, and Iwate Prefectures during the evening and into the night. In Fukushima Prefecture, the high-radionuclide concentration plume, due to the major release from FNPP1 in the afternoon, caused significant dry and wet radionuclide depositions in the area northwest of FNPP1 (Katata et al., 2012a,b).

Although a large amount of surface deposition of ^{137}Cs was observed by airborne monitoring in the north of Tochigi and Gunma Prefectures (MEXT, 2011a), in the calculation, it was mainly seen in the south of Niigata Prefecture and the west of Fukushima Prefecture. Because no comparable amount of surface deposition was calculated on other days in the north of Tochigi and Gunma Prefectures, the deposition on 15 March 2011 has the largest contribution to the total deposition for the area. From the observed precipitation and air dose rates (MEXT, 2011b), the discrepancy is due to the high-radionuclide concentration plume passing through Tochigi and Gunma Prefectures before the occurrence of precipitation in the area. In Tochigi Prefecture, additional surface deposition of ^{137}Cs from fog or cloud observed at Oku-Nikko also contributed to the observed surface deposition. In the calculation, a significant amount of wet deposition also appeared in the area from the north of the central part of Fukushima Prefecture, via Miyagi Prefecture, to the south of Iwate Prefecture, which was not the case from ground-level and airborne monitoring (MEXT, 2011a,b; Miyagi Prefecture, 2011). Considering that, in the calculation, the principal part of the plume was transported above clouds, a possible cause of the discrepancy between the calculations and observations is an overestimation of the wet deposition in the upper air by the model.

4.1.3 16–19 March 2011

The plume was calculated to have flowed mostly toward the ocean, due to the wind fluctuating from westerly to north-westerly from 16 to 19 March 2011. In the observations, however, air dose rate monitoring in the southeast of Fukushima Prefecture and in the east of Ibaraki Prefecture (Fukushima Prefecture, 2011; Ibaraki Prefecture, 2011; KEK, 2011; MEXT, 2011b) during the morning of 16 March 2011 suggests that the plume flowed over Ibaraki Prefecture and produced a lot of dry and wet radionuclide depositions on that morning. The model could not reproduce the above surface deposition because of errors in the calculated wind direction.

4.1.4 20 March 2011

The plume, which flowed to the east of FNPP1, started to turn clockwise before dawn on 20 March 2011. The plume passed over the Kanto region in the morning, and then changed its direction to the north-west of FNPP1 in the afternoon. The plume flowed to the west of Miyagi Prefecture, to the east of Yamagata Prefecture, and to the south of Iwate Prefecture. A large amount of wet deposition was produced in the area, due to precipitation in the evening. The plume started to flow to the south of FNPP1 again late at night.

4.1.5 21 March 2011

The plume flowed into Ibaraki Prefecture via the ocean on the morning of 21 March 2011, and then dispersed over the Kanto region and caused a large amount of the wet radionuclide deposition in Ibaraki, Chiba, Saitama, Tokyo, and Kanagawa Prefectures due to precipitation from the morning. Surface depositions were overestimated at the sampling sites of Kofu and Omaezaki because of the overestimation of daily precipitation.

4.1.6 22 March 2011

The plume released from FNPP1 on 22 March 2011 dispersed mainly in Fukushima and Tochigi Prefectures, and caused wet deposition in the east of Fukushima Prefecture due to precipitation from the night. Meanwhile, the radionuclides released on 21 March 2011 remained stagnant in the Kanto region, and they were deposited by precipitation on 22 March 2011.

4.1.7 23–24 March 2011

From the early morning on 23 March 2011, the plume was transported along the seaside of Ibaraki and Chiba Prefectures, due to the wind fluctuating from the northerly to the north-westerly directions. It flowed into the south-eastern part of the Kanto region from the afternoon, and produced wet deposition from the evening until late at night. The plume was mostly transported toward the ocean from the morning of 24 March 2011 to the morning of 25 March 2011.

4.1.8 25–29 March 2011

From the afternoon to the night of 25 March 2011, the plume was transported by the south-easterly wind in the northwest direction from FNPP1, and caused wet deposition mainly in Fukushima Prefecture and in the south of Yamagata and Miyagi Prefectures. Then, the plume changed its direction anticlockwise from northwest to southeast on the morning of 26 March 2011. The plume continuously flowed toward the ocean from the morning of 26 March 2011 to 29 March 2011.

4.1.9 30–31 March 2011

The plume turned clockwise from the east to the southwest of FNPP1 and flowed into the Kanto region on the morning of 30 March 2011, and caused wet deposition mainly in Tochigi and Ibaraki Prefectures during the afternoon. The plume then flowed in the northwest direction from FNPP1 from the afternoon until the night. On 31 March 2011, the plume mainly flowed toward the ocean.

4.2 Marine Releases of Radionuclides

Oceanic numerical simulations have been useful tools for estimating the rates of direct radionuclide releases to the ocean, and for representing and predicting the behavior of radionuclides in the marine environment. Tsumune et al. (2012, 2013) used a regional ocean model to estimate direct release rates until the end of February 2012. The analysis of $^{131}I/^{137}Cs$ activity ratios indicated that direct releases started on 26 March 2011 because $^{131}I/^{137}Cs$ activity ratios varied much more before 26 March 2011. The total amount of ^{137}Cs activity released was estimated to be 3.5 ± 0.7 PBq by the end of May, and 3.55 PBq by the end of February 2012.

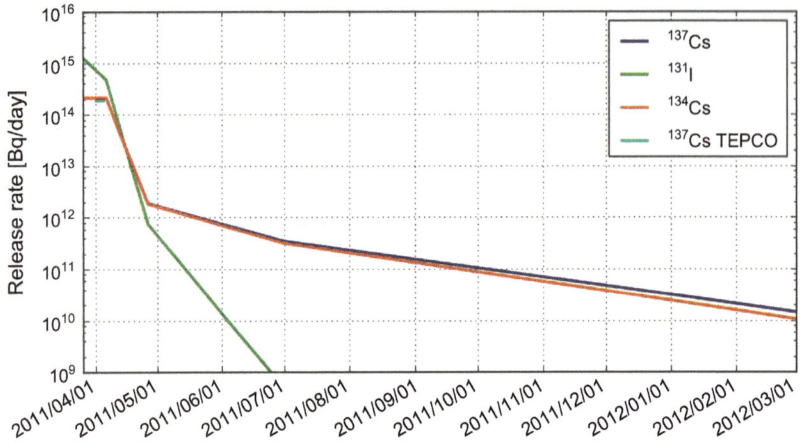

Figure 4.6 Estimated direct release rates (Bq/day) of ^{131}I, ^{134}Cs, and ^{137}Cs from the FNPP1. (Adapted from Tsumune et al., 2013)

An exponential model to describe the releases during each period was used to improve the agreement between the measured and simulated long-term trend of ^{137}Cs activity adjacent to the FNPP1, as follows from Fig. 4.6 (Tsumune et al., 2013):

26 March 2011–6 April 2011	1.1×10^7 Bq/m^3 (constant)
7 April 2011–25 April 2011	1.1×10^7 e $(-0.236t)$ Bq/m^3
26 April 2011–30 June 2011	1.2×10^5 e $(-0.026t)$ Bq/m^3
1 July 2011–29 February 2012	2.3×10^4 e $(-0.013t)$ Bq/m^3

where t is the elapsed time in days since the start of the time interval.

Near-shore simulated and measured ^{137}Cs activities were in good agreement except the period before 25 March 2011 (as shown in Fig. 4.7) because the atmospheric deposition was ignored in the simulation. Observed ^{137}Cs activities before 25 March 2011 were 10^5–10^7 Bq/m^3, while simulated ^{137}Cs activities were from 10^3–10^4 Bq/m^3. These larger differences could originate from additional radionuclide sources, e.g. due to the atmospheric deposition, different from the direct radionuclide releases from FNPP1 (Tsumune et al., 2012, 2013).

The direct release rates of ^{131}I and ^{134}Cs were also estimated from the activity ratios and half-lives of the isotopes. The $^{131}I/^{137}Cs$ and $^{134}Cs/^{137}Cs$ activity

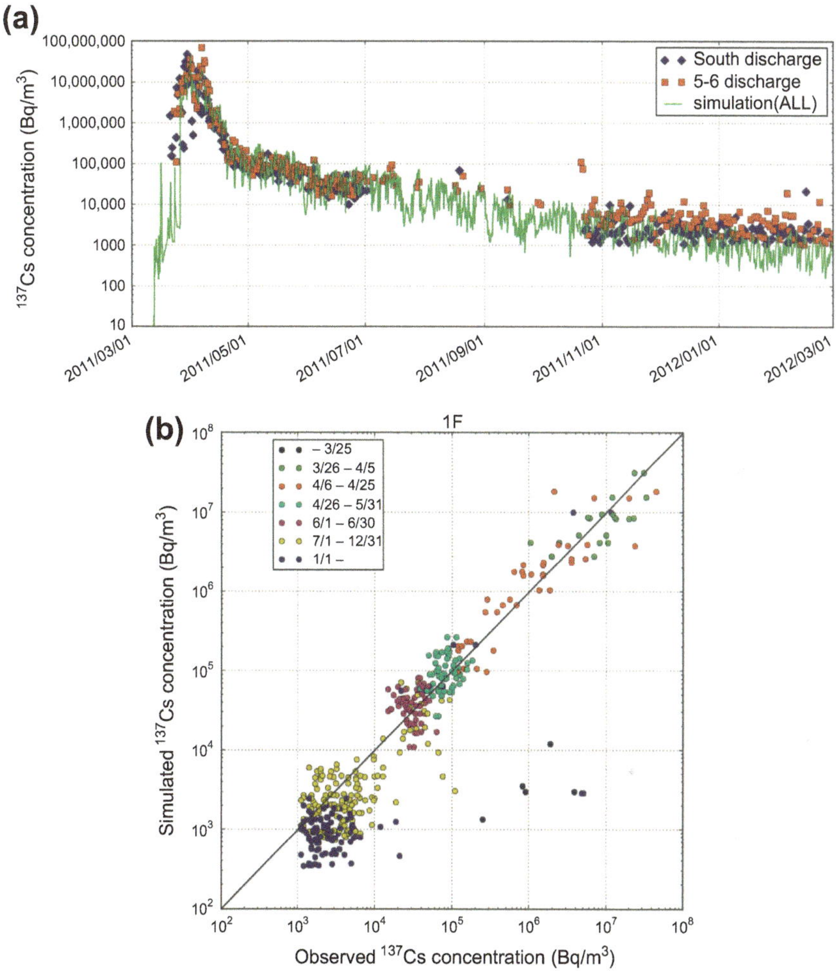

Figure 4.7 (a) Measured ^{137}Cs activities (Bq/m^3) at the 5–6 (north) and south discharge canals near the Fukushima Dai-ichi NPP, and the simulated ^{137}Cs activities in a grid adjacent to the FNPP1 site including direct discharge, atmospheric deposition and inflow at boundary as shown simulation (ALL) in this figure. (b) Scatter plot between daily mean measured activities (Bq/m^3) and the simulated activities. (Adapted from Tsumune et al., 2013)

ratios were changed in seawater only due to the decay of ^{131}I, but not due to the transport processes in seawater. The $^{131}I/^{137}Cs$ activity ratio was 5.7 on 26 March 2011 in a puddle of water in the basement of the FNPP1 reactor 2 turbine building; therefore, the $^{131}I/^{137}Cs$ activity ratio in the water directly discharged to the ocean should be 5.7 on 26 March 2011. The ratio decreased with time because the half-life of

^{131}I (8.02 days) is very short compared to the 30-year half-life of ^{137}Cs. The $^{134}Cs/^{137}Cs$ activity ratio was 0.99 ± 0.03 adjacent to the FNPP1 during the first month of radionuclide observations in coastal seawater (Buesseler et al., 2011). The $^{134}Cs/^{137}Cs$ activity ratio was estimated to be 1 on 26 March 2011. The ratio decreased with time because the half-life of ^{134}Cs (2 year) is shorter than the 30-year half-life of ^{137}Cs. The total amounts of ^{131}I, ^{134}Cs, and ^{137}Cs directly released to the ocean were estimated to be 11.1, 3.52, and 3.55 PBq until the end of February 2012, respectively. The uncertainties associated with these estimates are at least 20%, based on the estimate by Tsumune et al. (2012) that the total amount of ^{137}Cs activity released by the end of May had an uncertainty of 20% (3.5 ± 0.7 PBq). The radioactivity directly released by the end of May 2011 accounted for 98.8% of the total radioactivity directly released to the ocean. The rate at which ^{137}Cs was directly released to the ocean (outside of the port) was estimated to decrease exponentially and to be 10 GBq/day by the end of February 2012.

4.3 Total Amount of Radionuclides Released into the Atmosphere, Stagnant Water and the Ocean

Total amounts of released radionuclides from reactor cores of FNPP1 into the atmosphere were estimated and reported by Japanese Government for Cores 1, 2 and 3 (Japanese Government, 7 June 2011). After the severe accident in the FNPP1, a large amount of contaminated stagnant water was produced in turbine buildings and surrounding areas. Nishihara et al. (2012) reported activities of radionuclides in stagnant waters in turbine buildings and surrounding areas, and they also estimated inventories of radionuclides in stagnant water. It was revealed that the release ratios of iodine and caesium were several tens of percent, while those of strontium and barium were smaller by one or two orders of magnitude. These release ratios observed at the Fukushima accident were similar to those observed during the Three Miles Island

2 accident which occurred in USA in March 1979. Tsumune et al. (2012) reported for the total amount of direct discharge of ^{137}Cs the value of 3.5 ± 0.7 PBq as described in Section 4.2. Aoyama et al. (unpublished data) reported activity ratios of several radionuclides to ^{137}Cs in seawater collected on 13 April 2011 at 30 km distance south of FNPP1. The activity ratios in seawater sample measured might be similar with activity ratios in direct discharged contaminated water because sampling date, 13 April 2011, was close to a data when highest radiocaesium activities were observed on 6 April 2011 (as shown in Fig. 4.6). Therefore, amount of direct discharge of selected radionuclides could be calculated based on radioactivity ratios of measured radionuclides to ^{137}Cs (Table 4.1).

The releases of radioactive elements during the Fukushima accident might be controlled by the physical and chemical properties of the radioactive elements in the core. All the noble gases, including krypton and xenon, contained within the reactor cores were released immediately into the atmosphere by the first stage of the meltdowns at three cores of FNPP1. We can assume therefore that 100% of ^{85}K and ^{133}Xe were released from the three cores, Units 1–3 of FNPP1.

In the three reactor cores, Units 1–3 of FNPP1, the total inventory of ^{131}I was about 6000 PBq, and about 2.6% of ^{131}I (159 PBq) was released, as a mixture of vapor, aerosols, and organic iodine compounds. For comparison, about 2.2% of ^{137}Cs (15.3 PBq) was released in the aerosol form. For ^{134}Cs, about 2.4% of ^{134}Cs (17.5 PBq) was released in the aerosol form, similarly as in the case of ^{137}Cs. For iodine and caesium, the releases were not at once but they were following the development of the accident in each core as shown in Figs 4.3 and 4.4, respectively.

For radiostrontium isotopes, ^{89}Sr and ^{90}Sr, about 0.03% of them (1.96 PBq of ^{89}Sr and 0.14 PBq of ^{90}Sr) were released from the three cores. One millionth of plutonium isotopes—^{238}Pu, ^{240}Pu and ^{241}Pu—were estimated to be released from the three cores during the FNPP1 accident (Table 4.2).

In the stagnant water, the release ratios of iodine and caesium were several tens of percent, while that of strontium was smaller by one or two orders of

Table 4.2. Summary of the Radionuclide Reactor Core Inventories Released into the Atmosphere

			Core Inventory on 11 March 2011				Released Amount to the Atmosphere				
Radionuclide	Product	Half-life	Core 1 (Bq)	Core 2 (Bq)	Core 3 (Bq)	Total (Bq)	Core 1 (Bq)	Core 2 (Bq)	Core 3 (Bq)	Total (Bq)	Percent of Inventory (%)
^{85}Kr	Fission	10.756 years	2.31E + 16	3.11E + 16	2.95E + 16	8.37E + 16					100[1]
^{89}Sr	Fission	50.53 days	1.36E + 18	2.21E + 18	2.35E + 18	5.93E + 18	8.20E + 13	6.80E + 14	1.20E + 15	1.96E + 15	0.033[2]
^{90}Sr	Fission	28.74 years	1.50E + 17	1.91E + 17	1.81E + 17	5.22E + 17	6.10E + 12	4.80E + 13	8.50E + 13	1.39E + 14	0.027[2]
^{99}Mo	Fission	65.94 hour	2.57E + 18	4.43E + 18	4.42E + 18	1.14E + 19	2.60E + 09	1.20E + 09	2.90E + 09	6.70E + 09	
^{99m}Tc	Fission	6.01 hour	2.24E + 18	3.88E + 18	3.86E + 18	9.98E + 18					
^{110m}Ag	Fission	249.8 days	4.04E + 15	6.41E + 15	5.99E + 15	1.64E + 16					
^{125}Sb	Fission	2.758 years	1.09E + 16	1.65E + 16	1.57E + 16	4.31E + 16					
^{129m}Te	Fission	33.6 days	4.40E + 16	7.06E + 16	7.48E + 16	1.89E + 17	7.20E + 14	2.40E + 15	2.10E + 14	3.33E + 15	1.8[2]
^{131}I	Fission	8.02 days	1.35E + 18	2.34E + 18	2.33E + 18	6.01E + 18	1.20E + 16	1.40E + 17	7.00E + 15	1.59E + 17	2.6[2]
^{132}Te	Fission	3.2 days	1.95E + 18	3.36E + 18	3.37E + 18	8.69E + 18	2.50E + 16	5.70E + 16	6.40E + 15	8.84E + 16	1.0[2]
^{133}I	Fission	20.8 days	1.18E + 17	2.04E + 17	2.04E + 17	5.27E + 17	1.20E + 16	2.60E + 16	4.20E + 15	4.22E + 16	8.0[2]
^{133}Xe	Fission	5.24 days	2.71E + 18	4.67E + 18	4.67E + 18	1.20E + 19					100[1]
^{134}Cs	Fission	2.065 years	1.90E + 17	2.77E + 17	2.52E + 17	7.19E + 17	7.10E + 14	1.60E + 16	8.20E + 14	1.75E + 16	2.4[2]
^{136}Cs	Fission	13.16 days	5.42E + 16	8.17E + 16	8.18E + 16	2.18E + 17					
^{137}Cs	Fission	30.04 years	2.03E + 17	2.56E + 17	2.41E + 17	7.00E + 17	5.90E + 14	1.40E + 16	7.10E + 14	1.53E + 16	2.2[2]
^{144}Ce	Fission	284.9 days	1.33E + 18	2.31E + 18	2.28E + 18	5.92E + 18	3.10E + 11	1.10E + 13	1.40E + 11	1.15E + 13	0.00019[2]
^{238}Pu	Actinoid	87.74 years	4.63E + 15	4.58E + 15	5.54E + 15	1.47E + 16	5.80E + 08	1.80E + 10	2.50E + 08	1.88E + 10	0.00013[2]
^{239}Pu	Actinoid	24,100 years	7.00E + 14	8.83E + 14	1.04E + 15	2.62E + 15	8.60E + 07	3.10E + 09	4.00E + 07	3.23E + 09	0.00012[2]
^{240}Pu	Actinoid	6570 years	8.83E + 14	1.03E + 15	1.35E + 15	3.27E + 15	8.80E + 07	3.00E + 09	4.00E + 07	3.13E + 09	0.00010[2]
^{241}Pu	Actinoid	13.2 years					3.50E + 10	1.20E + 12	1.60E + 10	1.25E + 12	
^{241}Am	Actinoid	433 years	5.61E + 14	4.34E + 14	5.58E + 14	1.55E + 15					
^{242}Cm	Actinoid	162.8 days	8.94E + 16	8.97E + 16	1.04E + 17	2.83E + 17	1.10E + 10	7.70E + 10	1.40E + 10	1.02E + 11	0.00004[2]
^{244}Cm	Actinoid	18.11 years	2.71E + 15	3.22E + 15	2.71E + 15	8.64E + 15					
^{54}Mn	Activation	312.12 days	6.47E + 13	1.10E + 14	1.08E + 14	2.83E + 14					
^{60}Co	Activation	5.271 years	2.53E + 12	3.61E + 12	3.28E + 12	9.42E + 12					

[1]The 100% release is assumed due to the nature of the noble gas.
[2]Percentage of inventory was derived by "total release amount"/"total core inventory".
Source: Core inventory on 11 March 2011 by Nishihara et al., 2012. Released amount to the atmosphere by Japanese Government, 2011.

magnitude. We can neglect very small portion of dissolved noble gases in the stagnant water. About 32% of ^{131}I (1940 PBq) and 20% of ^{137}Cs (141 PBq) were dissolved in the stagnant water. For ^{134}Cs, 20% of ^{134}Cs (147 PBq) was dissolved in the stagnant water. For radiostrontium isotopes, ^{89}Sr and ^{90}Sr, about 1.2% of ^{89}Sr (70.6 PBq) and 1.6% of ^{90}Sr (8.6 PBq) were dissolved in the stagnant water from the three cores, and 0.015% of ^{125}Sb (6.6 TBq) was dissolved in the stagnant water (Table 4.3).

In the ocean, estimations of total amount of radionuclide releases were mainly done for ^{137}Cs (Japanese Government, 2011; Kawamura et al., 2011; Tsumune et al., 2012, 2013; Miyazawa et al., 2012, 2013; Bailly du Bois et al., 2012; Estournel et al., 2012). The total amount of directly released ^{137}Cs activity was estimated by various groups using numerical simulation models to be from 3.5 to 5.5–5.9 PBq. The upper bound of 5.5–5.9 PBq might be reduced if the estimated atmospheric deposition rates onto the ocean were more accurate. Much higher values in the range of 10–34 PBq were obtained by Bailly du Bois et al. (2012), but these were probably overestimates, as discussed by Dietze and Kriest (2012), who came with the range of 0.9–3.5 PBq.

Therefore, 3.6 ± 0.7 PBq recently reported by Tsumune et al. (2013) might be the best estimate. The releases of other radioactive elements were derived taking into account the total of released ^{137}Cs, and the activity ratios to ^{137}Cs of radionuclides as stated above and shown in Table 4.4. Using the estimated 0.5% of ^{137}Cs (3.6 PBq in 700 PBq of core inventories of three cores) released in the ocean, using the $^{134}Cs/^{137}Cs$ ratio close to one, we may estimate that also 0.5% of ^{134}Cs (3.6 PBq) was released in the ocean.

There are no data available on the ^{90}Sr levels in surface waters near the Fukushima Dai-ichi NPP just after the large releases of cooling water which occurred in early April 2011. With a few exceptions, the ^{90}Sr concentrations are by about a factor of 10 lower than the ^{137}Cs levels (Povinec et al., 2012). If we extrapolate the ^{90}Sr data to the beginning of the Fukushima accident with the average $^{90}Sr/^{137}Cs$ activity ratio of 0.02 observed on 18 April 2011, a total

Table 4.3. Summary of the Radionuclide Reactor Core Inventories Released to the Stagnant Water

Radionuclide	Product	Half-life	Total Core Inventory on 11 March 2011[2] (Bq)	Released Amount to the Stagnant Water				Percent of Inventory (%)
				Core 1 (Bq)	Core 2 (Bq)	Core 3 (Bq)	Total (Bq)	
^{85}Kr	Fission	10.756 years	8.37E + 16					
^{89}Sr	Fission	50.53 days	5.93E + 18	6.96E + 12	3.53E + 16	3.53E + 16	7.06E + 16	1.2[1]
^{90}Sr	Fission	28.74 years	5.22E + 17	1.65E + 12	4.59E + 15	3.98E + 15	8.56E + 15	1.6[1]
^{99}Mo	Fission	65.94 hour	1.14E + 19					
^{99m}Tc	Fission	6.01 hour	9.98E + 18		3.88E + 16	1.93E + 16	5.81E + 16	0.58[1]
^{110m}Ag	Fission	249.8 days	1.64E + 16					
^{125}Sb	Fission	2.758 years	4.31E + 16		6.61E + 12		6.61E + 12	0.015[1]
^{129m}Te	Fission	33.6 days	1.89E + 17					
^{131}I	Fission	8.02 days	6.01E + 18	9.30E + 16	1.22E + 18	6.28E + 17	1.94E + 18	32[1]
^{132}Te	Fission	3.2 days	8.69E + 18					
^{133}I	Fission	20.8 days	5.27E + 17					
^{133}Xe	Fission	5.24 days	1.20E + 19					
^{134}Cs	Fission	2.065 years	7.19E + 17	1.29E + 16	9.13E + 16	4.28E + 16	1.47E + 17	20[1]
^{136}Cs	Fission	13.16 days	2.18E + 17	3.79E + 15	2.21E + 16	1.15E + 16	3.73E + 16	17[1]
^{137}Cs	Fission	30.04 years	7.00E + 17	1.26E + 16	8.69E + 16	4.10E + 16	1.41E + 17	20[1]

^{144}Ce	Fission	284.9 days	5.92E + 18
^{238}Pu	Actinoid	87.74 years	1.47E + 16
^{239}Pu	Actinoid	24,100 years	2.62E + 15
^{240}Pu	Actinoid	6570 years	3.27E + 15
^{241}Pu	Actinoid	13.2 years	
^{241}Am	Actinoid	433 years	1.55E + 15
^{242}Cm	Actinoid	162.8 days	2.83E + 17
^{244}Cm	Actinoid	18.11 years	8.64E + 15
^{54}Mn	Activation	312.12 days	2.83E + 14
^{60}Co	Activation	5.271 years	9.42E + 12

[1]Percentage of inventory was derived by "total release amount to stagnant water"/"total core inventory".
[2]Total core inventory on 11 March 2011 are the same as in Table 4.2.
Source: Nishihara et al., 2012.

Table 4.4. Summary of the Radionuclide Reactor Core Inventories Directly Released to the Pacific Ocean

Radionuclide	Product	Half-life	Total Core Inventory on 11 March 2011[3] (Bq)	Total Amount of Direct Release to the Ocean (Bq)	Activity Ratio to ^{137}Cs in Seawater	Percentage of Released Amount to Total Core Inventory (%)
^{85}Kr	Fission	10.756 years	8.37E + 16			
^{89}Sr	Fission	50.53 days	5.93E + 18			
^{90}Sr	Fission	28.74 years	5.22E + 17	5.20E + 10	1.49E − 05	0.000010[1]
^{99}Mo	Fission	65.94 hour	1.14E + 19			
^{99m}Tc	Fission	6.01 hour	9.98E + 18			
^{110m}Ag	Fission	249.8 days	1.64E + 16		ND	
^{125}Sb	Fission	2.758 years	4.31E + 16	1.19E + 11	3.40E − 05	0.00028[1]
^{129m}Te	Fission	33.6 days	1.89E + 17			
^{131}I	Fission	8.02 days	6.01E + 18			
^{132}Te	Fission	3.2 days	8.69E + 18			
^{133}I	Fission	20.8 days	5.27E + 17			
^{133}Xe	Fission	5.24 days	1.20E + 19			
^{134}Cs	Fission	2.065 years	7.19E + 17	3.50E + 15	1.00E + 00	0.49[2]
^{136}Cs	Fission	13.16 days	2.18E + 17			

^{137}Cs	Fission	30.04 years	7.00E + 17	3.50E + 15	1.00E + 00	0.50[2]
^{144}Ce	Fission	284.9 days	5.92E + 18	1.88E + 12	5.38E − 04	0.00003[1]
^{238}Pu	Actinoid	87.74 years	1.47E + 16			
^{239}Pu	Actinoid	24,100 years	2.62E + 15			
^{240}Pu	Actinoid	6570 years	3.27E + 15			
^{241}Pu	Actinoid	13.2 years				
^{241}Am	Actinoid	433 years	1.55E + 15			
^{242}Cm	Actinoid	162.8 days	2.83E + 17			
^{244}Cm	Actinoid	18.11 years	8.64E + 15			
^{54}Mn	Activation	312.12 days	2.83E + 14	4.46E + 10	1.27E − 05	0.016[1]
^{60}Co	Activation	5.271 years	9.42E + 12	1.01E + 10	2.87E − 06	0.11[1]

[1]Data from Aoyama et al. (2013), unpublished data.
[2]Data from Tsumune et al. (2012).
[3]Total core inventory on 11 March 2011 are the same as in Table 4.2.

released amount of ^{90}Sr should be 0.07–0.08 PBq. A tentative total release of ^{89}Sr, corrected for radioactive decay at the time of cease of reactors, was estimated from the initial $^{89}Sr/^{90}Sr$ ratio to be 0.8–0.9 PBq. If the $^{90}Sr/^{137}Cs$ ratio in directly released water into the sea was equal to that in the stagnant water (0.08 as reported by Nishihara et al., 2012), then the total releases of ^{90}Sr and ^{89}Sr would be about 0.3 and 3.7 PBq, respectively (a conservative estimate based on the release of 4 PBq of ^{137}Cs). However, if we take a weighted average of the $^{90}Sr/^{137}Cs$ activity ratios observed in surface seawater offshore Fukushima Dai-ichi NPP equal to 0.24, then the estimated ^{90}Sr release rate would be around 1 PBq (Povinec et al., 2012).

The other radionuclides were released in much lower quantities, e.g. 0.00028% of ^{125}Sb (0.12 TBq), 0.016% of ^{54}Mn (45 GBq), and 0.11% of ^{60}Co (10 GBq).

It is important to compare magnitude of radionuclide releases to the environment during the Chernobyl accident in 1986 and the FNPP1 accident in 2011. The release of radionuclides from the nuclear fuel was largely controlled by their boiling points of the compounds. Radionuclides releases during the Chernobyl accident in 1986 are summarized in Table 4.5 (IAEA, 2003). All the noble gases, including krypton and xenon, contained within the reactor were released immediately into the atmosphere by the first steam explosion, which occurred during the Chernobyl accident, similarly as was the case during the Fukushima accident. For the volatile radionuclides, about 50–60% of ^{131}I, 20–40% of ^{134}Cs and ^{137}Cs were released, while several percentage of radiostrontium were also released. For the plutonium isotopes, about 3.5% of each plutonium isotope and 3.5% of ^{242}Cm were released as shown in Table 4.6.

A comparison of results in the percentage of releases during the Chernobyl and Fukushima accidents is summarized in Table 4.6. During the Chernobyl accident, 55% of the radioactive iodine in the reactor, containing about 1760 PBq of ^{131}I, was released into the atmosphere, as a mixture of vapor, solid particles, and organic iodine compounds, while only 2.6% of ^{131}I (159 PBq) was released into the

Table 4.5. Summary of Radionuclide Reactor Core Inventories and the Total Radionuclide Release during the Chernobyl Accident in 1986 (after IAEA, 2003)

	Core Inventory on 26 April 1986		Total Release during the Accident	
Radionuclide	Half-life	Activity (PBq)	Percentage of Inventory (%)	Activity (PBq)
^{33}Xe	5.3 days	6500	100	6500
^{131}I	8.0 days	3200	50–60	~1760
^{134}Cs	2.0 years	180	20–40	~54
^{137}Cs	30.0 years	280	20–40	~85
^{132}Te	78.0 h	2700	25–60	~1150
^{89}Sr	52.0 days	2300	4–6	~115
^{90}Sr	28.0 years	200	4–6	~10
^{140}Ba	12.8 days	4800	4–6	~240
^{95}Zr	65.0 days	5600	3.5	196
^{99}Mo	67.0 h	4800	>3.5	>168
^{103}Ru	39.6 days	4800	>3.5	>168
^{106}Ru	1.0 years	2100	>3.5	>73
^{141}Ce	33.0 days	5600	3.5	196
^{144}Ce	285.0 days	3300	3.5	~116
^{239}Np	2.4 days	27,000	3.5	~95
^{238}Pu	86.0 years	1	3.5	0.035
^{239}Pu	24,400.0 years	0.85	3.5	0.03
^{240}Pu	6580.0 years	1.2	3.5	0.042
^{241}Pu	13.2 years	170	3.5	~6
^{242}Cm	163.0 days	26	3.5	~0.9

Table 4.6. A Comparison of Radionuclide Releases during the Fukushima Accident in 2011 and the Chernobyl Accident in 1986

Radionuclide	Product	Half-life	Total Inventory of Three Cores[1] (Bq)	Release into the Atmosphere[2] (%)	Release in Stagnant Water[3] (%)	Direct Discharge to the Ocean[4] (%)	Chernobyl Accident[5] (%)
^{85}Kr	Fission	10.756 years	8.37E + 16	100			
^{133}Xe	Fission	5.24 days	1.20E + 19	100			100
^{133}I	Fission	20.8 days	5.27E + 17	8.0			
^{131}I	Fission	8.02 days	6.01E + 18	2.6	32		50–60
^{134}Cs	Fission	2.065 years	7.19E + 17	2.4	20	0.49	20–40
^{137}Cs	Fission	30.04 years	7.00E + 17	2.2	20	0.50	20–40
^{129m}Te	Fission	33.6 days	1.89E + 17	1.8			
^{132}Te	Fission	3.2 days	8.69E + 18	1.0			25–60
^{89}Sr	Fission	50.53 days	5.93E + 18	0.033	1.2		4–6
^{90}Sr	Fission	28.74 years	5.22E + 17	0.027	1.6	0.000010	4–6
^{144}Ce	Fission	284.9 days	5.92E + 18	0.00019		0.00003	3.5
^{238}Pu	Actinoid	87.74 years	1.47E + 16	0.00013			3.5
^{239}Pu	Actinoid	24,100 years	2.62E + 15	0.00012			3.5
^{240}Pu	Actinoid	6570 years	3.27E + 15	0.00010			3.5

^{242}Cm	Actinoid	162.8 days	2.83E + 17	0.00004			3.5
^{99}Mo	Fission	65.94 h	1.14E + 19				>3.5
^{99m}Tc	Fission	6.01 h	9.98E + 18		0.58		
^{110m}Ag	Fission	249.8 days	1.64E + 16				
^{125}Sb	Fission	2.758 years	4.31E + 16		0.015	0.00028	
^{136}Cs	Fission	13.16 days	2.18E + 17		17		
^{241}Pu	Actinoid	13.2 years					3.5
^{241}Am	Actinoid	433 years	1.55E + 15				
^{244}Cm	Actinoid	18.11 years	8.64E + 15				
^{54}Mn	Activation	312.12 days	2.83E + 14			0.016	
^{60}Co	Activation	5.271 years	9.42E + 12			0.11	

[1]Total inventory of three cores are the same as in Table 4.2.
[2]Release into the atmosphere are the same as in Table 4.2.
[3]Release into the stagnant water are the same as in Table 4.3.
[4]Direct discharge to the Ocean are the same as in Table 4.4.
[5]Chernobyl accident are the same as in Table 4.5.

atmosphere during the FNPP1 accident. However, 32% of ^{131}I was released in stagnant water. For ^{137}Cs, about 30% of ^{137}Cs (85 PBq) was released into the atmosphere during the Chernobyl accident, while 2.2% of ^{137}Cs (15.3 PBq) was released into the atmosphere, and 20% of ^{137}Cs was released in stagnant water.

References

Aoyama, M., Tsumune, D., Hamajima, Y., 2012. Distribution of ^{137}Cs and ^{134}Cs in the North Pacific Ocean: impacts of the TEPCO Fukushima-Daiichi NPP accident. J. Radioanal. Nucl. Chem., 1–5. http://dx.doi.org/10.1007/s10967-012-2033-2.

Aoyama, M., Uematsu, M., Tsumune, D., Hamajima, Y., 2013. Surface pathway of radioactive plume of TEPCO Fukushima NPP1 released ^{134}Cs and ^{37}Cs. Biogeosciences Discuss 10, 265–283. http://dx.doi.org/10.5194/bgd-10-265-2013.

Bailly du Bois, P., Laguionie, P., Boust, D., Korsakissok, I., Didier, D., Fievet, B., 2012. Estimation of marine source-term following Fukushima Dai-ichi accident. J. Environ. Radioact. 114, 2–9. http://dx.doi.org/10.1016/j.jenvrad.2011.11.015.

Buesseler, K., Aoyama, M., Fukasawa, M., 2011. Impacts of the Fukushima nuclear power plants on marine radioactivity. Environ. Sci. Technol. 45, 9931–9935. http://dx.doi.org/10.1021/es202816c.

Chino, M., Nakayama, H., Nagai, H., Terada, H., Katata, G., Yamazawa, H., 2011. Preliminary estimation of release amounts of ^{131}I and ^{137}Cs accidentally discharged from the Fukushima Daiichi nuclear power plant into the atmosphere. J. Nucl. Sci. Technol. 48, 1129–1134.

Dietze, H., Kriest, I., 2012. ^{137}Cs off Fukushima Dai-ichi, Japan: model based estimates of dilution and fate. Ocean Sci. 8, 319–332.

Estournel, C., Bosc, E., Bocquet, M., Ulses, C., Marsaleix, P., Winiarek, V., Osvath, I., Nguyen, C., Duhaut, T., Lyard, F., Michaud, H., Auclair, F., 2012. Assessment of the amount of cesium-137 released into the Pacific Ocean after the Fukushima accident and analysis of its dispersion in Japanese coastal waters. J. Geophys. Res. 117, C11014. http://dx.doi.org/10.1029/2012JC007933.

Furuta, S., Sumiya, S., Watanabe, H., Nakano, M., Imaizumi, K., Takeyasu, M., Nakada, A., Fujita, H., Mizutani, T., Morisawa, M., Kokubun, Y., Kono, T., Nagaoka, M., Yokoyama, H., Hokama, T., Isozaki, T., Nemoto, M., Hiyama, Y., Onuma, T., Kato, C., Kurachi, T., 2011. Results of the Environmental Radiation Monitoring Following the Accident at the Fukushima Daiichi Nuclear Power Plant. JAEA-Review 2011-035, Japan Atomic Energy Agency (in Japanese with English abstract, available online).

Fukushima Prefecture, 2011. http://www.pref.fukushima.jp/j/7houbu0311-0331.pdf (in Japanese, accessed 02.12.11.).

KEK, Kou Enerugii Butsurigaku Kenkyusho, 2011. http://rcwww.kek.jp/norm/index-e.html (in Japanese, accessed 02.12.11.).
METI, Minister of Economy Trade and Industry, 2011. http://www.meti.go.jp/press/2011/06/20110603019/20110603019.html (in Japanese, accessed 02.12.11.).
Ibaraki Prefecture, 2011. http://www.pref.ibaraki.jp/20110311eq/radiation.html (in Japanese, accessed 02.12.11.).
Miyagi Prefecture, 2011. http://www.pref.miyagi.jp/gentai/Press/PressH230315-3(sokutei).html (in Japanese, accessed 02.12.11.).
Tohoku Electric Power, 2011. http://www.tohoku-epco.co.jp/news/atom/topics/1183332_1984.html (in Japanese, accessed 02.12.11.).
IAEA, International Atomic Energy Agency, 2003. Chernobyĺs Legacy: Health, Environmental and Socio-economic Impacts. IAEA, Vienna.
JG, 2011. Japanese government report to IAEA Ministerial Conference on nuclear safety – accident at TEPCO's Fukushima nuclear power stations. http://www.iaea.org/newscenter/focus/fukushima/japan-report/ (accessed 31.01.11, 7.06.11.).
Katata, G., Ota, M., Terada, H., Chino, M., Nagai, H., 2012a. Atmospheric discharge and dispersion of radionuclides during the Fukushima Dai-ichi nuclear power plant accident. Part I: source term estimation and local-scale atmospheric dispersion in early phase of the accident. J. Environ. Radioact. 109, 103–113. http://dx.doi.org/10.1016/j.jenvrad.2012.02.006.
Katata, G., Terada, H., Nagai, H., Chino, M., 2012b. Numerical reconstruction of high dose rate zones due to the Fukushima Dai-ichi Nuclear Power Plant accident. J. Environ. Radioact. 111, 2–12. http://dx.doi.org/10.1016/j.jenvrad.2011.09.011.
Kawamura, H., Kobayashi, T., Furuno, A., In, T., Ishikawa, Y., Nakayama, T., Shima, S., Awaji, T., 2011. Preliminary numerical experiments on oceanic dispersion of ^{131}I and ^{137}Cs discharged into the ocean because of the Fukushima Daiichi nuclear power plant disaster. J. Nucl. Sci. Technol. 48, 1349–1356.
MEXT, Ministry of Education, Culture, Sports, Science and Technology, 2011a. MEXT and DOE Airborne Monitoring. http://radioactivity.mext.go.jp/en/list/203/list-1.html (accessed 04.05.11.).
MEXT, Ministry of Education, Culture, Sports, Science and Technology, 2011b. Reading of environmental radioactivity level by prefecture. http://radioactivity.mext.go.jp/en/list/192/list-1.html (accessed 04.05.11.).
Miyazawa, Y., Masumoto, Y., Varlamov, S. M., Miyama, T., (2012a). Transport simulation of the radionuclide from the shelf to open ocean around Fukushima, Cont. Shelf Res. 16–29, 50–51.
Miyazawa, Y., Masumoto, Y., Varlamov, S.M., Miyama, T., Takigawa, M., Honda, M., Saino, T., 2012b. Inverse estimation of source parameters of oceanic radioactivity dispersion models associated with the Fukushima accident. Biogeosciences 10, 2349–2363. http://dx.doi.org/10.5194/bgd-9-13783-2012.

Nishihara, K., Yamagishi, I., Yasuda, K., Ishimori, K., Tanaka, K., Kuno, T., Inada, S., Gotoh, Y., 2012. Radionuclide release to stagnant water in Fukushima-1 nuclear power plant. J. At. Energy Soc. Japan 11, 13–19 (in Japanese).

Povinec, P.P., Hirose, K., Aoyama, M., 2012. Radiostrontium in the western North Pacific: characteristics, behavior and impact of the Fukushima Dai-ichi nuclear power plant accident. Environ. Sci. Technol. 46, 10356–10363.

Povinec, P.P., Gera, M.K., Hirose, K., Lujanienė, G., Nakano, M., Plastino, W., Bartok, I., 2013. Dispersion of Fukushima radionuclides in the global atmosphere and the ocean. Appl. Radiat. Isot. http://dx.doi.org/10.1016/j.apradiso.2013.03.058.

Schöppner, M., Plastino, W., Povinec, P.P., Wotawa, G., Bella, F., Budano, A., De Vincenzi, M., a, b, Federico Ruggieri, F., 2012. Estimation of the time-dependent radioactive source-term from the Fukushima nuclear power plant accident using atmospheric transport modeling. J. Environ. Radioact 114, 10–14.

Schöppner, M., Plastino, W., Povinec, P.P., Nikkinen, M., Ruggieri, F., 2013. Estimation of the radioactive source dispersion from Fukushima nuclear power plant accident. Appl. Rad. Isot. http://dx.doi.org/10.1016/j.apradiso.2013.03.070.

Stohl, A., 2011. Interactive comment on "xenon-133 and caesium-137 releases into the atmosphere from the Fukushima Dai-ichi nuclear power plant: determination of the source term, atmospheric dispersion, and deposition". Atmos. Chem. Phys. Discuss. 11, C12298–C12304.

TEPCO, Tokyo Electric Power Company, 2011. Additional monitoring data at Fukushima Daiichi nuclear power station. http://www.tepco.co.jp/en/press/corp-com/release/11052811-e.html (accessed 02.12.11.).

Terada, H., Katata, G., Chino, M., Nagai, H., 2012. Atmospheric discharge and dispersion of radionuclides during the Fukushima Dai-ichi nuclear power plant accident. Part II: verification of the source term and analysis of regional-scale atmospheric dispersion. J. Environ. Radioact. 112, 141–154. http://dx.doi.org/10.1016/j.jenvrad.2012.05.023.

Tochigi Prefecture, 2011. http://www.pref.tochigi.lg.jp/kinkyu/documents/20110312-18.pdf (in Japanese, accessed 02.12.11).

Tokyo Metropolitan Government, 2011. http://www.sangyo-rodo.metro.tokyo.jp/whats-new/measurement-kako.html (in Japanese, accessed 02.12.11.).

Tsumune, D., Tsubono, T., Aoyama, M., Hirose, K., 2012. Distribution of oceanic ^{137}Cs from the Fukushima Dai-ichi nuclear power plant simulated numerically by a regional ocean model. J. Environ. Radioact. 111, 100–108. http://dx.doi.org/10.1016/j.jenvrad.2011.10.007.

Tsumune, D., Tsubono, T., Aoyama, M., Uematsu, M., Misumi, K., Maeda, Y., Yoshida, Y., Hayami, H., 2013. One-year, regional-scale simulation of 137Cs radioactivity in the ocean following the Fukushima Daiichi nuclear power plant accident, Biogeosciences Discuss. 10, 6259-6314. http://dx.doi.org/10.5194/bgd-10-6259-2013.

5

FUKUSHIMA RADIOACTIVITY IMPACT

CHAPTER OUTLINE

Fukushima Accident. http://dx.doi.org/10.1016/B978-0-12-408132-1.00005-X

Heavy environmental monitoring activities have been carried out by national Japan government, local governments, research institutes and universities in Japan and worldwide to assess the impact of the Fukushima accident on the terrestrial and marine environment. We shall discuss in detail an impact on the Japan terrestrial and the marine environment, as well as on the global atmosphere and the ocean.

5.1 Japan

5.1.1 Atmosphere

5.1.1.1 Emergency Monitoring of Airborne Radioactivity

Radioactivity measurement in surface air is one of the most important issues in emergency environmental monitoring. Japanese Government including Prefectural Governments started measurements of radioactivity in surface air using monitoring cars as emergency monitoring devices (Ministry of Education, Culture, Sports, Science and Technology (MEXT, 2012a,b,c)). Air sampling was carried out with a potable air sampler in the area from 20 to 60 km of the Fukushima Dai-ichi Nuclear Power Plant (FNPP1); sampling time was about 20 min in every day. Although ^{131}I, ^{132}I, ^{134}Cs, ^{136}Cs, ^{137}Cs, ^{132}Te, ^{99m}Tc and ^{140}Ba were detected in dust samples collected in March 2011, dominant radionuclides were ^{131}I, ^{134}Cs and ^{137}Cs. The temporal variations of ^{131}I and ^{137}Cs are shown in Fig. 5.1. The highest total ^{131}I concentration in surface air (5.6 kBq/m^3) occurred on 20 March 2011 at Hirono Town (23 km south of the FNPP1), whereas the highest total ^{137}Cs concentration in surface air (860 Bq/m^3) was observed at Futaba Country (32 km northwest of the FNPP1) on 21 March. The ^{131}I concentration in surface air rapidly decreased, although high ^{131}I concentrations were observed in early May. After June 2011, the ^{131}I concentrations in surface air were less than the detection limit (about 0.8 Bq/m^3). The ^{137}Cs concentration in surface air decreased in March 2011, however, it seems to be constant in April and May. It must be noted that actual temporal variations of the ^{131}I and ^{137}Cs concentrations in surface air as well as precious radionuclide compositions could not be obtained from the emergency monitoring because of a short sampling time and high detection limits.

5.1.1.2 Source Monitoring

Source monitoring, including activity concentrations of radionuclides in stack and surface air, is the most important issue in the emergency situations (IAEA, 2005), as does the environmental monitoring. Tokyo Electric Power Company (TEPCO) measured radioactivity in surface air at three points of the FDNPP site since March 19 2011. Air sampling was performed with the potable air sampler, consisting of gas and particulate collectors, using a monitoring car; sampling time was about 20 min in every day. The temporal variations of ^{131}I and ^{137}Cs in surface air are shown in Fig. 5.2.

The highest total ^{131}I concentration in surface air (5.94 kBq/m^3) occurred on March 19, whereas the highest total ^{137}Cs concentration in surface air

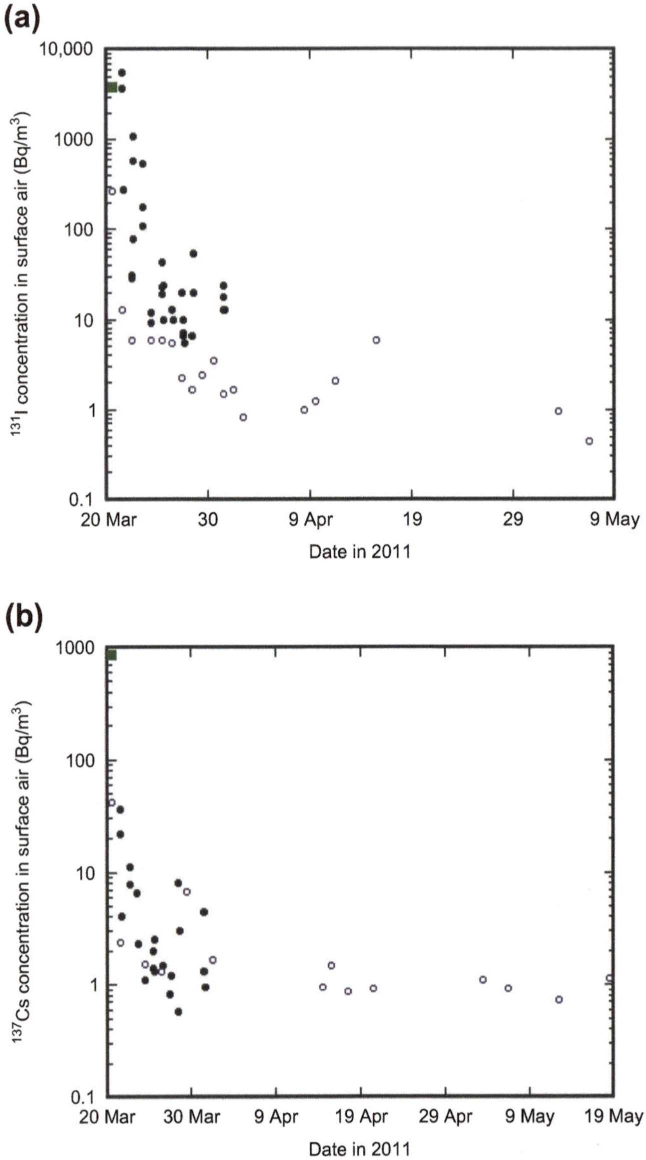

Figure 5.1 Temporal variations of ^{131}I and ^{137}Cs concentrations in surface air conducted by MEXT. Closed circles: Hirono Town (sampling period: 20–31 March), open circles: Iidate Village (sampling period: 20 March–31 May), and closed squares: Futaba Country (21 March). (Data from MEXT (2012a,b,c))

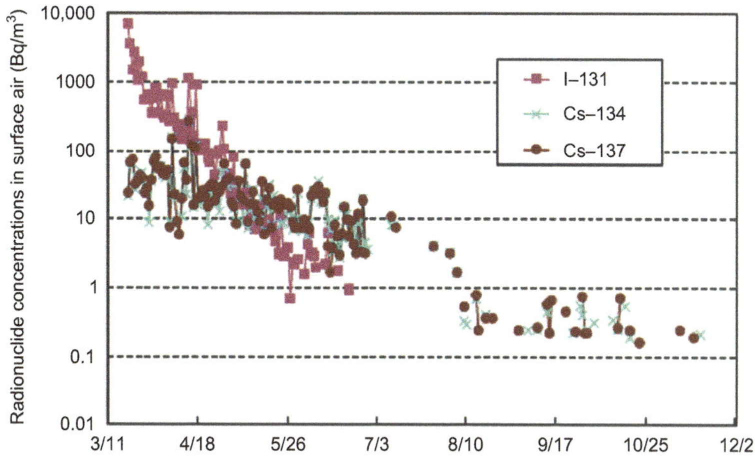

Figure 5.2 Temporal variations of ^{131}I and ^{137}Cs concentrations in surface air at the west gate of the Fukushima Dai-ichi NPP. (Data from TEPCO (2012a,b,c))

(430 Bq/m^3) was observed on March 25. The ^{131}I concentration in surface air exponentially decreased to about 1 Bq/m^3 on mid-June 2011 and was less than the detection limit (0.5 Bq/m^3) after late June. On the other hand, the ^{137}Cs concentration in surface air exhibited more gradual decrease with larger variability than ^{131}I; the ^{137}Cs concentration of about 10 Bq/m^3 was measured in July 19 and ^{137}Cs was sometimes detected in December 2011. This finding suggests that the ^{131}I and ^{137}Cs concentrations in surface air in the FNPP1 site reflect continuous emission of radionuclides from accidental reactors after the major release in March 2011, although there a possibility of resuspension of heavy contaminated soil particles around the FNPP1 cannot be excluded. The temporal change of the surface ^{131}I reveals that any significant recriticality could not occur in meltdown fuels of the reactors after April 2011. It is noteworthy that the monitoring data of TEPCO could not cover the whole day, as did loss of initial monitoring data at the FNPP1 accident.

5.1.1.3 Air Monitoring Carried out in Japan (Except MEXT)

Continuous sampling of air at both accidental FNPP1 site and the area from 20 to 60 km of the FDNPP was not conducted. On the other hand, some research institutes (MHLW, 2011; NERH, 2012; NSCJ, 2011;

JG, 2012; Amano et al., 2012; Doi et al., 2013; Haba et al., 2012; Kanai, 2012; Furuta et al., 2011) including Comprehensive Test-Ban Treaty Organization (CTBTO) (Yonezawa and Yamamoto, 2011) and universities (Momoshima et al., 2012) in Japan have performed continuous air sampling and measured radionuclides in surface air and detected many kinds of the FDNPP-derived radionuclides since late March 2011. Air monitoring results reveal that dominant detected radionuclides originating from the FDNPP were radioactive noble gases (^{133}Xe, ^{131m}Xe, ^{133m}Xe, and ^{135}Xe), radioiodine (^{132}I, and ^{131}I) and radiocesium (^{134}Cs, ^{136}Cs, and ^{137}Cs), although many short-lived radionuclides, such as ^{133}I, ^{132}I, ^{131}I, ^{134}Cs, ^{136}Cs, ^{137}Cs, ^{132}Te, ^{129m}Te (^{129}Te), ^{99}Mo(^{99m}Tc), ^{140}Ba(^{140}La), ^{95}Nb and ^{110m}Ag, were detected in air and airborne particles (Doi et al., 2013; Haba et al., 2012; Kanai, 2012; Yonezawa and Yamamoto, 2011). In contrast to the Chernobyl accident, significant amounts of refractory fission products such as ^{106}Ru, ^{103}Ru and ^{95}Zr have not been detected in environmental samples for the Fukushima accident. Surface air measurements documented that radioactive plume first arrived in Kanto Plain on March 15 2011, which coincided with abrupt increases of the ambient dose rate in the Kanto area (Fig. 5.3). After that, the Fukushima-derived radionuclides were transported on all Japanese Islands and on the northern hemisphere in late March and early April 2011. The ^{131}I and ^{137}Cs were detected in the Asia (Bolsunovsky and Dementyev, 2011; Hernández-Ceballos et al., 2012; Kim et al., 2012; Lee et al., 2012), North America (Biegalski et al., 2012; Bowyer et al., 2011; Zhang et al., 2011), Europe (Lozano et al., 2011; Masson et al., 2011; Beresford et al., 2012; Bikit et al., 2012; Bossew et al., 2012; Evrard et al., 2012; Loaiza et al., 2012; Lujaniené et al., 2012a,b, 2013; Pham et al., 2012; Perrot et al., 2012; Piñero García and Ferro García, 2012; Pittauerová et al., 2011; Povinec et al., 2012a, 2012b, 2013a), and Arctic (Paatero et al., 2012).

5.1.1.4 Radionuclides Measured in Regional and Local Monitoring Sites

Short-lived radioiodine (^{131}I, ^{132}I and ^{133}I) within the Fukushima-derived radionuclides is the important

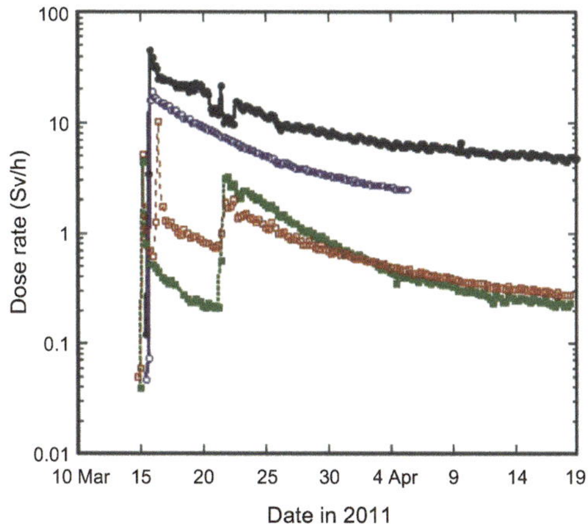

Figure 5.3 Ambient dose rates at Iidate (closed circles) and Fukushima City (open circles) Kitaibaraki (open square) observed by the Fukushima local government, and at Takahagi (closed squares) observed by Ibaraki local government. Abrupt peaks of the ambient dose rate observed in Kitaibaraki and Takahagi on 15 March resulted in passage of the radioactive plume.

radionuclides to monitor in early stage of nuclear reactor accidents because of higher health risk of radiation dose due to inhalation. In the initial stage of the FNPP1 accident, however, there was limited information on short-lived radioiodine concentrations in surface air including gaseous and particulate forms; in the sites near Fukushima, gaseous ^{131}I have been reported at Inage, Chiba (Amano et al., 2012), Oarai, Ibaraki (Furuta et al., 2011), and Tsukuba, Ibaraki (Doi et al., 2013). Sampling stations in the Kanto area are shown in Fig. 5.4. Although Kanai (2012) and Haba et al. (2012) reported the ^{131}I concentrations in aerosols at Tsukuba and Wako, respectively, there was no information on early stage data in March 2011 and gaseous ^{131}I. The gaseous ^{131}I is conventionally defined as radionuclides adsorbed onto activated carbon fiber filter.

The temporal variation of the total ^{131}I concentrations (total gas plus particulate) at Oarai (about 170 km of the FNPP1) and Tsukuba (about 100 km of the FNPP1) is shown in Fig. 5.5. The first peak of ^{131}I, being maximum concentrations at Tsukuba (33 Bq/m^3 (8 h mean value)) and Oarai (1.4×10^3 Bq/m^3 (3 h mean value)) during the observation period, respectively, occurred in 15 and 16 March 2011, which coincided

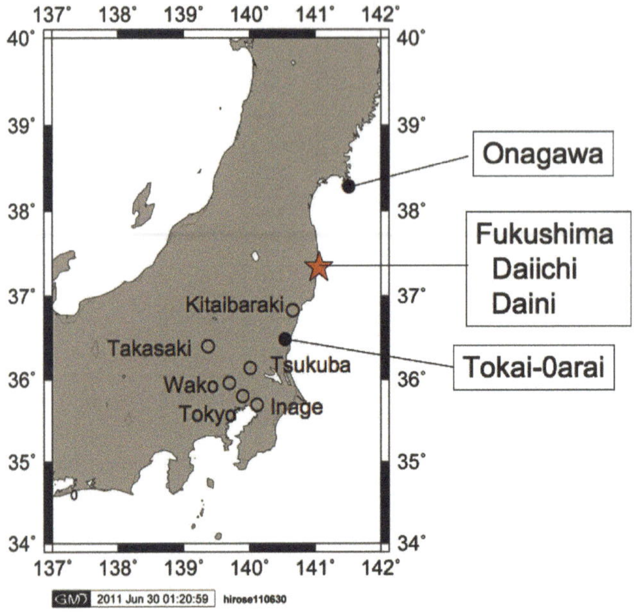

Figure 5.4 Monitoring sites of radionuclides in surface air in Kanto area. Closed circles: location of nuclear power plants. Open circles: Takasaki (CTBTO site), Tsukuba (National Institute of Environment Sciences), Wako (RIKEN), Inage (JCAC), Tokyo and Kitaibaraki (monitoring site of ambient dose rate).

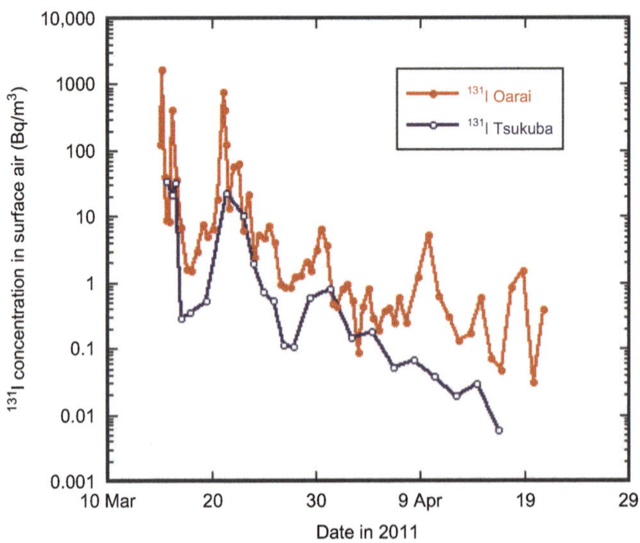

Figure 5.5 Temporal variations of total ^{131}I concentrations in surface air at Oarai and Tsukuba. (Data from Doi et al. (2013) and Furuta et al. (2011))

with that of the ambient dose rate of the corresponding site. In 15 and 16 March 2011, the high ^{131}I concentrations in surface air were observed in Setagaya (Tokyo) (about 240 km of the FNPP1) and Inage (about 220 km of the FNPP1), where the ^{131}I concentrations in surface air were 2.4×10^2 Bq/m^3 (1 h mean value) and 33 Bq/m^3 (27 h mean value) (Amano et al., 2012), respectively. The ^{131}I concentrations in surface air at four sites decreased with distance from the FNPP1. The second peak of the total ^{131}I at Tsukuba and Oarai appeared in 20–22 March, which also coincided with the second peak of the ambient dose rate in the corresponding sites. The similar peak of ^{131}I was observed in Setagaya and Inage. The total ^{131}I concentration in surface air at Tsukuba, which showed some peaks corresponding in 30 March–1 April, 17–19 April and 5–7 May, decreased at an apparent half time of 2.7 day (Doi et al., 2013), and it is difficult to detect ^{131}I in air samples sampled in Kanto Plain in June 2011 (detection limit: 0.1 mBq/m^3).

In order to have better understanding of atmospheric behaviors of ^{131}I, the temporal variation of gaseous ^{131}I/total ^{131}I ratio observed at Tsukuba was examined, in which the gaseous radioiodine was captured with the activated carbon fiber filter (Doi et al., 2013). The result is shown in Fig. 5.6. A lower portion of

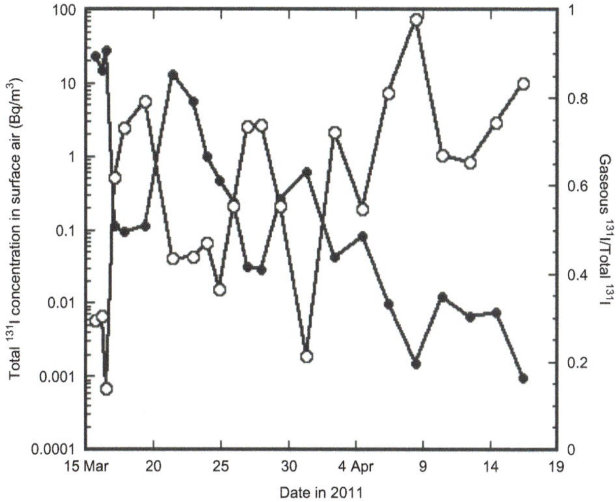

Figure 5.6 Temporal variations of the total ^{131}I concentration and (gaseous ^{131}I)/(total ^{131}I) ratio in surface air at Tsukuba. (Data from Doi et al. (2013))

the gaseous ^{131}I almost corresponded to the peak of the total ^{131}I concentration in surface air; the lowest percentage of the gaseous ^{131}I to the total ^{131}I (13%) appeared in occurrence of the maximum total ^{131}I concentration on 16 March. The percentage of the gaseous ^{131}I to the total ^{131}I in each peak increased with time. The similar trend of increasing percentage of the gaseous ^{131}I was observed in Oarai (Furuta et al., 2011). This finding suggests that as a possible process the gaseous ^{131}I remained in the atmosphere longer than the particulate ^{131}I, and as another possible process the gaseous ^{131}I was preferentially emitted into atmosphere comparing with the particulate ^{131}I due to change of the radioactivity-released conditions in the accidental reactors such as decline of temperature. Other possible cause of increasing gaseous ^{131}I may be volatilization of deposited ^{131}I due to biological processes (Amachi et al., 2001; Muramatsu and Yoshida, 1995).

As did ^{131}I, ^{133}I was detected in the surface air samples collected at Oarai, Tsukuba and Inage surface air samples collected on 15 March 2011 (Doi et al., 2013; Furuta et al., 2011; Amano et al., 2012). The maximum concentrations of ^{133}I in surface air at Oarai, Tsukuba and Inage on 15 March 2011 were 18, 2.6, and 1.9 Bq/m^3, respectively. The ^{133}I concentrations in surface air rapidly decreased and ^{133}I cannot be detected in the end of March 2011. The temporal change of $^{133}I/^{131}I$ ratio in surface air at each site, which was 0.1 on 15 March 2011, decreased exponentially as shown in Fig. 5.7. There was no geographical variation of the temporal changes of the $^{133}I/^{131}I$ ratios observed at Tsukuba, Oarai and Inage (Doi et al., 2013; Furuta et al., 2011; Amano et al., 2012), although lower $^{133}I/^{131}I$ ratios occurred at Tsukuba. A similar result has been obtained in the US CTBTO network data, in which the temporal change of $^{133}I/^{131}I$ ratios measured in the US sites excellently correlated with the ORIGEN model (Biegalski et al., 2011). The temporal change of the $^{133}I/^{131}I$ ratio coincided with a decay curve of a constant ($\lambda_{133}-\lambda_{131}$). This suggests that the $^{133}I/^{131}I$ ratio at the cessation of nuclear reactions in these reactors (on 11 March) was 1.4 ± 0.1, which coincides approximately with the value observed when radioactivity is released through pin holes and tight cracks of nuclear

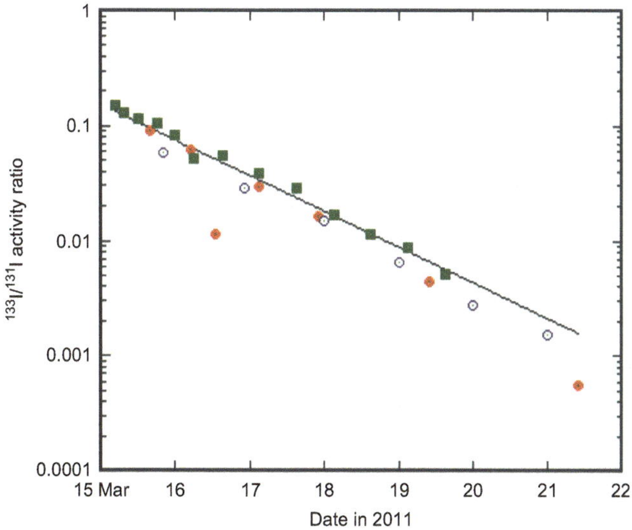

Figure 5.7 Temporal variations of $^{133}I/^{131}I$ activity ratios in surface air. Closed circles: Tsukuba, open circles: Inage, and closed squares: Oarai. (Data from Amano et al. (2012), Doi et al. (2013), and Furuta et al. (2011))

fuel rods in equilibrium with radioactivity inside the fuel cladding material (CPSMA, 1996).

The ^{134}Cs, ^{136}Cs (half-life: 13.16 days) and ^{137}Cs were detected in surface air at Oarai, Tsukuba, Wako and Inage (Doi et al., 2013; Furuta et al., 2011; Haba et al., 2012; Amano et al., 2012). The temporal variation of the ^{137}Cs concentrations in surface air at Oarai, Inage and Taksasaki is shown in Fig. 5.8. The first peak of the surface ^{137}Cs in Kanto Plain occurred in 15 and 16 March 2011, which coincided with that of the total ^{131}I peak and the ambient dose rate (Fig. 5.3). On 15 March 2011, the high ^{134}Cs and ^{137}Cs concentrations in surface air were observed in Oarai, Tsukuba, Wako and Inage, where the ^{137}Cs concentrations in surface air were 1.9×10^2 Bq/m^3 (3 h mean value) (Furuta et al., 2011), 3.8 Bq/m^3 (3 h mean value) (Doi et al., 2013), 8.8 Bq/m^3 (0.5 h mean value) and 0.87 Bq/m^3 (27 h mean value) (Amano et al., 2012), respectively. The ^{137}Cs concentrations in surface air at three sites decreased with distance from the FNPP1 except Wako, although a simple comparison between data cannot be carried out because of different air sampling time span. In 18–20 March, the ^{137}Cs concentrations in surface air rapidly

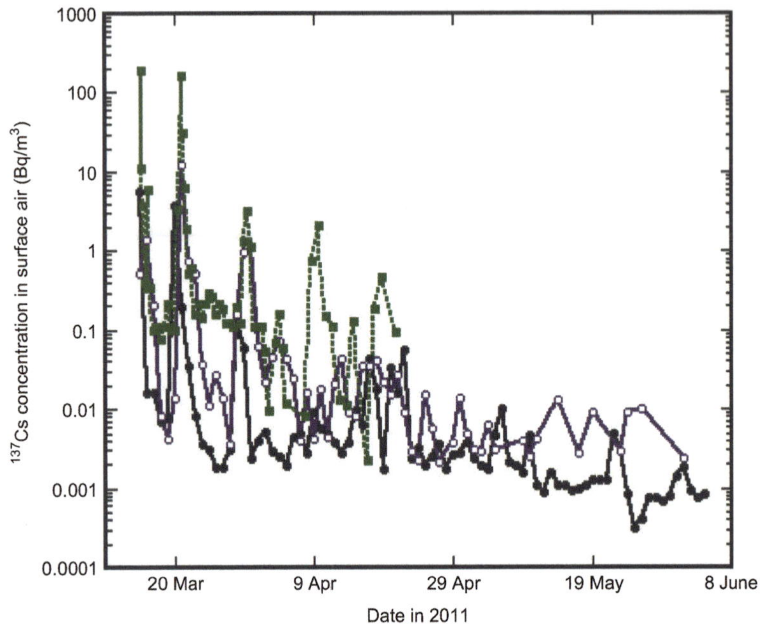

Figure 5.8 Temporal variations of the ^{137}Cs concentrations in surface air observed at Oarai (green squares), Tsukuba (closed circles) and Inage (open circles). (Data from Furuta et al. (2011), Doi et al. (2013), and Amano et al. (2012))

decreased to about three orders of magnitude lower than previous value in contrast of ^{131}I. The second peak of the surface ^{137}Cs in Kanto Plain appeared in 20–22 March. The observed second peak concentrations of ^{137}Cs in surface air at Oarai, Tsukuba and Inage were 160 Bq/m^3 (9 h mean value), 4.6 Bq/m^3 (48 h mean value) and 6.1 Bq/m^3 (26 h mean value), respectively. The third peak of the surface ^{137}Cs in Kanto Plain appeared in 29–31 March. The observed third peak concentrations of ^{137}Cs in surface air at Oarai, Tsukuba and Inage were 3.2 Bq/m^3 (12 h mean value), 0.23 Bq/m^3 (48 h mean value) and 0.69 Bq/m^3 (24 h mean value), respectively. After that, the ^{137}Cs concentration in surface air in Kanto Plain gradually decreased, although some ^{137}Cs peaks occurred in April–September, and ^{137}Cs can still be detected in air samples in early September 2011 (detection limit: 0.05 mBq/m^3). It is noteworthy that ^{134}Cs and ^{137}Cs in surface air at Fukushima City have still been observed in the early 2012, whose concentrations were 0.1 to several millibecquerel per cubic meter (MEXT, 2012a,b,c).

Isotope signature is an important tool to have better understanding of the environmental behavior of the FNPP1-derived radionuclides. As it is well known that the ^{134}Cs is not a direct fission product, ^{134}Cs is produced by neutron activation of ^{133}Cs, which is the decay product of ^{133}Xe present in nuclear reactors. $^{134}Cs/^{137}Cs$ ratios increase with burn-up time of nuclear fuel. Mutual relationships between ^{134}Cs and ^{137}Cs in surface air were examined (Amano et al., 2012; Doi et al., 2013; Haba et al., 2012; Biegalski et al., 2011). The $^{134}Cs/^{137}Cs$ activity ratio in surface air at all sites including Japanese and US stations, ranged from 0.9 to 1.2 as an average of 1.0 ± 0.1, was fairly stable over the time of observation, which coincides with that of deposition (Amano et al., 2012; Hirose, 2012). It must be noted that the $^{134}Cs/^{137}Cs$ activity ratio in the FNPP1-derived radionuclides is higher than that in the Chernobyl fallout ($^{134}Cs/^{137}Cs$ ratio: 0.5) (IAEA, 1986; UNSCEAR, 2000). Temporal change of the $^{136}Cs/^{137}Cs$ activity ratio is shown in Fig. 5.9. In contrast to the

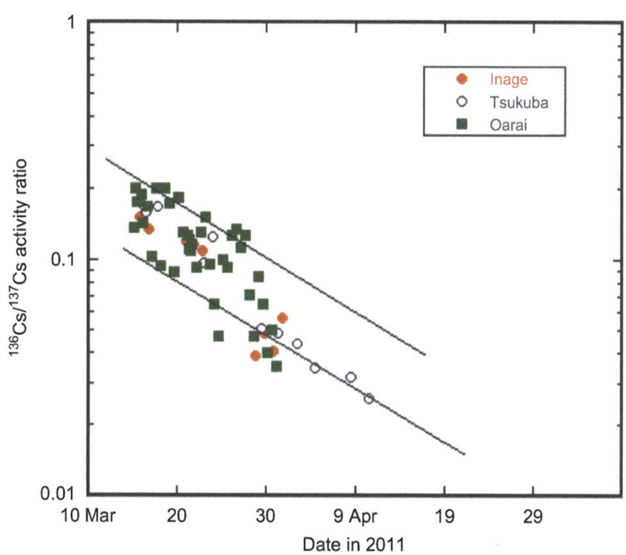

Figure 5.9 Temporal variations of $^{136}Cs/^{137}Cs$ activity ratios in surface air observed at Oarai, Tsukuba and Inage. (Data from Furuta et al. (2011), Doi et al. (2013), and Amano et al. (2012))

^{134}Cs/^{137}Cs ratios, there was rather large variability of the ^{136}Cs/^{137}Cs ratio in surface air. The similar different nature of the ^{136}Cs/^{137}Cs ratio from the ^{134}Cs/^{137}Cs ratio was observed in the US monitoring sites (Biegalski et al., 2011). Although the ^{136}Cs/^{137}Cs ratio decreased exponentially, the ^{136}Cs/^{137}Cs ratios at three sites in Japan seem not to fit to a simple decay curve of ^{136}Cs; the ^{136}Cs/^{137}Cs ratios in late March–April 2011 were slightly lower than that in the period of the first and second peaks (15–22 March). The ^{136}Cs/^{137}Cs ratio was calculated to extrapolate at cease of reactors (11 March 2011) from the decay curve of ^{136}Cs. The initial ^{136}Cs/^{137}Cs ratio ranged from 0.09 to 0.27. These findings suggest that there is no isotopic fractionation between ^{134}Cs and ^{137}Cs, whereas the ^{136}Cs/^{137}Cs ratio may be changed between major emission and following minor emission, although its cause is unknown. In this connection, Biegalski et al. (2012) mentioned that excess ^{137}Cs could be present in the samples due to a release from damaged fuel in the FNPP1 spent nuclear fuel pools, taking into account the fact that temporal change of the ^{136}Cs/^{137}Cs ratios appear to be generally below the ORIGEN model (ORNL, 2004). Although there is no report of the radioactivity release from the FNPP1 spent nuclear fuel pools, a possibility of radioactivity release from the spent fuel ponds (Units 2 and 3) could not be excluded (Chapter 3). On the other hand, it is noteworthy that determination of ^{136}Cs by gamma-spectrometry is rather complicated in the initial stage because of interference of short-lived radionuclides.

As other radionuclides, ^{132}Te (^{132}I), ^{129m}Te (^{129}Te), and ^{99}Mo (^{99m}Tc) were detected in the surface air samples at Tsukuba (Doi et al., 2013; Haba et al., 2012; Kanai, 2012). Temporal variations of ^{132}Te, ^{129m}Te, and ^{99}Mo (^{99m}Tc) in surface air at Tsukuba are shown in Fig. 5.10. ^{132}Te, ^{129m}Te, and ^{99}Mo(^{99m}Tc) were detected on 15 March 2011, which coincided with the first detection of radioiodine and radiocesium. The maximum concentrations of ^{132}Te and ^{129m}Te, which occurred on 15 March, were 23 and 3.8 Bq/m^3, respectively, which corresponded to the maximum of radioiodine, whereas the maximum ^{99}Mo (^{99m}Tc) concentration (0.46 Bq/m^3) occurred in

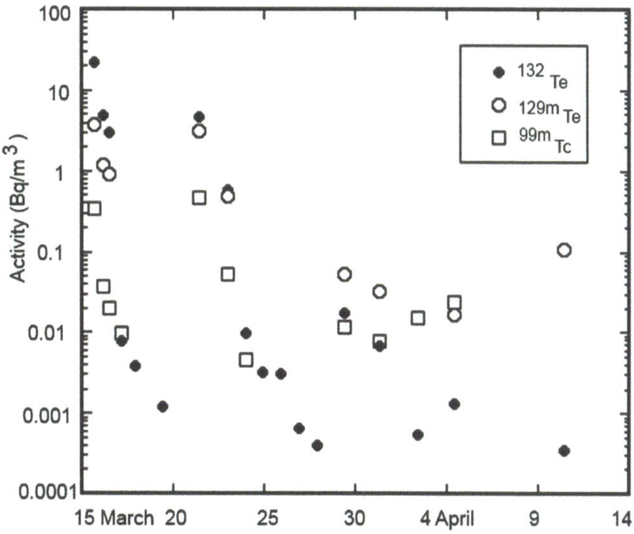

Figure 5.10 Temporal variations of ^{132}Te, ^{129m}Te and ^{99m}Tc concentrations in surface air at Tsukuba. (Data from Doi et al. (2013))

20–22 March corresponding to the maximum of radiocesium. The surface air concentrations of ^{132}Te, ^{129m}Te, and ^{99}Mo (^{99m}Tc) exhibited three major peaks, which corresponded with that of ^{131}I and ^{137}Cs, and decreased to less than detection limit after mid-April 2011.

In order to elucidate difference of atmospheric behavior between radionuclides, mutual relationships between ^{132}Te, ^{129m}Te, and ^{99}Mo (^{99m}Tc) were examined, in which activity concentrations of short-lived radionuclides were corrected radioactive decay on 11 March 2011 (Doi et al., 2013). The decay-corrected ^{132}Te concentration in surface air well correlated with that of the decay-corrected ^{129m}Te, in which the ^{129m}Te/^{132}Te ratio was 0.086 ± 0.005. It is noteworthy that PMJC (2011) documented the ^{129m}Te/^{132}Te ratio of 0.04. The ratios of the decay-corrected ^{132}Te, decay-corrected ^{129m}Te, and decay-corrected ^{99m}Tc concentrations to the ^{137}Cs concentration in surface air were calculated. The decay-corrected ^{132}Te/^{137}Cs and ^{129m}Te/^{137}Cs ratios were in the ranges of 4–65 and 0.3–5.4, respectively. Peaks of the decay-corrected ^{132}Te/^{137}Cs and ^{129m}Te/^{137}Cs ratios appeared in

22–23 March after occurrence of the maximum concentrations of ^{137}Cs in 20–22 March. There was rather large variation of $^{132}Te/^{137}Cs$ and $^{129m}Te/^{137}Cs$ ratios, suggesting that atmospheric behavior of radiotellurium differed from that of the radiocesium. The decay-corrected ^{99}Mo (^{99m}Tc)/^{137}Cs, ranging from 0.16 to 22, showed larger variation than that of the decay-corrected $^{132}Te/^{137}Cs$ and $^{129m}Te/^{137}Cs$. The decay-corrected ^{99}Mo (^{99m}Tc)/^{137}Cs increased with time. These findings suggest that the atmospheric behavior of ^{99}Mo (^{99m}Tc) was significantly different from that of ^{137}Cs and ^{132}Te.

Atmospheric behaviors of the Fukushima-derived radionuclides depend on physical and chemical properties of radionuclide-bearing particles. Particle size distributions of the Fukishima-derived radionuclide-bearing particles collected at Tsukuba were determined during the periods of 4–11 April and 14–21 April 2011 (Doi et al., 2013) and 29 April–12 May and 12–26 May 2011 (Kaneyasu et al., 2012). The activity median aerodynamic diameters (AMAD) of ^{131}I-bearing particles were calculated to be 0.7 and 0.7 μm in 4–11 April 2011 and in 14–21 April 2011, respectively, whereas the AMADs of ^{134}Cs and ^{137}Cs-bearing particles were 1.8 and 1.5 μm in 4–11 April 2011, and 1.0 and 1.0 μm in 14–21 April 2011, respectively. Measurements of the particle-size distributions of radiocesium-bearing particles in May (Kaneyasu et al., 2012) revealed that ^{137}Cs as did ^{134}Cs attached on submicrometer particles, typically sulfate particles. These findings suggest that the particle size of the ^{131}I-bearing particles differed from those of radiocesium, which implies that the dispersion and deposition behaviors of ^{131}I differed from that of ^{134}Cs and ^{137}Cs and that the particle size of the radiocesium-bearing particles varied with time. The result reveals that the FNPP1-derived ^{134}Cs and ^{137}Cs-bearing particles observed in April 2011 were larger than that of the Chernobyl ^{137}Cs found in remote sites, which existed as submicrometer particles observed in Japan (Aoyama et al., 1992). This finding suggests that the Fukushima-derived radiocesium observed in April was preferentially removed from atmosphere by dry and wet deposition processes, and that as a result, the Fukushima-derived radiocesium observed in May, existing as submicrometer particles, remained in the atmosphere (Aoyama et al., 1992; Hirose et al., 1993; Hirose, 1995). Paatero et al. (2012) estimated that a significant part of the Fukushima-derived radioactivity is in hot particles from autoradiogram of a filter sample from 1 to 4 April 2011 at Mt. Zeppelin, Ny-Ålesund, Svalbard. It must be noted that there is no information about particle size of radionuclide-bearing particles in major radioactive plume occurring in the late March 2011.

The surface air concentrations of the Fukushima-derived radionuclide decreased with distance from the FNPP1. The model simulation (Morino et al., 2011) revealed that the radioactive plume originating from the FNPP1 flowed southwestward in 20–21 March according to northeast wind. In order to elucidate difference of the atmospheric behavior between ^{131}I and ^{137}Cs, the surface air concentrations of ^{131}I and ^{137}Cs at four sites corresponding to downstream of the FNPP1 were potted as a function of distance from the FNPP1 as shown in Fig. 5.11. The result reveals that the surface air concentrations of ^{131}I and ^{137}Cs decreased exponentially with distance shown in Fig. 5.11. Apparent half decrease distances (AHDDs) of ^{131}I and ^{137}Cs can be calculated to be 32 and

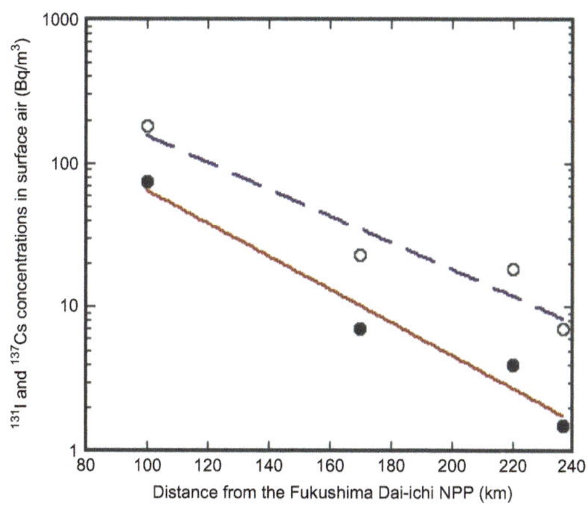

Figure 5.11 Relationship between ^{131}I and ^{137}Cs concentrations in surface air and distance of the monitoring sites from the Fukushima Dai-ichi NPP.

26 km from the fitting curves, respectively. The AHDD of ^{131}I is longer than that of ^{137}Cs, which implies that ^{137}Cs emitted from the FNPP1 is easily removed from atmosphere comparing with ^{131}I.

5.1.1.5 Air Monitoring in Remote Sites in Japan

The concentrations of the FNPP1-derived radionuclides in surface air were determined in the western part of Japan (Momoshima et al., 2012). Figure 5.12 shows temporal variations of the ^{131}I and ^{137}Cs concentrations in surface air (dust) in Fukuoka (32° 27′ N, 130° 25′ E, 1050 km west of the FNPP1) and Matsue (35° 28′ N, 133° 3′ E, 750 km west of the FDNPP). Iodine-131 was first detected in dust

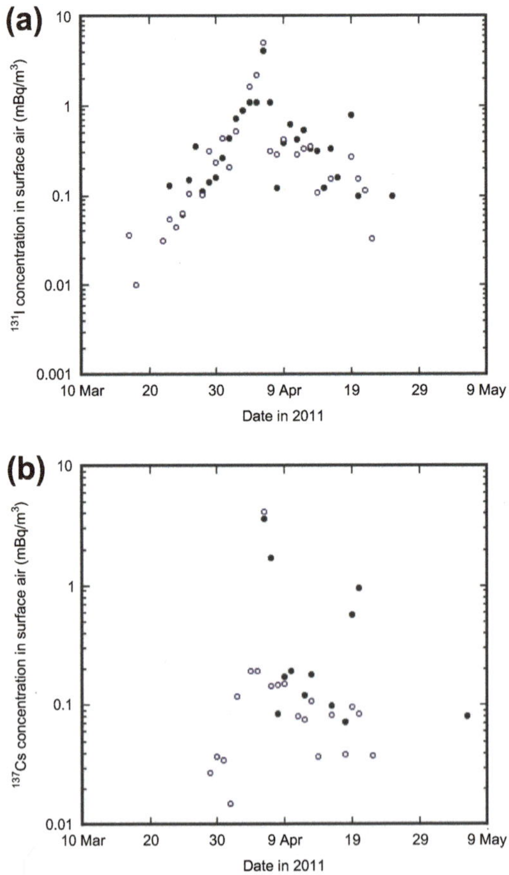

Figure 5.12 Temporal variations of ^{131}I and ^{137}Cs concentrations in surface air. Open circles: Fukuoka and closed circles: Matsue. (Data from Momoshima et al. (2012))

samples on 17 March 2011 at Fukuoka, although its level was very low (0.036 mBq/m^3) in contrast of the stations in Kanto Plain. At Fukuoka and Matsue, the ^{131}I concentrations in surface air exponentially increased and reached a maximum on 6 April. After occurrence of a first peak, the ^{131}I concentration in surface air decreased. On 20 April, a weak second peak of the surface ^{131}I occurred, in which the second peak was more remarkable at Matsue. After that, the ^{131}I concentration decreased and was less than detection limit until end of April.

Cesium-137 was first detected in dust samples on 29 March 2011 at Fukuoka, although its level was very low (0.027 mBq/m^3) as did ^{131}I. At Fukuoka, the ^{137}Cs concentrations in surface air rapidly increased and reached a maximum on 6 April, when coincided with the maximum of ^{131}I. After occurrence of a peak, the ^{137}Cs concentration in surface air in Fukuoka decreased and was less than the detection limit until the end of April, whereas a marked second peak appeared on 20 April. A similar peak of the ^{137}Cs concentrations in surface air was observed in South Korea (Busan; 35° 27′ N, 129° 06′ E) and North Taiwan (Pengchiayu Islet; 25° 37′ 12″ N, 122° 4′ 12″ E) on 7 April (Huh et al., 2012; Kim et al., 2012).

5.1.2 Dry Deposition and Precipitation

5.1.2.1 Daily Deposition at Local and Regional Stations

Measurements of radionuclides in daily deposition samples, which include wet and dry depositions, started on 18 March as emergency monitoring at Japanese Government monitoring stations (MEXT, 2012a,b,c). Daily radioactive deposition rates in 44 stations in Japan were recorded until the end of 2011. High radioactive deposition derived from the FNPP1 accident occurred in a wide area of Kanto Plain and South Tohoku from 21–23 March 2011. On the other hand, the record of the ambient dose rate in the area (Fukushima Prefecture) adjacent to the FNPP1 (Fig. 5.3) suggests that high radioactive deposition occurred in 15–16 March accompanied with snowfall, although there was no measurement of the daily radioactivity deposition during the period of March 2011.

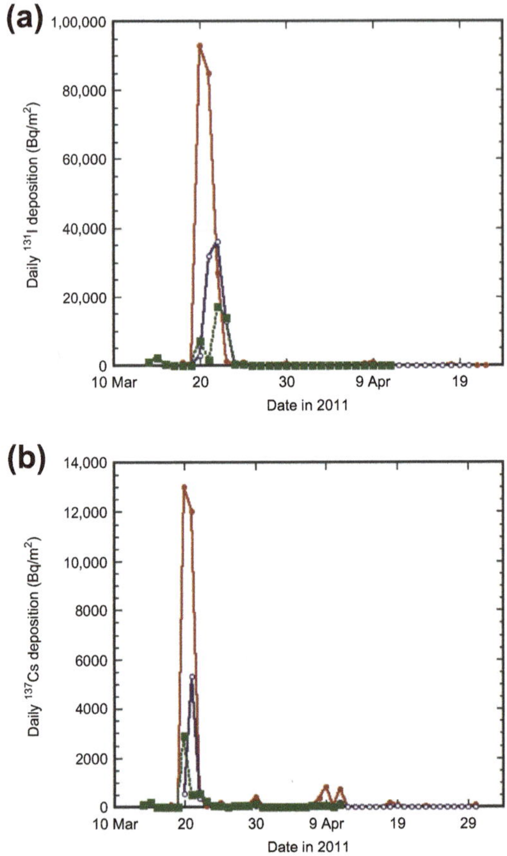

Figure 5.13 Daily ^{131}I and ^{137}Cs deposition rates at Hitachinaka, Tokyo and Inage. (Data from MEXT (2012a,b,c) and Amano et al. (2012))

The temporal changes of the daily ^{131}I and ^{137}Cs deposition rates during the period of March–April 2011 are shown in Fig. 5.13. Earliest daily deposition sampling in Kanto Plain was performed by Japan Chemical Analysis Center (JCAC) (Inage, Chiba Prefecture) (Amano et al., 2012). The first deposition of the Fukushima-derived radionuclides appeared on 15 March, which coincided with the first arrival of the radioactive plume as seen in Figs 5.1 and 5.5. The first peaks in ^{131}I and ^{137}Cs depositions (2400 and 76 Bq/m^2, respectively) were observed at Inage on 16 March. It must be noted that at the emergency stage of the accident, Government Monitoring documented only ^{131}I and ^{137}Cs as dominant radionuclides in deposition samples from the FDNPP accident by MEXT (MEXT, 2012a,b,c). On 20 March, the level of ^{131}I, a dominant radionuclide, at Hitachinaka (36.40° N, 140.54° E), which is located about 120 km south of the FNPP1, was 91 kBq/m^2 day as a maximum value, and ^{137}Cs deposition at Hitachinaka was 13 kBq/m^2 day.

The high radionuclide deposition rates (^{131}I: 58 kBq/m^2 day and ^{137}Cs: 4.3 kBq/m^2 day) were observed at Yamagata (38.25° N, 140.33° E), which is located about 110 km northwest of the FNPP1. On 21 March, the high deposition rates were observed in all the monitoring stations of Kanto Plain. In Inage, the highest daily deposition rates of ^{131}I and ^{137}Cs were recorded on 23 March (^{131}I: 17 kBq/m^2 day and ^{137}Cs: 2.9 kBq/m^2 day) (Amano et al., 2012). The high radioactive deposition rates were accompanied with rainfall continued until 23 March. On 24 March, the radioactive deposition rates decreased dramatically due to fine weather with coverage of high-pressure system. Most of the FNPP1-derived radionuclides, which were transported from the FDNPP by northeast wind, were deposited on land surface by rainfall during 20–23 March. It must be noted that the enhanced ambient dose rates in the Kanto and Tohoku areas were governed by high radioactive deposition rates in 20–23 March as seen in Fig. 5.3. The daily radioactive deposition derived from the FNPP1 accident was observed in April 2011. In May 2011, the daily deposition rate of the Fukushima-derived radionuclides in most of the Japanese monitoring stations decreased below the detection limit except in the Fukushima City (37.75° N, 140.47° E), which is located about 60 km northwest of the FNPP1. In Fukushima City, in which the radioactivity measurement in daily deposition samples started on 1 April, the daily ^{137}Cs deposition decreased from April 2011 to October 2011, although some high peaks occurred in the period of June–August 2011, which corresponded to heavy rainfall events. However, the enhanced daily ^{137}Cs deposition (100 Bq/m^2 day) appeared in the period of December 2011–April 2012. The significant amount of ^{137}Cs was still detected in the daily deposition samples in June 2012 (Fig. 5.14).

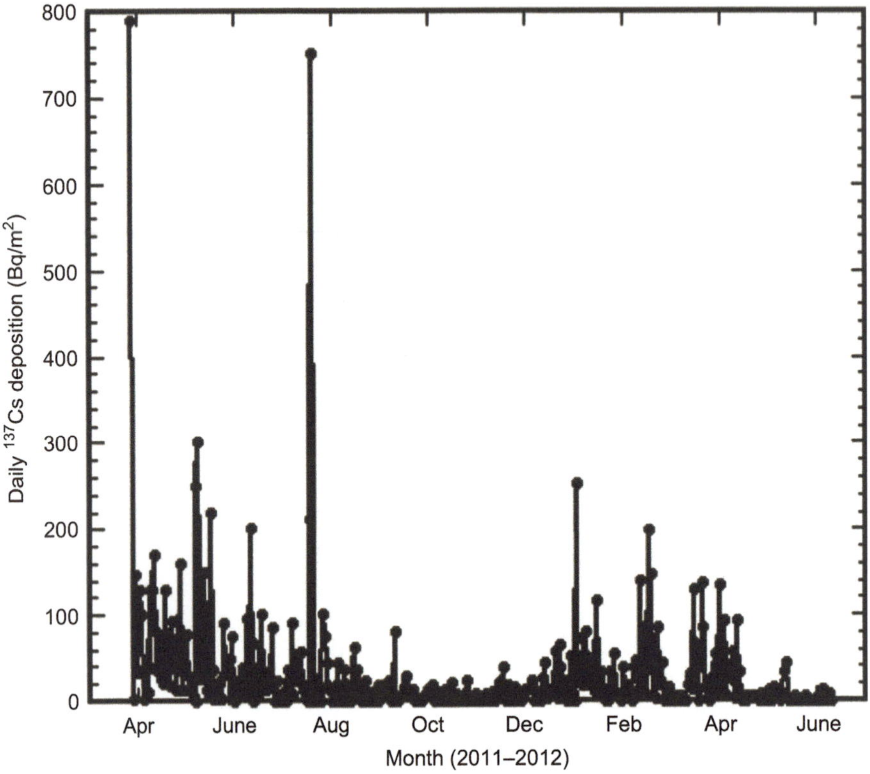

Figure 5.14 Daily ^{137}Cs depositions observed in Fukushima City. (Data from MEXT (2012a,b,c))

The $^{131}I/^{137}Cs$ activity ratio in the daily deposition samples, ranged from 0.3 to 230, varied temporally and spatially. The $^{131}I/^{137}Cs$ activity ratio in total atmospheric release was estimated to be 11 (NSC, 2011). To compare with the $^{131}I/^{137}Cs$ activity ratio at the initial release of radionuclides from the FDNPP, the radioactive decay of ^{131}I was corrected on 11 March when shutdown of the nuclear reactors was conducted. The decay-corrected $^{131}I/^{137}Cs$ activity ratios ranged from 3.5 to 550, with a median value of 15. The high decay-corrected $^{131}I/^{137}Cs$ activity ratios occurred in 22–23 March, which corresponded to the time just after occurrence of the maximum deposition rates (20–21 March). Higher $^{131}I/^{137}Cs$ activity ratios appeared at the inland site (Utsunomiya: 36.60° N, 139.94° E), which means that the depositional behavior between ^{131}I and ^{137}Cs largely differed from each other. The $^{134}Cs/^{137}Cs$ activity ratios in the daily deposition samples (=1.0) were constant (Amano et al., 2012).

5.1.2.2 Spatial Distributions of Monthly Deposition

Environmental Monitoring Centers in Local Governments have determined the Fukushima-derived radionuclides in monthly deposition samples. ^{131}I, ^{134}Cs and ^{137}Cs were measured as dominant radionuclides in the monthly deposition samples in Japan, whereas ^{129m}Te, ^{129}Te, ^{136}Cs, ^{110m}Ag, ^{95}Zr, ^{95}Nb (half-life: 35 days), ^{140}Ba and ^{140}La (half-life: 1.68 days) were detected as minor radionuclides in the monthly deposition samples at the monitoring stations within 300 km of the FNPP1. For the monthly deposition, long-lived radionuclides (^{134}Cs, ^{137}Cs and ^{110m}Ag) are meaningful because the monthly deposition rates of the short-lived radionuclides are not well defined. Activity ratios of $^{134}Cs/^{137}Cs$ observed on March, ranged from 1.0 to 1.1 as an average of 1.03, were constant in cases where the monthly ^{137}Cs deposition exceeded 100 Bq/m^2. The $^{134}Cs/^{137}Cs$ activity ratios on April and May at MEXT monitoring sites were from 0.99 to 1.09 with an average of 1.03 and from 0.93 to 1.04 with an average of 0.99, respectively. The result suggests that there were no spatial and temporal variations of the $^{134}Cs/^{137}Cs$ activity ratios in the monthly deposition samples during the period from March to May 2011. Activity ratios of $^{110m}Ag/^{137}Cs$ observed on March ranged from 0.0009 to 0.006 as an average of 0.0019.

The spatial distributions of the monthly ^{137}Cs deposition in March–May are depicted in Fig. 5.15(a), (b) and (c), respectively. The highest monthly ^{137}Cs deposition (3340 kBq/m^2) was observed in March at Futaba (Fukushima Prefecture), about 5 km of the FDNPP. The monthly ^{137}Cs depositions at the stations within 300 km of the FNPP1 except Kofu (35.65° N, 138.57° E; inland site) and the Japan Sea side sites (Niigata: 37.91° N, 139.04° E and Akita: 39.72° N, 140.10° E) were in the range from 1.1 to 17 kBq/m^2, which are greater than the maximum monthly ^{137}Cs deposition (0.55 kBq/m^2) originating

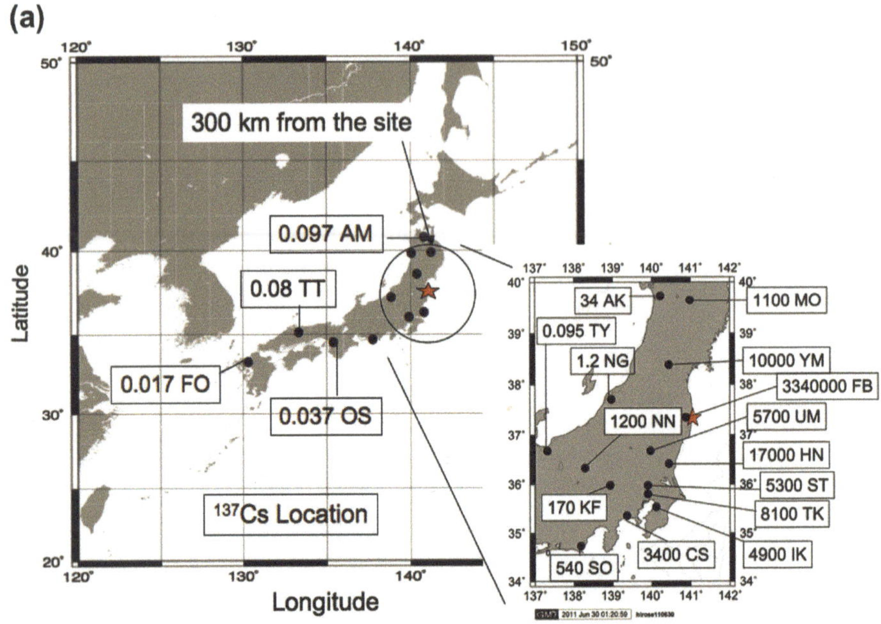

Figure 5.15 Spatial distributions of monthly ^{137}Cs deposition rates in Japan (unit: Bq/m^2). a: March 2011, b: April 2011, c: May 2011. Monitoring stations: SP (Sapporo), TM (Tomari), AM (Aomori), HT (Higashi-dori), AK (Akita), MO (Morioka), YM (Yamagata), NG (Niigata), FB (Futaba), UM (Utsunomiya), HN (Hitachinaka), ST (Saitama), TK (Tokyo), IK (Ichihara), CS (Chigasaki), SO (Shizuoka), KF (Kofu), NN (Nagano), TY (Toyama), FK (Fukui), OS (Osaka), FO (Fukuoka), KC (Kochi), and SS (Satsuma-sendai). (Data from MEXT (2012a,b,c))

from the 1961–1962 large-scale atmospheric nuclear testing observed at Koenji (Tokyo) in 1963 (Hirose et al., 2008; Katsuragi, 1983).

The results reveal that the high ^{137}Cs-deposited areas, comparable to the cumulative amount of the ^{137}Cs deposition at Tokyo until mid-1960 (about 7 kBq/m^2), appeared within a region band from 100 to 300 km from the FNPP1. The spatial distribution of the monthly ^{137}Cs deposition on March 2011 revealed that the major deposition of the FDNPP-derived radionuclides occurred in the North Pacific coast and inland area of the east Honshu Island, whereas there was less contribution of the Fukushima-derived radionuclides in the Japan Sea side sites of the east Honshu Island. These findings suggest that the transport of the radioactive plume is strongly affected by land topography and that most of the FDNPP-derived radionuclides might be injected in

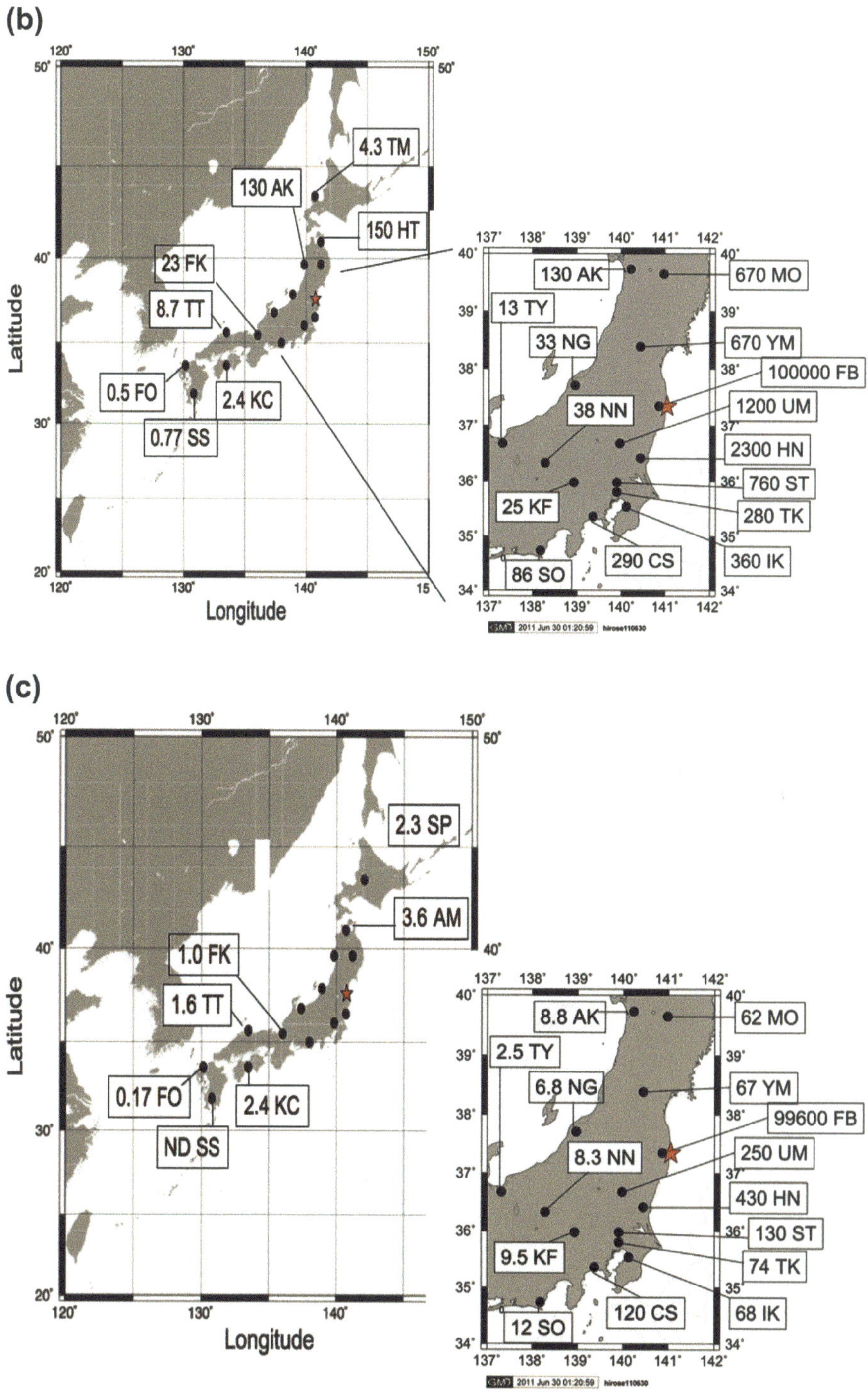

Figure 5.15 Continued

the boundary layer (about 1000 m). Model simulation (Morino et al., 2011) revealed that land topography controls the transport of the Fukushima-derived radioactive plume.

The higher monthly ^{137}Cs deposition (0.17 Bq/m^2), which was one order of magnitude higher than pre-Fukushima levels (Hirose et al., 2008; Igarashi et al., 2003, 2005), was observed on March at Fukuoka (33.51° N, 130.50° E) and Uruma (26.31° N, 127.90° E, Okinawa), which are located about 1050 and 1750 km southwest apart from the FDNPP, respectively. Detection of ^{131}I and ^{134}Cs in the same sample revealed that the FDNPP-derived radionuclides were transported to Fukuoka and Uruma in the late March, which is consistent with the results of the Fukushima-derived radionuclides in surface air (Momoshima et al., 2012). The highest monthly deposition rates of radiocesium occurred at Fukuoka and Uruma in April 2011; the monthly ^{137}Cs depositions at Fukuoka and Uruma were 0.5 and 3.7 Bq/m^2, respectively. It must be noted that the ^{137}Cs deposition at Uruma is higher than that at Fukuoka. Model simulation (Takemura et al., 2011) suggested that the FDNPP-derived radioactivity plume spread far eastern Siberia on 24 March 2011. Another model simulation (Huh et al., 2012) revealed that the FDNPP-derived radioactive plume was predominantly transported toward the southwest under phases of northern-easterly winds in the first week of April (6–7). It is likely that the Fukushima-derived radionuclides observed at Fukuoka and Uruma were transported via the east Siberia or the North Pacific southwest of Honshu Islands rather than the round globe.

In April, higher monthly ^{137}Cs depositions were observed at the North Pacific side stations and East Japan inland stations, although its levels decreased markedly. This suggests that the atmospheric emission of radionuclides from the FNPP1 at least continued within April 2011, although the release rate of radioactive materials dramatically decreased. On the other hand, the monthly ^{137}Cs deposition increased at southwest sites in Japan and at the Japan Sea side sites comparing with that in March, suggesting that the Fukushima-derived radioactive cloud dominantly affected north part of the Northern

Hemisphere atmosphere. In May, the monthly ^{137}Cs depositions decreased at all the monitoring stations of Japan (Fig. 5.15(c)), although the higher ^{137}Cs depositions were observed within 300 km of the FNPP1. In August, the monthly ^{137}Cs depositions at many monitoring stations in southwest Japan decreased below detection limit.

5.1.2.3 Temporal Variations of the Fukushima-Derived Radionuclides

In order to elucidate the depositional behavior of the Fukushima-derived radionuclides, it is important to examine the temporal variation of the monthly ^{137}Cs deposition in each monitoring station. The monthly ^{137}Cs deposition at the sites within 300 km of the FNPP1 except Futaba (Fukushima) exhibited exponential decreases during the period of March–June as shown in Fig. 5.16. Apparent half-lives of the Fukushima-derived atmospheric ^{137}Cs ranged from 8.8 days for Tokyo to 14 days for Morioka (39.70° N, 141.16° E) (Hirose et al., 2012). The Chernobyl-derived ^{137}Cs in surface air observed at Chilton, Gibraltar, Tromso and Hong Kong declined a half-life of 6.3 days

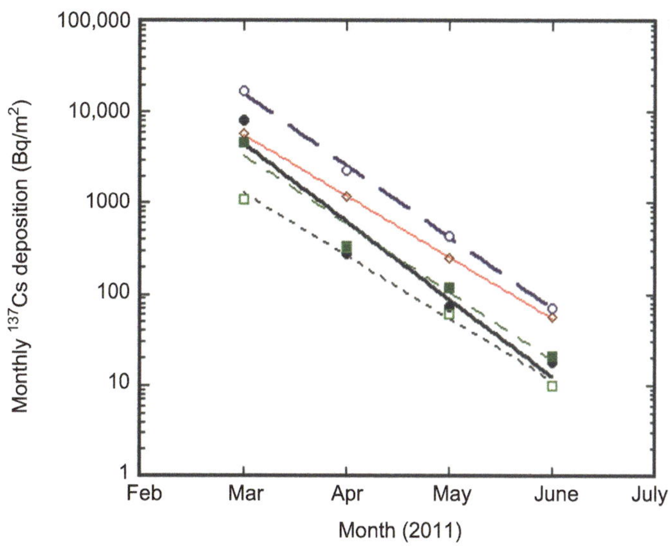

Figure 5.16 Temporal variations of monthly ^{137}Cs deposition rates. Closed circles: Tokyo, open circles: Hitachinaka, closed squares: Maebashi, and open squares: Morioka, rhombics: Utsunomiya. (Data from MEXT (2012a,b,c))

during 2 months after the Chernobyl accident (Chamberlain, 1991).

The apparent atmospheric half-life of the Fukushima-derived ^{137}Cs is similar to the atmospheric half-residence time (8 days) estimated for ^{210}Pb over the west central United States, based on the radioactive equilibrium of ^{210}Pb with its short-lived progenies (Moore et al., 1973). Lambert et al. (1982) estimated a global mean aerosol half-residence time of 6.5 days by using atmospheric inventories of ^{222}Rn and ^{210}Pb extrapolated from observations and computing the ^{210}Pb deposition sink to balance ^{222}Rn decay. The half-residence times of tropospheric aerosols inferred from a global three-dimensional simulation of ^{210}Pb were 5–10 days, which depends on season and latitude (Balkanski et al., 1993). On the other hand, longer troposphere residence time of aerosols (about 30 days) was obtained from radioactive debris of atmospheric nuclear explosions (Katsuragi, 1983). A similar long residence time was estimated for the Chernobyl ^{137}Cs (25 days) (Aoyama, 1988), which was emitted as submicrometer particles (Hirose, 1995). The residence times of aerosols in troposphere, which are in the range of 5–30 days, have been determined by natural and anthropogenic radionuclides, which depend on particles size and altitude (Ehhalt, 1973). The shorter half-life of the Fukushima-derived radionuclides suggests that most of the Fukushima-derived radionuclides were injected into the lower layer of the troposphere and/or emitted as larger particles in the atmosphere. As another possible process, the apparent half-life may partly reflect a time course of the atmospheric emission because the radioactive emission from the FNPP1 continued still now (the end of 2011) although the emission rate of radioactivity dramatically decreased (TEPCO, 2011a,b).

In Futaba, the monthly ^{137}Cs deposition showed different deposition history from other monitoring stations as shown in Fig. 5.17. After the record of the highest monthly deposition (3.34 MBq/m^2) on March 2011, rapid decrease of the monthly ^{137}Cs deposition occurred on April, after that, the ^{137}Cs monthly deposition decreased until October 2011. The decrease rate of the monthly ^{137}Cs deposition was slower than that of

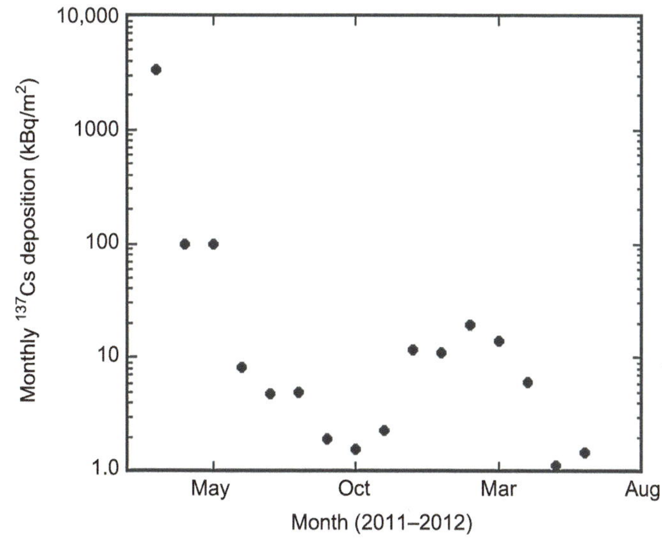

Figure 5.17 Temporal variation of monthly ^{137}Cs deposition rates at Futaba, Fukushima Prefecture. (Data cited from MEXT (2012a,b,c))

other stations in the corresponding period. The monthly ^{137}Cs deposition increased on November 2011 and continued a higher level of >10 kBq/m^2 during the period from December 2011 to March 2012. The minimum value of the monthly ^{137}Cs deposition (1.12 kBq/m^2) at Futaba occurred on May 2012 during the period from March 2011 to June 2012, which was more than 70 times greater than that observed in the Kanto area. It is likely that the monthly ^{137}Cs deposition at Futaba was strongly affected by the atmospheric emission from the FNPP1.

The monthly ^{137}Cs deposition after July 2011 at the monitoring stations in Kanto Plain exhibited slower decrease rates as shown in Fig. 5.18, although ^{137}Cs was below detection limit in the monthly deposition samples for most of the monitoring stations in southwestern region of Japan. In most of the monitoring stations in Kanto Plain, meaningful amounts of the monthly radiocesium deposition have been measured on June 2012 (MEXT, 2012a,b,c). The decrease rate of the monthly ^{137}Cs deposition changed during the period from July to November 2011. The monthly ^{137}Cs deposition showed increasing tendency after occurrence of a minimum deposition on October or November. The second

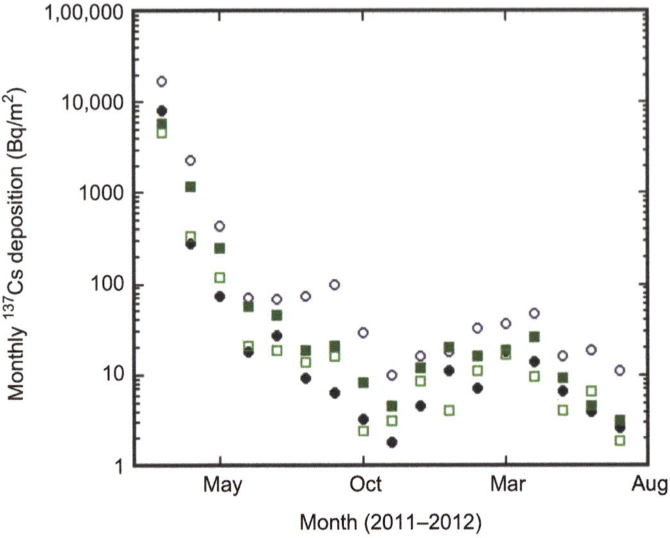

Figure 5.18 Temporal variations of monthly ^{137}Cs deposition rates in the Kanto Plain during the period from March 2011 to July 2012. Closed circles: Tokyo, open circles: Hitachinaka, closed squares: Utsunomiya, and open squares: Maebashi. (Data from MEXT (2012a,b,c))

peak of the monthly ^{137}Cs deposition was observed in March or April 2012 (Hirose, 2013). The relatively high levels of the radiocesium deposition rates in late fall 2011–spring 2012 may be supported by resuspension of radiocesium deposited on land surface (Hirose et al., 2012; Igarashi et al., 2005). Another possible source may be subsequent emission of radiocesium from the FNPP1.

5.1.2.4 Wet Deposition Processes

The major pathway of deposition of the natural radioactivity in the atmosphere is precipitation scavenging. In fact, the areas with enhanced ambient dose rate occurred within 300 km of the FDNPP were formed by snowfall and rainfall in 15–16 and 20–23 March as seen in Figs 5.3 and 5.13. In order to elucidate the wet removal processes of radionuclides, wet deposition velocity (V_{wet}, R) of the radionuclides is a useful tool as does dry deposition velocity, which is calculated from wet deposition flux (D_R: Bq/m^2 s) and surface air concentration ($C_{a,R}$: Bq/m^3) of radionuclides using an equation (V_{wet}, $R = D_R/C_{a,R}$). The wet deposition velocities of radioactive aerosols varied considerably and depended on the type of radionuclides and rainfall events (Hirose, 2000). Wet deposition velocity is a useful indicator to have better understanding of the atmospheric behavior of the Fukushima-derived radionuclides. Amano et al. (2012) have determined the wet deposition velocities of the Fukushima-derived ^{131}I and ^{137}Cs during the periods of 21–24 March, corresponding to the second arrival (heavier rainfall: 38 mm), respectively. The wet deposition velocities, corresponding to the second arrival of the radioactive plume, were around 0.01–0.14 m/s for ^{134}Cs and ^{137}Cs and 0.004–0.03 m/s for ^{131}I. The wet deposition velocities of the Chernobyl fallout were determined to be 0.0026–0.11 m/s for ^{137}Cs (Hirose et al., 1993). The results suggest that radiocesium is slightly easily removed from the atmosphere by wet deposition processes comparing with radioiodine. The wet deposition velocities of the Fukushima-derived radiocesium are the same order of magnitude as those of the Chernobyl fallout, although the wet deposition velocities depend on size and chemical characteristics of radionuclide-bearing particles and meteorological conditions such as rainfall intensity (Harvey and Matthews, 1989; Hirose, 2000; Hirose et al., 1993; McNeary and Baskaran, 2003; Slinn, 1978).

5.1.2.5 Dry Deposition Processes

Dry deposition is one of the important pathways for the removal of radioactive particles from atmosphere. Theoretical and experimental studies of dry deposition processes for particles and gases were carried out and reviewed by many researchers (Chamberlain, 1953; Schmel, 1980; Slinn, 1978; Underwood, 2001). Dry deposition, which depends on conditions of land surface such as grass, plat plain and others and on meteorological conditions, is controlled by gravitational settling, Brownian diffusion and compaction, as does physicochemical form of radioactive aerosols. Radioactive aerosols are an important tool to have better understanding of dry deposition processes of aerosols in environments.

The dry deposition velocity has been applied in evaluation of aerosol transfer to the earth's surface.

The dry deposition velocity, $V_{d,r}$, is defined as the ratio of dry deposition flux to land to surface air concentration.

$$V_{d,r} = F_r / C_{air,r} \tag{5.1}$$

where F_r and $C_{air,r}$ are the deposition flux of radioactive aerosols (Bq/m^2 s) and the surface air concentration of radioactive aerosols (Bq/m^3), respectively. The dry deposition velocity is a function of particle size and wind velocity; the dry deposition velocity increases with decreasing particle diameter in the particle size range of <0.1 μm, which may be controlled by the Brownian diffusion, exhibits a minimum in the particle size range of 0.1–1 μm, and increases with increasing particle diameter in the particle size range of more than 1 μm, where impaction and gravitational settling govern shape of increasing curve. The gravitational settling rate of soil particles is calculated to be around 1×10^{-3} m/s for a mean particle size of 7 μm and a particle density of about 1.5 g/cm^3, which is the minimum value because the dry deposition velocity increases due to friction velocity and surface conditions (Schmel, 1980).

The dry deposition velocities of radioactive aerosols have been determined by field observations. The dry deposition velocities depend on the kind of radionuclides and isotopes. For ^{7}Be-bearing particles, the dry deposition velocity varied from 1.0×10^{-3} to 3.4×10^{-2} m/s (Chamberlain, 1953; Small, 1960; Peirson et al., 1973; Young and Silker, 1980; Turekian et al., 1983; Todd et al., 1989; Papastefanou et al., 1995; Rosner et al., 1997; McNeary and Baskaran, 2003). For the ^{210}Pb-bearing particles, the dry deposition velocity ranged from 0.7×10^{-2} to 1.1×10^{-2} m/s (Turekian et al., 1983; Todd et al., 1989; McNeary and Baskaran, 2003). For the ^{137}Cs-bearing particles, the dry deposition velocity varied from 3.8×10^{-4} to 6.3×10^{-2} m/s (Aoyama et al., 1992; Papastefanou et al., 1995; Rosner et al., 1997). Amano et al. (2012) have determined the dry deposition velocities of the Fukushima-derived ^{131}I and ^{137}Cs during the periods of 14–17 March, which were around 2×10^{-3}–3×10^{-3} m/s for ^{134}Cs and ^{137}Cs and 1×10^{-3}–3×10^{-3} m/s for ^{131}I. The dry deposition velocities of thorium isotopes, which are a typical indicator of soil dust, ranged from 3.8×10^{-3} to 4.0×10^{-2} m/s (Hirose, 2000) and from 2.0×10^{-3} to 6.4×10^{-2} m/s (Crecelius et al., 1978). The dry deposition velocities of the Fukushima-derived radionuclides were the same order of magnitude as that of previous, although they were lower than that of soil particles.

5.1.2.6 Resuspension Processes

Relatively large amounts of the ^{137}Cs and ^{134}Cs depositions within 300 km of the FDNPP have been measured until June 2012. Resuspension is an important process to sustain a level of anthropogenic radionuclides in the surface air and deposition. The radionuclides deposited onto ground and/or vegetation are adsorbed onto fine organic or mineral particles. Some meteorological conditions such as dry and strong wind may blow off fragments of dried soil and vegetation. Initially resuspension process was concerned about inhalation of resuspended radioactivity, especially plutonium, for radioactivity-contaminated area by nuclear tests and accidental release from nuclear facilities. To characterize resuspension of radioactive aerosols, resuspension factor, k_r, and resuspension ratio, Λ, have been introduced, which is defined as a ratio of the radioactivity concentration in the air (C_{air}: Bq/m^3) to the radioactivity deposited on the ground (D_s: Bq/m^2) and the ratio of vertical flux (Q: Bq/m^2 s) to the radioactivity deposited on the ground (D_s: Bq/m^2), respectively.

$$k_r = C_{air}/D_s \tag{5.2}$$

$$\Lambda = Q/D_s \tag{5.3}$$

The resuspension factor is a measure of resuspension phenomenon. The resuspension factors have been calculated for aerosols associated with ^{7}Be and ^{137}Cs. For the ^{7}Be-bearing particles, the resuspension factor varied from 1.4×10^{-4} to 4.2×10^{-4}/m (average 2.3×10^{-4}; Papastefanou et al., 1995). For the ^{137}Cs-bearing particles, the resuspension factor, ranging from 10^{-8} to 1.2×10^{-4}/m (Stewart, 1966; Garland and Cambray, 1988; Papastefanou et al., 1995), showed large variability. Shinn et al. (1983) deduced the lower resuspension factor (a mean value: 3×10^{-10}/m) from measurements of aerosol Pu over a bare field near the Savannah River Processing Plant. The resuspension factor depends on meteorological factors such as wind velocity and humidity, land conditions (climatologically factors, industrial and agricultural activities) and chemical and physical properties of radionuclide-bearing particles.

After the 1990s, the anthropogenic radionuclides observed in surface air and rainwater over the globe except the Chernobyl fallout are considered to have been derived from the resuspension of fallout radionuclides deposited on the land surface (Arimoto et al., 2005; Karlsson et al., 2008; Nicholson, 1988; Rosner and Winkler, 2001). After more than 10 years since the cessation of atmospheric nuclear testing, the atmospheric deposition of ^{137}Cs, ^{90}Sr and ^{239,240}Pu derived from atmospheric nuclear tests is negligible. It is likely that the anthropogenic radionuclides and their activity ratios in surface air and rainwater reflect their redistribution processes on the land surface. Especially, regional transport of soil dust from the desert and arid region (e.g. the Sahara dust in Europe and the Asian dust in East Asia) is an important factor for resuspension of the anthropogenic radionuclides (Fujiwara et al., 2007; Igarashi et al., 2001, 2003, 2005, 2009; Hirose et al., 2003, 2004; Lee et al., 2002; Masson et al., 2010; Pham et al., 2011, 2013; Povinec et al., 2012d; Sýkora et al., 2012).

Although levels of dominant anthropogenic radionuclides (^{137}Cs, ^{90}Sr and ^{239,240}Pu) in surface air in the 1990s and 2000s were supported by resuspension, atmospheric behavior differs among the anthropogenic radionuclides. In fact, ^{90}Sr and ^{137}Cs deposition fluxes observed in Japan since 1990 were decreasing slowly, whereas ^{239,240}Pu deposition has been almost constant since 1985, although there is interannual variability of the ^{239,240}Pu deposition. The monthly anthropogenic radionuclide depositions in East Asia exhibited seasonal changes, with a maximum in spring (Hirose et al., 2003, 2008; Igarashi et al., 2001, 2003); especially, ^{239,240}Pu deposition showed the most typical seasonal variation among ^{137}Cs, ^{90}Sr, and ^{239,240}Pu (Hirose et al.,

2003, 2008). The annual and seasonal changes in $^{239,240}Pu$ deposition coincide with those of the occurrence of the Kosa (Asian dust) event observed in spring (typically, March and April) in Japan (Hirose et al., 2007). These findings suggest that the major origin of resuspension of the anthropogenic radioactive aerosols in Japan in the 1990s and 2000s is aeolian dust produced in the East Asian deserts and arid areas (Hirose et al., 2003; Igarashi et al., 2001).

Natural radionuclides, especially lithogenic elements, are a useful tool to elucidate resuspension processes. Most of the thorium in aerosols originates from soil particles (Hirose, 2000). The variation of thorium in surface air, therefore, provides a key to solve factors controlling resuspension. The temporal variation of ^{232}Th deposition in the 1990s and 2000s shows marked increase in spring coinciding with the Asian dust (Kosa) events. Residual materials in deposition samples consist of soil dust, fly ash, sea salt and others. Soil dust in residual materials comprises local soil and long-range transported soil particles so-called "Kosa". The specific activities of thorium in deposition samples collected at Tsukuba ranged from 1.5 to 23 mBq/g (Hirose et al., 2012). The thorium concentration in surface soil at Tsukuba is 17 mBq/g in average (range: 4.4–32 mBq/g, unpublished data). The ^{232}Th-specific activities in deposition are roughly lower than the mean ^{232}Th concentration in surface soil because soil particles occupy a part of residual materials in deposition. The $^{230}Th/^{232}Th$ activity ratios vary according to sources, in which its ratio is high in the local cultivated fields due to fertilization and low in arid and desert soils, and can therefore be used to differentiate between locally derived and remotely derived ^{232}Th (Hirose et al., 2010). The $^{230}Th/^{232}Th$ activity ratios in aerosol and deposition samples showed large variability with high ratios occurring in early spring. These high $^{230}Th/^{232}Th$ ratios in deposition can be attributed to local dust storms that cause resuspension of soils from cultivated fields. The $^{230}Th/^{232}Th$ ratios in deposition allow us to separate locally derived and remotely derived ^{232}Th fractions. The results reveal that both locally and remotely derived ^{232}Th depositions showed seasonal variations with maxima in spring although the remotely derived fraction is dominant rather than the locally derived one. The locally derived ^{232}Th deposition showed a peak in early spring, which can be attributed to local dust storm events. The ^{232}Th deposition maximum later in spring is attributable to the remotely derived fraction, corresponding to the Kosa events. Thorium isotopes in aerosols clearly show coexistence of two components of resuspension, i.e. locally and remotely derived components.

Although, since July 2011, the decrease rate of the monthly ^{137}Cs deposition in each sampling station of Kanto Plain was declined, the level of the monthly ^{137}Cs deposition in Fukushima Prefecture and Kanto Plain is more than two orders of magnitude higher than the pre-Fukushima level (Hirose, 2013; Hirose et al., 2008). These findings and the detection of ^{134}Cs reveal that the emission of the Fukushima-derived radionuclides has affected the atmospheric environment even in 2012. It is considered that there were possible processes supporting the monthly ^{137}Cs deposition after July 2011 such as resuspension of the Fukushima-derived ^{137}Cs deposited on land surface and continuous emission of radionuclides from the FNPP1. In order to elucidate atmospheric effects of the Fukushima-derived radioactivity, it is important to estimate levels of resuspended ^{137}Cs (Hirose, 2013).

In order to evaluate contribution of resuspension of the Fukushima-derived radionuclides, it is needed to estimate amounts of locally derived soil dust in deposition samples, which comprises locally derived soil and long-range transported soil particles ("Kosa"). Thorium in deposition samples is a good indicator of soil particles because it is one of the typical lithogenic elements (Hirose et al., 2010). The massic activities of thorium in deposition samples collected at Tsukuba ranged from 1.5 to 23 mBq/g (Hirose et al., 2012). The thorium concentration in surface soils at Tsukuba is 17 mBq/g as an average (range: 4.4–32 mBq/g, unpublished data). The ^{232}Th massic activities in deposition are roughly lower than the mean ^{232}Th concentration in surface soil because soil particles occupy a part of residual materials. The ^{137}Cs massic activities in deposition samples at

Tsukuba are calculated to range from 2.4 to 16 mBq/g and 1.3–8.5 mBq/g for dust season (January–May) and nondust season (June–December) in 2000–2007. The ^{137}Cs concentration in surface soil at Tsukuba is 10 mBq/g as an average (range: 0.92–20 mBq/g, unpublished data). As did thorium, the ^{137}Cs massic activities in deposition are roughly lower than the mean ^{137}Cs concentration in surface soil. If amounts of resuspended materials derived from locally derived soil particles are estimated, a portion of the resupended FDNPP-derived ^{137}Cs to the total ^{137}Cs deposition can be calculated. Hirose et al. (2012) have estimated the locally and remotely derived thorium fractions for thorium deposition. The amount of locally derived soil in residual materials is estimated from the locally derived thorium. Unfortunately, weights of residual materials in monthly deposition samples in monitoring stations as done in the case of locally derived thorium deposition have not been recorded in the MEXT database (MEXT, 2012a,b,c). Therefore, the data set of the weight of residual materials in the monthly deposition measured by Meteorological Research Institute (MRI) during the period 1990–2007 was used. The amount of the locally derived soils in monthly deposition at Tsukuba may be representative of that in monitoring stations located in the Kanto Plain. Seasonal variation of locally derived soil contributions observed at Tsukuba is shown in Fig. 5.19. The locally derived soil amount in the monthly deposition showed seasonal variation with high deposition in January–May and low in October–December, in which soil dust including filaments of withered plants is blown by strong wind of dry spring in Kanto Plain. The temporal variation of the FDNPP-derived ^{137}Cs deposition at several stations of Kanto Plain in 2012, having a maximum in March or April, is similar to that of the resupension of the local soil. Hirose (2013) estimated contribution of resuspension of radionuclide-bearing soil particles to the radioactivity deposition.

According to the monitoring data of ^{137}Cs deposition and airborne monitoring (Yoshida and Takahashi, 2012), the Fukushima-derived ^{137}Cs deposition in Kanto Plain was $<$60 kBq/m^2 (20 kBq/m^2

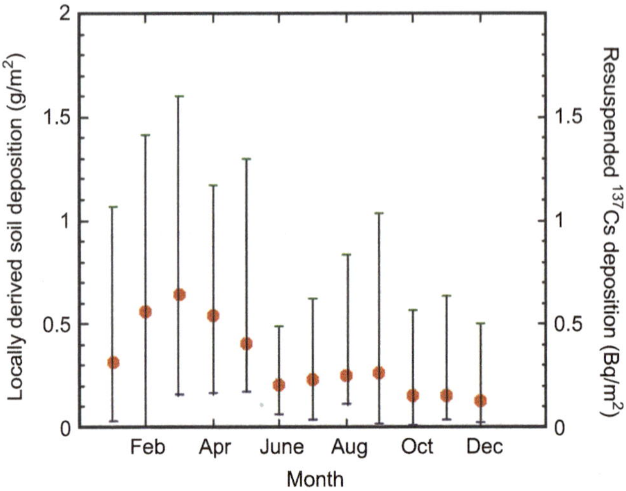

Figure 5.19 Seasonal changes of locally derived soil deposition and estimated monthly ^{137}Cs deposition rates derived from resuspension of locally derived ^{137}Cs-bearing soil particles. (Source: Cited from Hirose (2013))

for Hitachinaka, 8.5 kBq/m^2 for Tokyo, 7.2 kBq/m^2 Utsunomiya, 5.2 kBq/m^2 for Maebashi), which is corresponding to 2.5 kBq/kg as a ^{137}Cs concentration in surface soil, assuming that most of the Fukushima-derived ^{137}Cs exists in the upper 2 cm of soils (Kato et al., 2012) and soil density is 1.2 g/cm^3. The contribution of the Fukushima-derived ^{137}Cs due to resuspension, which is in the range of 0.33–1.6 Bq/m^2 as monthly average values, is <4 Bq/m^2 in Kanto Plain. For the monitoring stations in Kanto Plain, minimum monthly ^{137}Cs deposition (2.4–8.4 Bq/m^2) was observed in October after the Fukushima accident. These estimates, therefore, suggest that all the ^{137}Cs deposition at the monitoring stations in Kanto Plain after June 2011 is hardly supported by any resuspension of local soil particles. However, taken into account the hypothesis that ^{137}Cs-enriched fine particles, so-called "plant opal", are produced on land surface by degradation of radioactivity-contaminated plants (Katayama et al., 2012), resupension is one of the important processes supporting the higher ^{137}Cs deposition.

Resuspension of the Fukushima-derived radionuclide-bearing soil particles is a potential important process supporting the ^{137}Cs deposition.

However, it is inadequate to disregard the possibility of other processes to enhance the atmospheric ^{137}Cs. Small-scale biomass burning, which occurred in November–December in Kanto Plain, is considered to be a potential source to the ^{137}Cs deposition as another process, which has been recognized for the Chernobyl fallout (Lujanienė et al., 2009). It must be noted that the direct emission of radionuclides from the FNPP1 has continued at least up to October 2011. The amounts of released radionuclides were, however, more than three orders of magnitude lower than the initial release as a monthly emission (TEPCO, 2011a,b).

5.1.3 Radionuclides in Grass, Crop, Livestock and Soil

5.1.3.1 Radionuclides in Soil

MEXT started the emergency monitoring of environmental samples on 18 March, including radioactivity measurements of soil and grass samples at the area from 20 km to about 50 km of the FNPP1 (MEXT, 2012a,b,c). The radioactivity monitoring of soil samples, which can provide information about total ground deposition and kinds of emitted radionuclides, is important to find highly contaminated areas so-called "hot spot". This is important to implement early protective actions to mediate radiological effects. In the emergency monitoring, radioactivity concentrations in surface soil is defined as a mean value of five points in the sampling site, in which samples collected with a cylinder sampler with 5 cm diameter and 5 cm depth were homogenized and subjected by gamma-spectrometry. The number of soil sampling sites was about 60. Although measured radionuclides included ^{131}I, ^{134}Cs, ^{136}Cs, ^{137}Cs, ^{129m}Te, ^{132}Te, and ^{90}Sr, major target radionuclides were ^{131}I, ^{134}Cs, and ^{137}Cs. The highest concentration of ^{131}I in soil (340 kBq/kg) collected on 30 March was observed at the site of 34 km northwest of the FNPP1. The early radioactivity monitoring revealed that the spatial distribution of radionuclide concentrations in soil was very variable, depending on sampling and local conditions.

Besides MEXT emergency monitoring, some university teams (Endo et al., 2012; Kinoshita et al., 2011; Fujiwara et al., 2012; Kato et al., 2012) conducted environmental radioactivity monitoring of surface soil. Endo et al. (2012) collected undisturbed soil samples (6.5 cm diameter and 0–5 cm and 5–10 cm depths) at a site (Futaba-Yamada), 5 km of the FNPP1, and 15 sites in the area of 20–40 km of the FNPP1 in the period of 15–30 March and measured activity concentrations in soil samples. The result reveals that following gamma-emitting radionuclides were detected in soil samples: ^{131}I, ^{134}Cs, ^{136}Cs, ^{137}Cs, ^{129m}Te, ^{129}Te, ^{132}Te, ^{132}I, ^{140}Ba, ^{140}La, ^{99m}Tc and ^{110m}Ag. The highest concentrations of gamma-emitting radionuclides in soil samples, which were 2.38 MBq/kg (^{131}I), 1.43 MBq/kg (^{132}Te), 1.33 MBq/kg (^{132}I), 0.186 MBq/kg (^{134}Cs), 0.04 MBq/kg (^{136}Cs) and 0.193 MBq/kg (^{137}Cs), were found at Futaba-Yamada on 15 March. The higher concentrations of the radionuclides in soil were observed in the area northwest of the FNPP1, which coincided with the airborne monitoring (Lyons and Colton, 2012). The result suggests that dominant radionuclides emitted from the FNPP1 were volatile ones such as ^{131}I, ^{134}Cs, ^{137}Cs, ^{132}Te and ^{132}I, whereas refractory radionuclides such as ^{95}Zr, ^{103}Ru and ^{106}Ru, which were detected in the Chernobyl accident, could not be detected. Kinoshita et al. (2011) collected a lot of soil samples in Fukushima and Ibaraki prefectures in the period of late March–April in order to construct geographical maps of the Fukushima-derived radionuclides. Kinoshita et al. (2011) depicted spatial distributions of ^{131}I, ^{134}Cs, ^{137}Cs, ^{129m}Te and ^{136}Cs in Fukushima and Ibaraki prefectures except the evacuation area. Yamamoto et al. (2012) documented concentrations of gamma-emitting radionuclides (^{131}I, ^{134}Cs, ^{136}Cs, ^{137}Cs, ^{129m}Te, ^{129}Te, ^{132}I, and ^{140}La) in soil samples collected in 20–60 zones of the FNPP1 on 27 March and in Okuma Town (1.7 km to about 5 km west and south of the FNPP1) on 25 May 2011. Watanabe et al. (2012) reported concentrations of gamma-emitting radionuclides (^{110m}Ag, ^{134}Cs, ^{137}Cs, and ^{129m}Te) in 35 soil samples collected in Miyagi prefecture located just north of Fukushima prefecture on 16–29 April 2011. They detected ^{110m}Ag in 16 soil samples, in which the ^{110m}Ag concentrations in soil ranged from 2.6 to 48.4 Bq/kg and $^{110m}Ag/^{137}Cs$ activity ratios were in a range of 0.002–0.011, which were roughly higher than values (0.0009–0.006) in deposition samples in the regional sites (Hirose et al., 2012). Fujiwara et al. (2012) measured concentrations of gamma-emitting radionuclides in soil samples collected in 50 sites in the eastern Fukushima prefecture (within 60 km of the FNPP1) on 20 April. For most of the sampling sites, only one soil sample was collected with a cylinder sampler of 5 cm diameter and 5 cm depth and at several sites, soil core samples were collected. The results reveal that higher concentrations of the Fukushima-derived radionuclides spread northwest direction and large spatial variation of $^{131}I/^{137}Cs$ ratio occurred, whereas there was a small spatial variation of $^{136}Cs/^{137}Cs$ ratios (0.17–0.25, radioactive decay corrected on 11 March), which is consistent with the ratio (0.27) observed in surface air (Doi et al., 2013). The vertical profiles of radionuclides in the soil core samples indicated that most of the ^{131}I and ^{137}Cs exist in the surface layer, although ^{131}I migrates deeper layer than ^{137}Cs. Tagami et al. (2011) reported concentrations of radionuclides in soil samples collected at J Village, about 20 km south of the FNPP1 site, on 20 and 28 April and 4 May 2011, in which radioiodine and radiocesium were detected as major nuclides and trace amounts of ^{95}Nb, ^{110m}Ag and ^{140}La were found. The average activity ratios of $^{131}I/^{137}Cs$, $^{134}Cs/^{137}Cs$, $^{136}Cs/^{137}Cs$ and $^{129m}Te/^{137}Cs$ were 55, 0.90, 0.22 and 4.0 (radioactive decay corrected on 11 March) in the surface soil (0–2 cm). Measurements for three soil layers (0–2, 5–7, and 10–12 cm depth) revealed that most of the radionuclides deposited on soil surface existed a surface layer (0–2 cm). Ohno et al. (2012) measured radioactivity in soil core samples collected in a wheat field, a rice paddy, an orchard, and a cedar forest in Koriyama, which is located about 60 km west of the FNPP1. Soil sampling was conducted using a stainless core sampler (30 cm in length and 5 cm in diameter) on 23 and 24 April 2011. The measurement of radioiodine and radiocesium was performed for six layers of soil cores

(0–2, 2–4, 4–6, 6–10, 10–15, and 15–20 cm). The results demonstrated that more than 90% of the radionuclides exited in the upper 6 cm of the wheat field soil column and within 4 cm of the rice paddy, orchard, and cedar forest soil surface. The cumulative ^{137}Cs deposition depended on field type; higher value occurred in the wheat field, whereas lower values appeared in rice paddy. In order to elucidate migration behaviors of radionuclides in soil layers, detail profiles of ^{131}I and ^{137}Cs in soil were determined by using a scraper plate with a sampling area of 450 cm^2 (15 × 30 cm) (Kato et al., 2012), in which samples were collected with 0.5 cm thicknesses for the depth of 0–5 cm, 1.0 cm thicknesses for the depth of 5–10 cm and 2.0 cm thicknesses for the depth of 10–30 cm. Cultivated soil samples were collected in Kawamata Town (37° 38′ 37″ N, 140° 39′ 3″ E) on 28 April 2011. Typical vertical profiles of ^{131}I and ^{137}Cs are shown in Fig. 5.20. The results reveal that more than 86% of

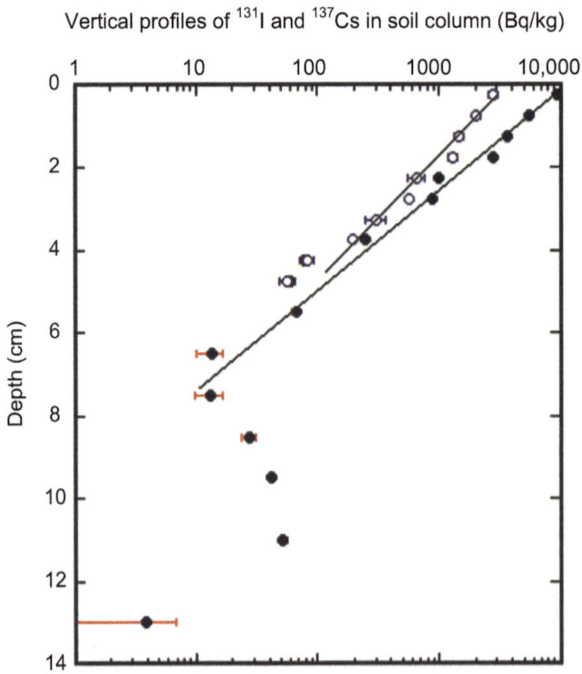

Figure 5.20 Vertical profiles of ^{131}I and ^{137}Cs concentrations in soil column. (Data from Kato et al. (2012))

the total radiocesium and 79% of the total ^{131}I were absorbed onto soil particles in the upper 2.0 cm layer of the soil profile. The ^{131}I and ^{137}Cs concentrations in soil layers exponentially decreased with increasing depth. These findings suggest that most of the radiocesium and ^{131}I were tightly absorbed onto soil particles in surface layer just after the deposition, although the vertical migration of radionuclides depends on the field type, soil type and clay content.

To construct maps of the Fukushima-derived radionuclides with a 2 km mesh within a 100 km radius of the FNPP1, the MEXT team, in cooperation with Japan Atomic Energy Agency (JAEA), JCAC and Japan universities, collected soil samples at 2200 sites throughout Fukushima prefecture and northern Ibaraki and southern Miyagi prefectures during the period of 4–29 June 2011. The soil sampling was the same as the Emergency sampling method. The result of radionuclide profiles in soil (Endo et al., 2012; Fujiwara et al., 2012; Kato et al., 2012; Ohno et al., 2012; Tagami et al., 2011) suggested that more than 90% of the Fukushima-derived radionuclides exist in the upper 5 cm layer. Therefore, the radionuclide concentrations in the upper 5 cm soil layer can be converted to the cumulative ground deposition (unit: Bq/m^2). MEXT documented the maps of the cumulative depositions of ^{134}Cs, ^{137}Cs, ^{129m}Te and ^{110m}Ag, whereas the fine-mesh mapping of the ^{131}I deposition failed due to decay out of ^{131}I in most of the sampling sites because of delay of sampling (meaningful values of ^{131}I in soil samples were obtained only for 20% of the sites).

The spatial distribution of the cumulative ^{137}Cs deposition calculated in soil samples as shown in Fig. 5.21 revealed that the highly contaminated area covered the area northwest of the FNPP1, which is consistent with the airborne monitoring results and preliminary soil surveys (Kinoshita et al., 2011; Fujiwara et al., 2012). The highest cumulative ^{137}Cs deposition (15.5 MBq/m^2) was measured in Okuma Town, close to the NPP (<3 km), where an annual radiation dose at 1 m height was 480 mSv. Higher cumulative ^{137}Cs deposition (7.9 MBq/m^2) occurred around the border between Namie Town and Iidate Village, about 30 km

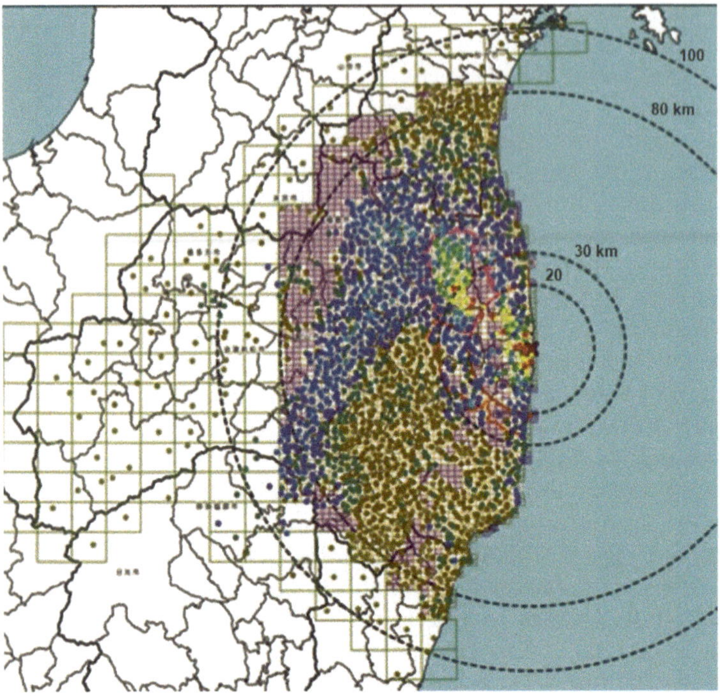

Figure 5.21 Mapping of ^{137}Cs deposition in Fukushima Prefecture. (Source: Cited from MEXT (2012a,b,c))

northwest of the FNPP1. Moderate radioactivity-contaminated area (0.1–0.6 MBq/m^2) covered in the Nakadori region (central part of Fukushima prefecture). The 2 km mesh sampling by the MEXT team revealed that there was a heterogeneous distribution of soil radioactivity at a smaller scale, as recognized in the emergency monitoring, which was not evident in the airborne monitoring.

Tellurium-129m and ^{110m}Ag—The distribution of the cumulative ^{129m}Te deposition was approximately similar to that of the ^{137}Cs. However, there was a partial spatial difference between ^{129m}Te and ^{137}Cs distributions, which was supported by the preliminary study (Kinoshita et al., 2011). The larger cumulative deposition of ^{129m}Te occurred south of the FNPP1, which suggests that radioactive plumes with different $^{129m}Te/^{137}Cs$ ratios were emitted in the atmosphere and their deposition differed spatially, depending on meteorological conditions in the time course of the accident. On the other hand, the distribution of the cumulative ^{110m}Ag deposition did not coincide with that of the ^{137}Cs. This finding suggests that the emission processes of ^{110m}Ag from the FNPP1 were significantly different from those of ^{129m}Te and ^{137}Cs because of refractory property of Ag in contrast of ^{129m}Te and ^{137}Cs.

Iodine-131—in order to assess stochastic effects due to radiation exposure from radioiodine, it is important to elucidate the spatial distribution of the cumulative ^{131}I deposition. Unfortunately, MEXT team failed to reconstruct precious mapping of the cumulative ^{131}I deposition due to decay out of most of ^{131}I. Kinoshita et al. (2011) conducted early mapping of the cumulative ^{131}I deposition using soil measurements. The high ^{131}I deposition region spread northwest of the FDNPP, which was similar to that of radiocesium, whereas the relatively high ^{131}I deposition comparing with ^{137}Cs occurred in the south region of the FDNPP, including Ibaraki prefecture. The result suggested that the depositional behaviors of ^{131}I differed from that of the ^{137}Cs. This finding reveals that it is difficult to reconstruct the fine-mesh map of the ^{131}I deposition from the fine-mesh mapping of ^{137}Cs and $^{131}I/^{137}Cs$ ratio.

Iodine-129 is a minor activity component in a reactor accident. If the $^{129}I/^{131}I$ ratio is constant in radionuclides emitted throughout the accident, the distribution of ^{129}I, determined by accelerator mass spectrometry (AMS), allows us to depict more reliable distribution of the ^{131}I deposition. The ^{129}I concentrations in soil samples, in which the ^{131}I concentrations had been determined (Fujiwara et al., 2012), were measured by AMS (Miyake et al., 2012). The cumulative ^{129}I deposition estimated from soil data was in the range of 0.0156–6.06 Bq/m^2. Soil samples in Japan were generally contaminated with ^{129}I derived from global fallout due to the atmospheric nuclear weapons testing (Matsuzaki et al., 2007; Muramatsu et al., 2008). In order to distinguish between the Fukushima-derived ^{129}I and global fallout-derived ^{129}I, $^{129}I/^{127}I$ atom ratios were plotted to ^{127}I (stable isotope). Most of the plots of $^{129}I/^{127}I$ vs ^{127}I in the Fukushima soil samples were deviated from the relationship between $^{129}I/^{127}I$ and ^{127}I for global fallout. This implies that ^{129}I in the Fukushima soil samples was derived from the

Fukushima accident. The ^{129}I deposition linearly correlated with the ^{131}I deposition and the average $^{129}I/^{131}I$ atom ratio was 31.6 ± 8.9. The measurements of ^{129}I, therefore, are desirable to depict the detailed ^{131}I map (Yoshida and Takahashi, 2012).

Radiostrontium (^{89}Sr, ^{90}Sr) attracted public concerns because radiostrontium has a higher risk for internal radiation exposure via ingestions of contaminated agricultural crops. The data of radiostrontium in soil samples were limited because radiochemical analysis with longer chemical separation/purification processes and resulting longer measurement time is required for measurements of radiostrontium because of their beta-emitter. TEPCO and MEXT documented preliminary results of ^{89}Sr and ^{90}Sr concentrations in surface soils collected at three points in the FNPP1 site, five points within 20 km of the FNPP1 and three points in the area of 30–40 km from the FNPP1, which are summarized in Table 5.1. The ^{90}Sr concentrations in surface soil samples ranged from 2.5 to 570 Bq/kg. The $^{89}Sr/^{90}Sr$ ratios in surface soils (^{89}Sr radioactive decay corrected at 11 March 2011) were 12–13. The $^{90}Sr/^{137}Cs$ ratios were almost <0.001. In order to depict the spatial distribution of radiostrontium, MEXT in cooperation with JCAC measured concentrations of radiostrontium in 100 soil samples, which were selected from 2200 samples of the 2 km mesh sampling. A significant amount of radiostrontium was released from the FNPP1 in the area within 80 km since ^{89}Sr was detected at about 70% of the sites. Maximum cumulative ^{89}Sr and ^{90}Sr depositions, which were 22 and 4.8 kBq/m^2, occurred in Namie Town. The $^{89}Sr/^{90}Sr$ activity ratios (sample number: 15 ($^{90}Sr > 500$ Bq/m^2), ^{89}Sr radioactive decay corrected at 11 March 2011) ranged from 6.5 to 15.7 with an average of 13.3. The result suggests that the high radiostrontium deposition was localized to a narrow area northwest of the FNPP1.

Plutonium attracted public concerns because plutonium isotopes have a higher risk for internal radiation exposure via ingestions of contaminated agricultural crops. Major plutonium isotopes are ^{238}Pu (half-life: 87.74 years), ^{239}Pu (half-life: 24,110 years), ^{240}Pu (half-life: 6563 years), and ^{241}Pu

Table 5.1. Concentrations of Radiostrontium in Soil Samples in the FDNPP Site and Fukushima Prefecture

Location	Sampling Date (MM/DD/YY)	^{89}Sr	^{90}Sr (kBq kg^{-1})	^{137}Cs	$^{89}Sr^{*}/^{90}Sr$ (kBq kg^{-1})	$^{90}Sr/^{137}Cs$ x 10^{-3}
In site						
Ss1	5/8/2011	2.9	0.4	500	12.9	0.8
	5/31/2011	2.8	0.48	500	12.2	0.96
Ss2	5/8/2011	0.023	0.003	1.6	12.4	2.1
	5/31/2011	1.9	0.34	40	11.4	8.5
Ss3	5/8/2011	4.4	0.57	2800	12.5	0.2
	5/31/2011	1.7	0.30	1100	13.0	0.27
Within 20 km						
S6	4/30/2011	0.018	0.0041	17	8.7	0.24
S41	4/29/2011	0.063	0.012	49	10.3	0.24
A13	5/1/2011	0.43	0.068	270	12.7	0.25
A14	5/1/2011	0.013	0.0025	5	10.5	0.5
Out of 30 km						
S31	3/17/2011	0.013	0.0033	2.3	4.3	1.4
S32	3/16/2011	0.081	0.0094	19	9.2	0.49
S33	3/16/2011	0.26	0.032	51	8.7	0.63

Ss 1-3: TEPCO data, S6: 4 km SSW, S41: 3 km WSW, A13: 2 km WSW, A13: 7 km W, S31: 30 km NW , S32: 32 km NW , S33: 33 km NW .
$^{89}Sr^{*}$: radioactivity decay corrected on March 11.

(half-life: 14.36 years). Plutonium isotope ratios such as ^{238}Pu/^{239,240}Pu, ^{241}Pu/^{239,240}Pu activity ratios and ^{240}Pu/^{239}Pu atom ratio depend on sources and burn-up time of nuclear fuel. Therefore, the accurate measurements of Pu isotopic compositions provide information on the extent of destruction of the damaged reactors including nuclear fuels and release processes. TEPCO reported that trace amounts of ^{239,240}Pu, ranging from 0.018 to 0.75 Bq/kg with an average of 0.078 Bq/kg, are detected in soil samples collected in the FNPP1 site, in which radiochemical analysis and alpha spectrometry were conducted by the JCAC and JAEA. The Fukushima-derived plutonium was evident from observed higher ^{238}Pu/^{239,240}Pu activity ratios (range: 1.2–3.3, mean: 2.2), which are about two orders of magnitude higher than those of global fallout (0.026). A median of ^{137}Cs/^{239,240}Pu ratios in the Fukushima soil was 7.5×10^6. Yamamoto et al. (2012) documented plutonium isotope measurements in soil samples collected at 11 sites outside the evacuation zone, four sites in Iidate Village and 10 sites in Okuma Town. Although high ^{239,240}Pu concentrations in soil (0–5 cm), ranged from 0.0022 to 0.27 Bq/kg, were not observed, evidently high ^{238}Pu/^{239,240}Pu activity ratios (0.05–1.2) occurred in the Iidate soil samples. Zheng et al. (2012) measured ^{241}Pu/^{239,240}Pu activity ratios and ^{240}Pu/^{239}Pu atom ratio and ^{239,240}Pu concentrations in litter and soil samples collected in Iidate Village and J Village using a sector-field inductively coupled plasma mass spectrometry (ICP-MS). The ^{239,240}Pu concentrations in soil (0–2 cm) ranged from 0.019 to 1.4 Bq/kg. These findings suggest that the ^{239,240}Pu concentrations in soil collected in Fukushima were generally in a concentration range (0.15–4.31 Bq/kg) of the global fallout-derived plutonium in Japanese soils before the Fukushima accident (Muramatsu et al., 2003). On the other hand, Zheng et al. (2012) found higher ^{241}Pu/^{239}Pu (0.103–0.135) and ^{240}Pu/^{239}Pu atom ratios (0.303–0.330) in the soils and litters than those of the global fallout, in which mean ^{241}Pu/^{239,240}Pu and ^{240}Pu/^{239}Pu atom ratios are 0.00194 ± 0.00014 and 0.180 ± 0.007, respectively (Kelley et al., 1999). The typical ^{241}Pu/^{239,240}Pu and ^{137}Cs/^{239,240}Pu activity ratios for the Fukushima-derived radionuclides were 108 and 1.48×10^7, respectively. In order to depict the spatial distribution of plutonium isotopes, MEXT in cooperation with JCAC measured concentrations of plutonium isotopes in 100 soil samples, which were selected from about 2200 samples of the 2 km mesh sampling. Significant amounts of the $^{239,240}Pu$ deposition, ranging from 0.44 to 15 Bq/m^2, were observed within 80 km of the FNPP1, whereas the high $^{238}Pu/^{239,240}Pu$ ratios in soil were observed at only five sites. The result suggests that the high $^{238}Pu/^{239,240}Pu$ deposition was localized to a narrow area northwest of the FNPP1. The plutonium isotope ratios revealed that the Fukushima-derived plutonium was dominantly deposited in a narrow area northwest of the FNPP1.

Americium and curium were key radionuclides to provide information on the extend of destroy of accidental reactors including nuclear fuels in ponds and release processes. TEPCO, in cooperation with the JACA and JAEA, measured americium and curium in soil samples collected in the FDNPP site. The ^{241}Am (half-life: 432.7 years), ^{242}Cm (half-life: 162.8 days) and $^{243,244}Cm$ (^{243}Cm (half-life: 28.5 years), ^{244}Cm (half-life: 18.1 years)), whose concentrations were 0.012–0.053 Bq/kg, 0.64–4.0 Bq/kg and 0.022–0.2 Bq/kg, respectively, were detected in the soil samples. Mean $^{242}Cm/^{238}Pu$ and $^{243,244}Cm/^{238}Pu$ activity ratios in soil (radioactive decay corrected on 11 March 2011) were 18 and 0.67, respectively.

To have better understanding of migration of the radionuclides in soil is important to assess the radiological and ecological effects of the Fukushima-derived radionuclides in the terrestrial environment. Migration of the radionuclides in soil is closely related to their chemical and mineralogical properties in soil, as do meteorological and ecological conditions. Current knowledge about radiocesium is that clay minerals in soil and sediment are major sorbents of cesium (Anderson and Sposito, 1991; Kim et al., 1996; Kim and Kirkpatrick, 1997; Choi et al., 2005). However, chemical and mineralogical knowledge of the FDNPP-derived radionuclides in soil is limited. Tanaka et al. (2012) applied leaching experiments to soil samples. The partitioning of ^{137}Cs and

^{131}I between soil (5.0 g) and solution (15 mL, pH: 1–10) revealed that most of the ^{137}Cs were tightly attached to soil particles, whereas ^{131}I was leachable for acidic solution and dissolved fraction of ^{131}I increased in alkaline solution. Qin et al. (2012) documented results of a sequential leaching experiment of radionuclides in Fukushima soil and speciation of Cs species in natural samples using stable Cs and extended X-ray absorption fine structure spectroscopy. The result suggested that clay mineral or mica, which can adsorb Cs, significantly contributes migration behaviors of the Fukushima-derived ^{137}Cs.

5.1.3.2 Radionuclides in Grass

Radioactivity measurements of grass samples are involved in major targets for the emergency monitoring, which was conducted since 18 March 2011. The grass samples were collected at the area from 20 km to about 50 km of the FNPP1. The radioactivity monitoring of grass is important to assess radioactivity contamination of vegetables and feeds. According to the MEXT results (MEXT, 2012a,b,c), maximum concentrations of ^{137}Cs and ^{131}I in Yagisawa (Iidate Village) grass samples, which were 2.65 and 3.87 MBq/kg, respectively, occurred on 20 and 26 March, respectively. The high concentrations of ^{137}Cs and ^{131}I in grass samples were observed just after the rainfall (snowfall). The temporal variation of the ^{131}I concentrations in grass samples is shown in Fig. 5.22. Although the ^{131}I concentrations in grass showed large variation at a short time scale of several days, the ^{131}I concentrations in grass exponentially decreased and were less than detection limit (about 30 Bq/kg) on June 2011. The apparent half-lives of the ^{131}I and decay-corrected ^{131}I concentrations in grass were 3.9–6.2 days and 7.4–28 days, respectively (Table 5.2). The apparent half-lives of the decay-corrected ^{131}I concentration in grass were relatively short in the in-land sites and long in the coastal sites. The temporal variation of the ^{137}Cs concentrations in grass samples is shown in Fig. 5.23. The ^{137}Cs concentrations in grass showed larger variation than ^{131}I at a short time scale. However, the ^{137}Cs concentrations in grass samples exponentially

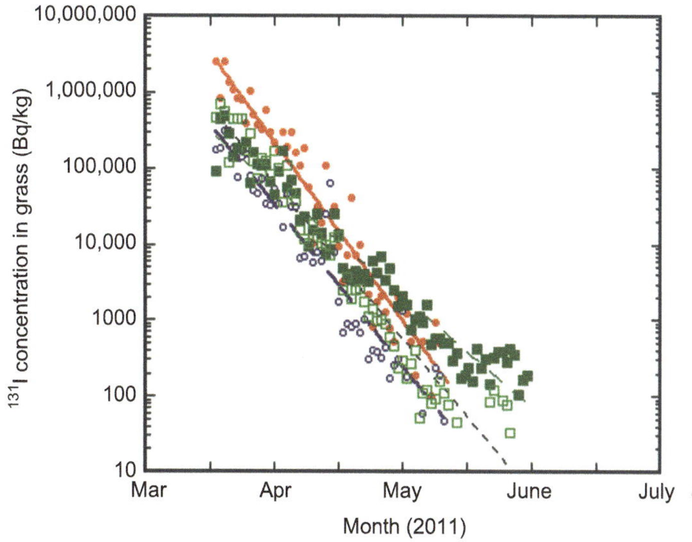

Figure 5.22 Temporal variations of ^{131}I concentrations in grass samples. Closed circles: Iidate (Yagisawa), open circles: Kawamata, closed squares: Minamisoma, and open squares: Iwaki. (Data from MEXT (2012a,b,c))

Table 5.2. Apparent Half-lives of Radionuclides in Grass Samples (unit: days)

Location	^{131}I	^{131}I*	^{137}Cs
Yagisawa	3.9	7.4	9.3
Kawamata	4.3	6.7	12
Minamisoma	6.2	28	13
Iwaki	4.5	13	10

Iidate (Yagisawa): 36 km northwest of the NPP, Kawamata: 45 km northwest of the FDNPP, Minamisouma: 25 km north of the NPP, Iwaki: 43 km south southwest of the FDNPP.
^{131}I*: radioactive decay was corrected on March 11, 2011.

decreased during the period of March–June 2011. The apparent half-lives of the ^{137}Cs concentrations in grass were 9.3–14 days during the period of March–June 2011. This half-life is similar to that of the monthly ^{137}Cs deposition observed in Kanto Plain (Hirose et al., 2012). In the period of July–November, the ^{137}Cs concentrations in the grass samples were

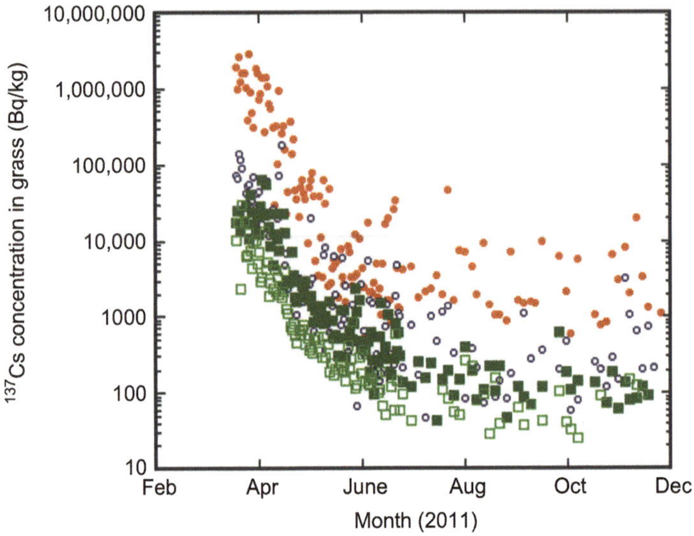

Figure 5.23 Temporal variations of ^{137}Cs concentrations in grass samples. (Data from MEXT (2012a,b,c))

approximately at constant level, although its levels spatially varied at about one order of magnitude.

5.1.3.3 Radionuclides in Crop

Ministry of Agriculture, Forestry and Fisheries (MAFF) conducted the emergency monitoring of crops including rice, vegetables, fruits, milk and beef produced in Japan. Rice is the most important staple food in Japan. The radioactivity measurements using gamma spectrometer were conducted for 1276 rice samples harvested in fall 2011. The result as shown in Fig. 5.24 revealed that most of the ^{137}Cs concentrations in rice (99.9%) were at a low level (<50 Bq/kg). However, two samples exceeded the provisional limit value for restriction of intake (radiocesium ($^{134}Cs + ^{137}Cs$): >500 Bq/kg), although the restriction of rice planting has been conducted for paddy field with a radiocesium concentration of more than 5000 Bq/kg in soil. Most of the radiocesium concentrations in vegetables (96.2% during the period of March–June and 99.9% during the period of July–December) were <100 Bq/kg as shown in Fig. 5.25. However, just after the radioactive deposition, significant amounts of vegetable samples

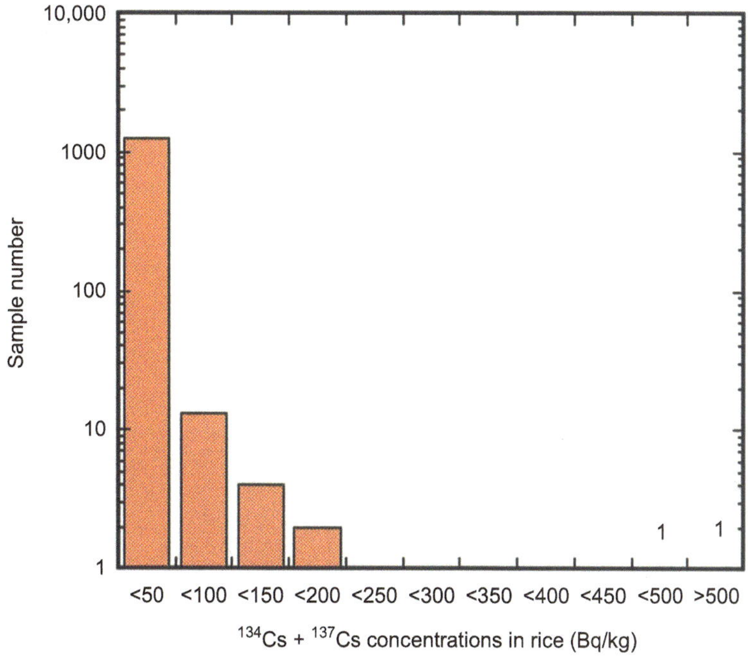

Figure 5.24 Frequency distribution of radiocesium ($^{134}Cs + ^{137}Cs$) concentration in rice. (Data from MAFF (2012a,b))

collected in Fukushima, Ibaraki, Chiba, Gunma, Tochigi and Tokyo exceeded the provisional limit values for the restriction of intake; for example, maximum concentrations of ^{131}I and radiocesium in vegetables collected on 21 March 2011 were 22 and 82 kBq/kg, respectively, just after the large deposition of the FDNPP-derived radionuclides (Hamada and Ogino, 2012). In contrast to rice and vegetables, high ^{137}Cs concentrations (>200 Bq/kg) were 6% of fruit samples, 8.6% of tea leaves and 6.6% of mushroom samples as shown in Fig. 5.26. It is noteworthy that high radiocesium concentrations were found in tea leaves sampled at a distance of more than 300 km from the FDNPP; for example, radiocesium concentrations of dried tea leaf (Kanagawa) sampled on 12 May and processed tea (Shizuoka) sampled on 9 June were 3000 and 879 Bq/kg, respectively (Hamada and Ogino, 2012).

The contamination processes of radiocesium differed between agricultural crops. For rice, major uptake process of radiocesium is considered to be

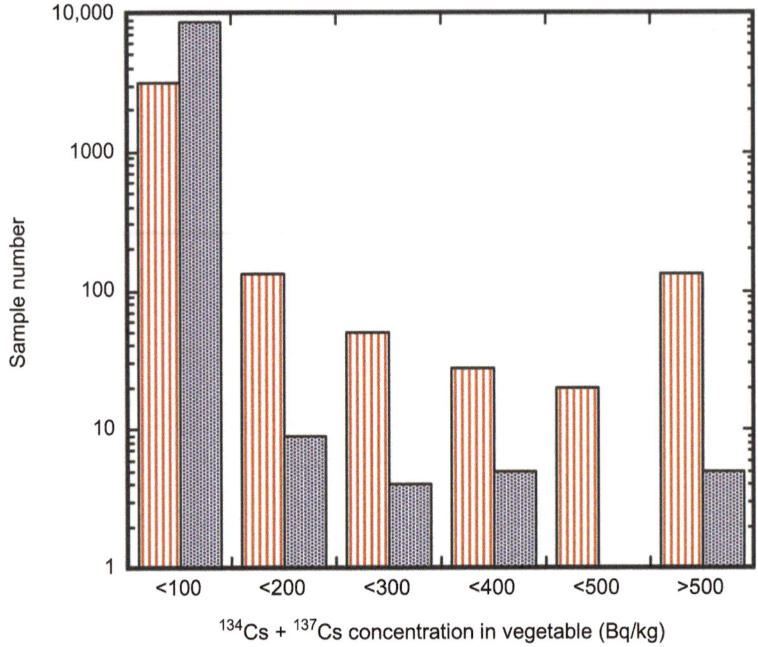

Figure 5.25 Frequency distribution of radiocesium ($^{134}Cs + ^{137}Cs$) concentration in vegetable. Red bars: March–June 2011 and blue bars: July–December 2011. (Data from MAFF (2012a,b))

via radioactive-contaminated soil because of no planting of rice during heavy Fukushima fallout (March–April); dissolved Cs is easily ingested by rice, whereas Cs absorbed in soil minerals is hardly ingested by rice. Transfer coefficient (TF) of radiocesium to crops from soil is an important factor to assess radioactive contamination. The TF value of radiocesium for rice was determined for paddy fields contaminated by the Fukushima fallout; the TF for rice roots was highest at 0.48. The TF values for chaff, rice bran, brown rice and polished rice were estimated to be 0.15, 0.10–0.16, 0.013–0.017 and 0.005–0.013, respectively (Endo et al., 2013). The TF value of cesium from soil to polished rice is generally low (0.0016) (Tsukada et al., 2002). The primary process of radioactive contamination of vegetables is considered to be direct dry or wet interception of airborne radionuclides by leaves. On the other hand, it was suggested that translocation of radiocesium from stems and leaves of plants is a major factor to

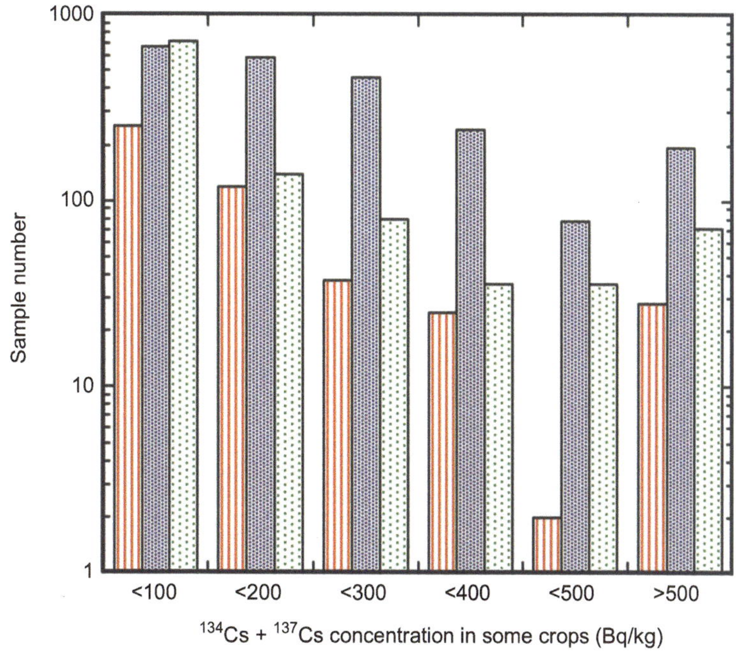

Figure 5.26 Frequency distribution of radiocesium ($^{134}Cs + ^{137}Cs$) concentration in fruits (red bars), tea leaves (blue bars) and mushroom (gray bars). (Data from MAFF (2012a,b))

result in high radiocesium contamination of newly emerged tea leaves because root uptake of stable Cs for trees was not high (Tagami and Uchida, 2010). Topcuoglu et al. (1997) documented that ^{137}Cs from the Chernobyl accident was stored into the stem of tea trees from old contaminated leaves and then translocated to new tea leaves. Recent study suggests that the translocation of radiocesium is an important process to transfer radiocesium to newly emerged plant tissues, although effects of translocation depended on kind of trees (Tagami et al., 2012).

Radionuclides in milk are one of the major concerns because of larger health effects to infants. High concentrations of ^{131}I and radiocesium, which were 1700 and 64 Bq/kg, respectively, were detected in raw milk sampled in Ibaraki prefecture on 19 March 2011 (Hamada and Ogino, 2012). MAFF has monitored radioactivity in raw milk. The result is shown in Fig. 5.27. The high radiocesium concentration (>200 Bq/kg) was detected in one raw milk sample within 170 samples measured in March 2011,

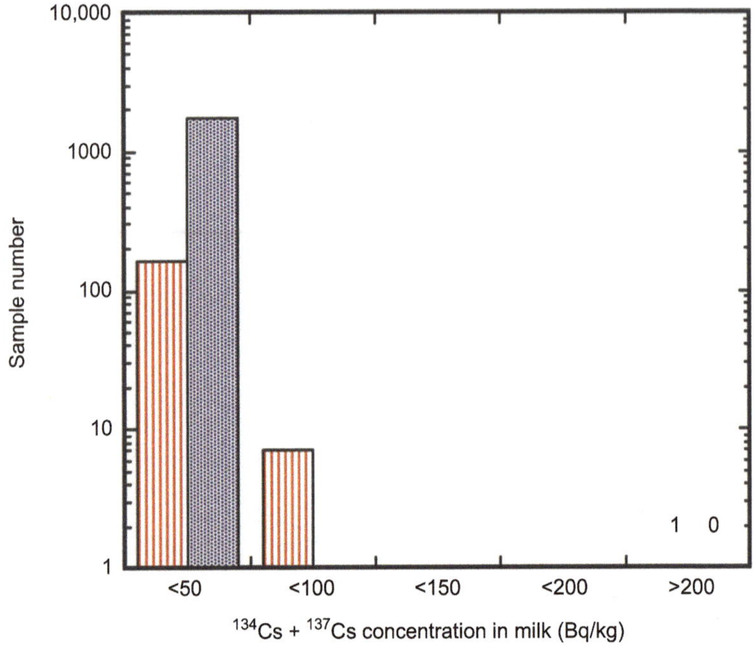

Figure 5.27 Frequency distribution of radiocesium ($^{134}Cs + ^{137}Cs$) concentration in milk. Red bars: March 2011 and blue bars: April–December 2011. (Data from MAFF (2012a,b))

whereas it was a low level (<50 Bq/kg) for 1744 samples measured since April 2011.

The radioactivity monitoring was performed for beef produced in northeast Honshu Island. The frequency graph is shown in Fig. 5.28. High radiocesium concentrations (>500 Bq/kg) were detected in 157 beef samples within 90,661 samples measured in March 2011. A major cause of radioactivity-contaminated beef was that cattle took highly radiocesium-contaminated rice straw as feeds.

5.1.4 Radionuclides in River and Lake Waters

In the emergency monitoring, MEXT has measured radionuclide concentrations in pond water in Iidate Village. In this case, unfiltered water was subjected to nondestructive gamma-spectrometry. Maximum concentrations of ^{131}I and ^{137}Cs in pond water, which were 2450 and 940 Bq/l, respectively, occurred on 19 March 2011. Both ^{131}I and ^{137}Cs

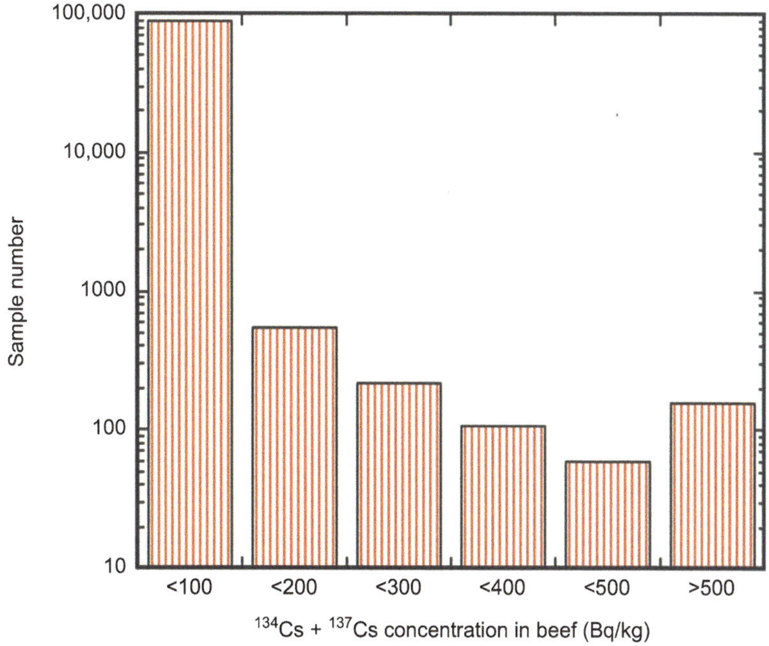

Figure 5.28 Frequency distribution of radiocesium ($^{134}Cs + ^{137}Cs$) concentration in beef. (Data from MAFF (2012a,b))

concentrations in pond water exponentially decreased and were less than the detection limit in late May (Fig. 5.29). Apparent half-lives of decay-corrected ^{131}I and ^{137}Cs in bond water were 45 and 17 days, respectively, which are longer than atmospheric half-life of ^{137}Cs. The apparent half-life of ^{137}Cs was shorter than that of the decay-corrected ^{131}I, which may mean that ^{137}Cs is preferentially removed from pond water by adsorption of pond sediments and suspended particles.

Ministry of Environment has conducted the emergency monitoring of river, lake, reservoir and pond out of the evacuation area in Fukushima prefecture. The preliminary measurements of radionuclides (^{131}I, ^{134}Cs, and ^{137}Cs) in water and sediment were performed to samples collected in the period of 24–29 May 2011. There was no radionuclide detected in river water of 29 sampling sites because of higher detection limit of ^{137}Cs measurement (10 Bq/l). On the other hand, radiocesium was found in all the river sediment samples, in which the ^{137}Cs concentrations

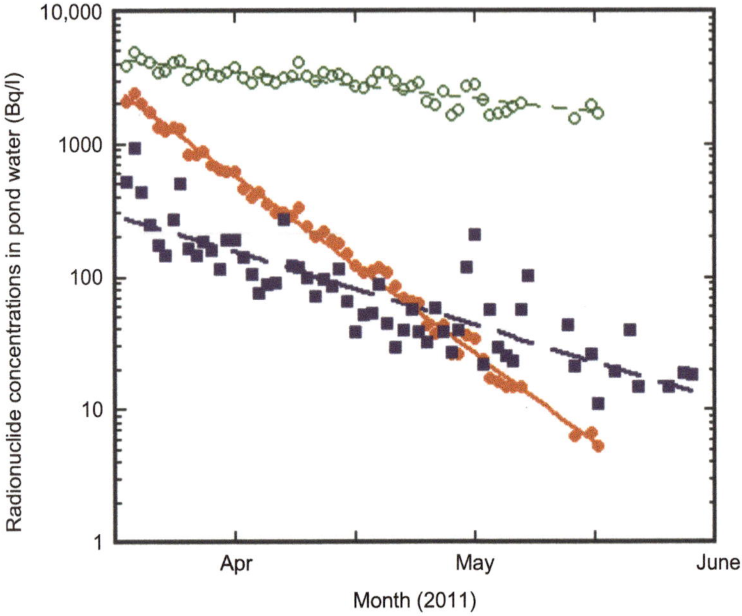

Figure 5.29 Temporal variations of ^{131}I and ^{137}Cs concentrations in pond water. Closed circles: ^{131}I, open circles: decay-corrected ^{131}I, and closed squares: ^{137}Cs. (Data from MEXT (2012a,b,c))

ranged from 51 to 16,000 Bq/kg dry weight (dw). The second measurements of radionuclides in water and sediment were performed to samples collected on 1 July 2011. Radiocesium (^{137}Cs: 17 Bq/l) was detected in river water of only one site within 29 sampling sites because of higher detection limit of ^{137}Cs measurement (10 Bq/l). Radiocesium was found in all the river sediment samples, in which the ^{137}Cs concentrations ranged from 49 to 14,000 Bq/kg dw. The first campaign of comprehensive water and sediment sampling was conducted in the period of 15 September–14 October 2011. In this campaign, water and sediment samples were collected at 113 river sites and 80 lake (including pond and reservoirs) sites and detection limit of radionuclide measurements was lowered to 1 Bq/l. The radiocesium was detected in river water of 13 sites, in which the ^{137}Cs concentrations were 1 Bq/l. The ^{137}Cs concentrations in river sediments were from 43 to 24,000 Bq/kg dw. The second campaign of comprehensive water and sediment sampling was conducted in the period of 15–30 November 2011. In this campaign, water and sediment samples were collected at 113 river sites and 46 lake (including pond and reservoirs) sites. The radiocesium was detected in river water of only one site, in which the ^{137}Cs concentration was 4 Bq/l. The ^{137}Cs concentrations in river sediments were from <10 to 49,000 Bq/kg dw. The radiocesium was detected in lake water of eight sites, in which the ^{137}Cs concentrations were in the range of 1–15 Bq/l. The ^{137}Cs concentrations in lake sediments were from <10 to 38,000 Bq/kg dw. The third campaign of comprehensive water and sediment sampling was conducted in the period of 15–30 January 2012. In this campaign, water and sediment samples were collected at 113 river sites and 32 lake (including pond and reservoirs) sites. The radiocesium concentrations in river waters were less than the detection limit (1 Bq/l) at all sampling sites. The ^{137}Cs concentrations in river sediments were from <10 to 30,000 Bq/kg dw. The radiocesium was detected in the lake water of two sites, in which the ^{137}Cs concentrations were in the range of 1–5 Bq/l. The ^{137}Cs concentrations in lake sediments were from <10 to 45,000 Bq/kg dw. The fourth campaign of comprehensive water and sediment sampling was conducted in the period of 25 February–14 March 2012. In this campaign, water and sediment samples were collected at 113 river sites and 25 lake (including pond and reservoirs) sites. The radiocesium was detected in the river water of three sites, in which the ^{137}Cs concentrations were 1 Bq/l. The ^{137}Cs concentrations in river sediments were from <10 to 54,000 Bq/kg dw. The radiocesium was detected in the lake water of two sites, in which the ^{137}Cs concentrations were 1 Bq/l. The ^{137}Cs concentrations in lake sediments were from <10 to 150,000 Bq/kg dw. The results revealed that the concentrations of the Fukushima-derived radionuclides in freshwaters were generally lower than the provisional limit value of drinking water since June 2011. The bottom sediments in river, lake, reservoir and pond were contaminated by the Fukushima-derived radiocesium, although the concentrations of radiocesium in sediments varied largely, depending on chemical and mineralogical properties of sediments and location. Unfortunately, there is little information about behaviors of the Fukushima-derived radionuclides in terrestrial environments including river, lake and ponds. It is noteworthy that the detection limit of radiocesium in freshwater is too high to assess the effects of radiocesium to the freshwater ecosystem because of high concentration factors (CFs) (6000: Saxèn et al., 2010) of biota in freshwater.

The precious measurements of radionuclides in river water (Kuchibuto River and Abukuma River), and paddy inflow and outflow were performed (Sakaguchi et al., 2012). Uranium isotopes, plutonium and radiocesium in filtered river water collected in June–July 2011 were determined by AMS, ICP-MS and gamma-spectrometry, respectively. ^{236}U among uranium isotopes is a good indicator to contamination of reactor-uranium because ^{236}U is enriched in nuclear reactors (Boulyga and Heumann, 2006). The ^{137}Cs concentrations dissolved in river waters were 0.02–0.46 Bq/kg, which are more than three orders of magnitude greater than that collected in Kuji River (Ibaraki prefecture) in 1988 (0.05–0.12 mBq/kg) (Matsunaga et al., 1991) and about two orders of magnitude larger than that collected in Tone River (Kanto Plain) just after the Chernobyl fallout (1.4 mBq/l) (Hirose et al., 1990). The ^{137}Cs concentrations in inflow and outflow waters of the Fukushima paddy fields were 0.10 Bq/kg. The $^{236}U/^{238}U$ atom ratios in river waters were in the range of 2.57×10^{-9} to 8.2×10^{-9} and the ^{236}U concentrations were $(0.82–4.48) \times 10^5$ atoms/kg. The results indicate that there was no evidence of enriched ^{236}U in river waters and therefore there was no indication of release of nuclear fuels.

The Fukushima-derived radionuclides in river waters in Kanto Plain were determined in late March and early April 2011 (Oura and Ebihara, 2012). A high concentration of ^{131}I in Edogawa River (31 Bq/l) occurred on 28 March, which corresponded to the first sampling time, whereas a peak of ^{137}Cs concentration (2.5 Bq/l) appeared on 31 March.

The available monitoring data of the Fukushima-derived radionuclides in freshwaters were limited now. The further monitoring data in freshwaters would be needed to have better understanding of migration of the radionuclides in the environment and transfer processes of the radionuclides to plants.

5.1.5 Radionuclides in Biota from Freshwater

Since 23 March 2011, MAFF has started emergency monitoring of radionuclides in freshwater food products in Fukushima and adjacent prefectures (Chiba, Ibaraki, Tochigi, Gunma, Miyagi, Iwate and Aomori). The temporal variations of radiocesium concentrations in fish caught in river are shown in Fig. 5.30. On May 2011, radiocesium ($^{134}Cs + {}^{137}Cs$) and ^{131}I concentrations, being 460 and 19 Bq/kg wet weight (ww), respectively, were detected in ayu sweetfish (*Plocoglossus altivelis*), which is an amphidromos fish and a typical 1-year life species, collected in rivers of the Fukushima prefecture. The highest concentration of radiocesium (4.4 kBq/kg ww) was measured in ayu on June 2011.

The radiocesium concentrations in ayu gradually decreased during the period from July to October 2011. In 2012, most of the radiocesium concentrations

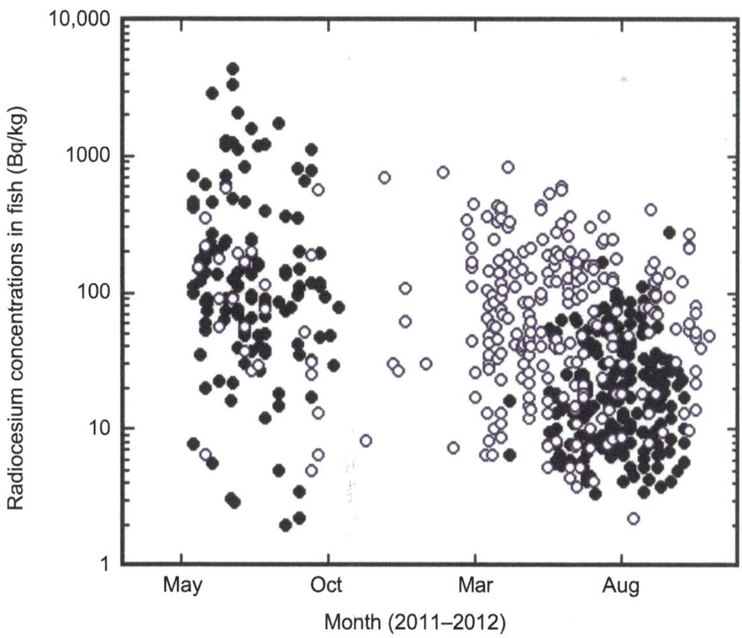

Figure 5.30 Temporal variations of radiocesium ($^{134}Cs + {}^{137}Cs$) concentrations in fish caught in river water. Closed circles: ayu sweetfish and open circles: Japanese trout. (Data from MEXT (2012a,b,c))

in ayu decreased to be <100 Bq/kg ww. On the other hand, the radiocesium concentrations in white-spotted char (Japanese trout: *Salvelinus leucomaenis*), whose mean lifetime is about 6 years in river, showed different time course from ayu. Although the Fukushima-derived radiocesium was detected in char in May 2011 as did it in ayu, the radiocesium concentrations in char in 2011 were almost lower than that in ayu. Since the radiocesium concentrations in char showed increased tendency, these were higher than that in ayu in 2012. The highest radiocesium concentration in char (840 Bq/kg ww) occurred in April 2012. The difference of the temporal variation of the radiocesium concentrations between ayu and char depends on their physiological conditions.

The temporal variation of the radiocesium concentrations in common carp (*Cyprinus carpio*) collected in ponds and reservoirs is shown in Fig. 5.31. The radiocesium was first detected in carp in May 2011. The radiocesium concentrations in carp, most of which were <100 Bq/kg ww, were generally lower than that in river-living fishes. The highest radiocesium concentration in carp (330 Bq/kg ww) occurred in April 2012. The result suggests that there is no clear trend of the radiocesium concentration in carp during the period 2011–2012.

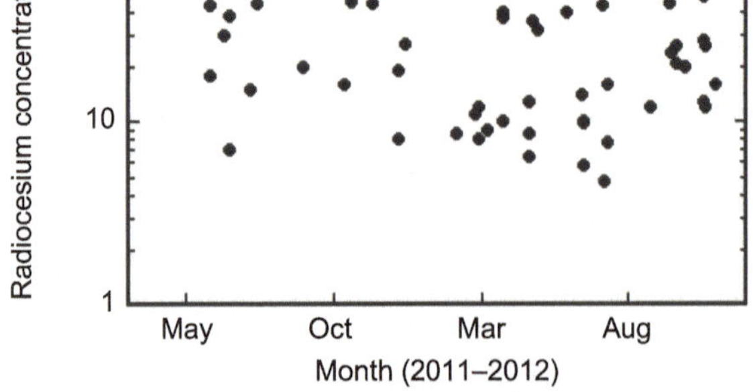

Figure 5.31 Temporal variations of radiocesium ($^{134}Cs + ^{137}Cs$) concentrations in common carp collected in ponds and reservoirs. (Data from MEXT (2012a,b,c))

5.1.6 Radionuclides in Groundwater

The MEXT conducted emergency monitoring of groundwater in Fukushima prefecture during the period of June–July 2011. All the measurement data for groundwater was less than the detection limit (10 Bq/l). These data suggest that levels of the Fukushima-derived radionuclides in groundwater in Fukushima prefecture were less than the provisional limit value for restriction of drinking water, although actual levels of the Fukushima-derived radionuclides are still unknown.

Radionuclides in 41 spring samples collected in the period 21–24 June were measured. Both ^{131}I and radiocesium concentrations were <10 Bq/l. The data suggest that levels of the Fukushima-derived radionuclides in spring in Fukushima prefecture were less than the provisional limit value for restrict of drinking water, although actual levels of the Fukushima-derived radionuclides are still unknown.

5.1.7 Radionuclides in Coastal Waters

In this chapter, we show trends of ^{134}Cs and ^{137}Cs close to the FNPP1 site, Onahama and Hasaki, located south of the FNPP1. ^{134}Cs and ^{137}Cs in seawater were observed very frequently at coastal stations located south or north of FNPP1, along the east coast of Honshu Island (Fig. 5.32).

5.1.7.1 Trends in ^{134}Cs and ^{137}Cs Concentrations Close to the Accident Site

We shall compile radionuclide monitoring data of MEXT and TEPCO and discuss observed trends at the accident site. A part of the data was already discussed in Chapter 4.2 and used to estimate total amount of direct discharge of ^{134}Cs and ^{137}Cs. The measured ^{137}Cs concentration in a seawater sample near the FNPP1 site reached 68 MBq/m^3 on 7 April (Buesseler et al., 2011). An analysis of ^{137}Cs concentrations and $^{131}I/^{137}Cs$ activity ratios suggest that major direct release of ^{137}Cs from the FNPP1 reactors occurred for 12 days, from 26 March to 6 April 2011 (Tsumune et al., 2012) and then it decreased much but it was still continuing until July 2011 (Buesseler et al., 2011) and thereafter. During the period from August 2011

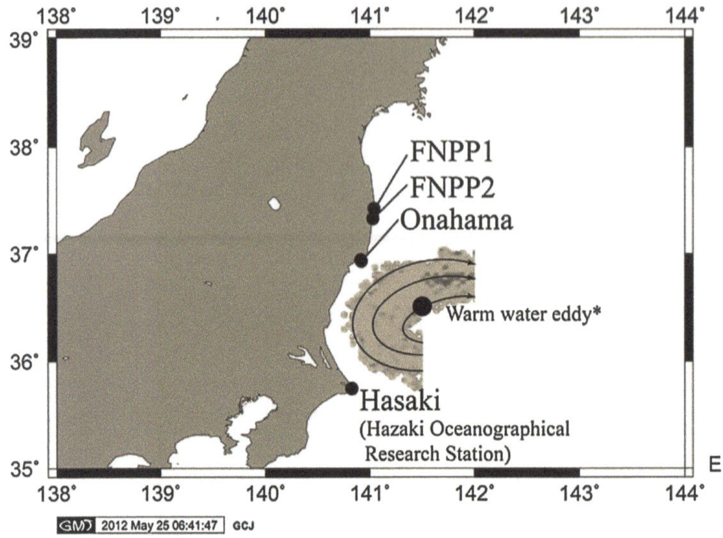

Figure 5.32 Sampling locations for ^{137}Cs and ^{134}Cs in coastal surface waters (*: Center of warm water eddy in the mid of May 2011).

to July 2012, the activities of ^{134}Cs and ^{137}Cs at near FNPP1 site were kept around 1000–10,000 Bq/m^3, which means that direct discharge becomes very small but still continues until July 2012 as shown in Figs 5.33 and 5.34.

5.1.7.2 Trends in ^{134}Cs and ^{137}Cs Concentrations at Hasaki and Onahama (South of FNPP1)

^{134}Cs activity in surface water at Hasaki was around 40–110 Bq/m^3 until the end of May 2011, there after it suddenly increased and reached 2080 $\pm$ 150 Bq/m^3 on 6 June 2011. Then ^{134}Cs activity decreased steadily with an apparent half-residence time of about 21 days until the end of August 2011. After that ^{134}Cs activity decreased with an apparent half-residence time of about 60 days until December 2011, at which time ^{134}Cs activity in surface water was 40–50 Bq/m^3 (Aoyama et al., 2012a,b). During the period from January 2012 to March 2012, ^{134}Cs activity in surface water stayed at levels of 10–30 Bq/m^3.

$^{134}Cs/^{137}Cs$ activity ratios observed at Hasaki were close to 1 (when we take into account the ^{134}Cs half-life of 2.1 years). This ratio is consistent with the $^{134}Cs/^{137}Cs$ activity ratio of 0.99 $\pm$ 0.02 observed at

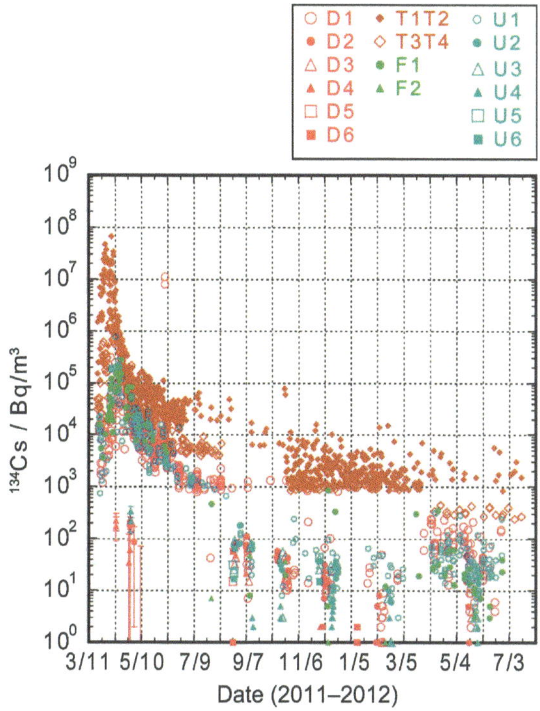

Figure 5.33 ^{134}Cs activity in surface water during the period from 3 March 2011 to July 2012 off Fukushima Dai-ichi NPP. (Modified from Aoyama et al. (2013))

FNPP1 in March and April 2011 (Buesseler et al., 2011) and in the western North Pacific in April and May 2011 (Honda et al., 2012). This ratio also indicates that the ^{134}Cs and ^{137}Cs observed at Hasaki originated from the FNPP1 accident.

^{137}Cs activity showed the same trend as ^{134}Cs. Before the FNPP1 accident, ^{137}Cs activity in the surface water of the western North Pacific Ocean was around 1–2 Bq/m^3 (Aoyama et al., 2008; Aoyama et al., 2011); therefore, ^{137}Cs activity in June 2011 was 1000 times higher than that before the FNPP1 accident.

The sudden increase of radiocesium activity in surface water at Hasaki on 6 June 2011 came 2 months after the maximum of radiocesium activity in surface water at FNPP1, which was observed on 6 April 2011 (upper arrow in Fig. 5.35). At Onahama, 30 km south of FNPP1 (Fig. 5.32), a delayed maximum of ^{134}Cs activity in surface water was also observed at the beginning of June 2011 (lower arrow in Fig. 5.35) as

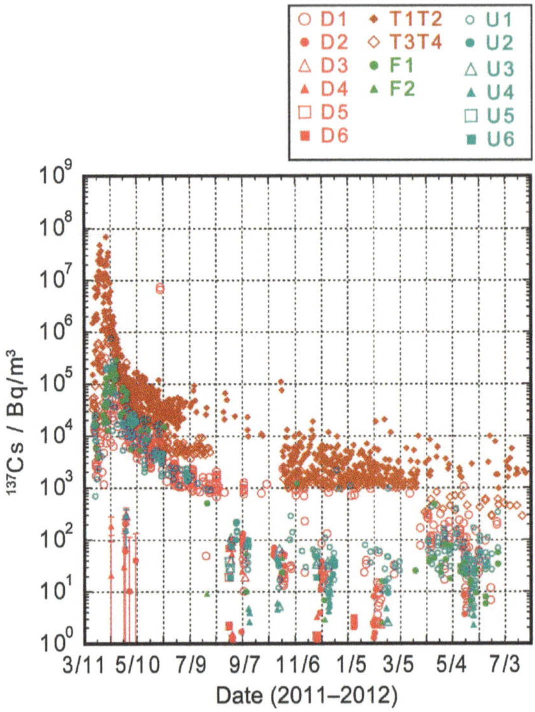

Figure 5.34 ^{137}Cs activity in surface water during the period from 3 March 2011 to July 2012 off Fukushima Dai-ichi NPP. (Modified from Aoyama et al. (2013))

was the case at Hasaki. In contrast, Inoue et al. (2012a,b,c) showed little increase of ^{134}Cs activity in surface water north of Fukushima at 10 coastal stations, shown as "Aomori" in Fig. 5.35. The ^{134}Cs activity in May and June 2011 at these stations north of FNPP1 was only a few Becquerels per cubic meter, and three orders of magnitude lower than those observations at Hasaki and Onahama south of FNPP1.

The trend of radiocesium activity in surface water north and south of FNPP1 might be regulated by characteristics of direct discharge of ^{134}Cs and ^{137}Cs at the FNPP1 site, transport processes in the coastal zone, and characteristics of atmospheric deposition of ^{134}Cs and ^{137}Cs released from FNPP1 into the atmosphere. The main source of the ^{134}Cs and ^{137}Cs measured at these coastal stations was the variable flux of radiocesium from FNPP1. Buesseler et al. (2011) reported a peak in ocean discharge in early April, 1 month after the earthquake and a decrease by

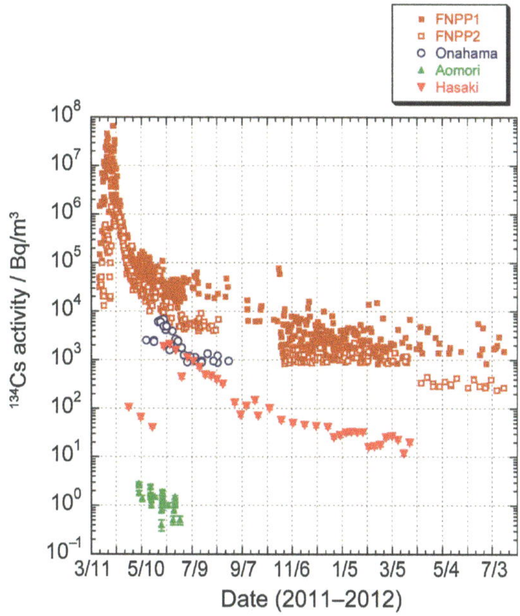

Figure 5.35 Temporal variations of ^{134}Cs at the FNPP1 and FNPP2 sites, Onahama, and Aomori stations.

a factor of 1000 by the following month. Concentrations through the end of July at FNPP1 remained several orders of magnitude higher than the levels in coastal waters measured in 2010, implying continuing releases from the reactors or other sources. ^{134}Cs activity in surface water at the FNPP1 and FNPP2 sites remained at 103–104 Bq/m^3 in December 2011 (Fig. 5.33). ^{134}Cs activity at Hasaki (Fig. 5.35) was about two orders of magnitude lower than that at FNPP1 and FNPP2 from June 2011 to March 2012 and similar to that at Onahama, whereas ^{134}Cs activity in Aomori stations, these stations locate north of FNPP1, was four orders of magnitude lower than that at FNPP1 and FNPP2. This pronounced difference in activity to the south and north of FNPP1 shows that transport of directly discharged ^{134}Cs and ^{137}Cs was dominantly southward, at least in May and June 2011 off north eastern Honshu.

Sudden increase of ^{134}Cs and ^{137}Cs activities in surface water at Hasaki occurred between 23 May 2011 and 6 June 2011 as shown in Fig. 5.35. Before this sudden change, ^{134}Cs and ^{137}Cs activities at Onahama already exceeded 2000 Bq/m^3 while those at Hasaki were only around 50–100 Bq/m^3 in mid of May 2011 (Fig. 5.35). This indicates that southward transport of ^{134}Cs and ^{137}Cs released from FNPP1 to Hasaki were relatively limited rather than southward transport to Oanahama until the end of May 2011. Coast transport processes are very complex in this sea area and these might be controlled by Kuroshio, mesoscale eddies associated with Kuroshio and freshwater flux from land. Therefore, it is interesting and important to discuss about the sudden increase of ^{134}Cs and ^{137}Cs activities at Hasaki regarding with hydrographic conditions near the coast of this region. In fact, there was a warm water eddy of which center located at 36.5° N, 141.4° E off Iwaki between Onahama and Hasaki in the middle of May 2011 as shown in Fig. 5.32. Clockwise current associated with this warm water eddy, which means northward current east of this warm water eddy, might be able to prevent southward transport of ^{134}Cs and ^{137}Cs as we stated previously. This warm water eddy disappeared on 30 May 2011 and as a consequence, sudden increase of ^{134}Cs and ^{137}Cs activities in surface water at Hasaki and Onahama were observed. Just before these changes, a difference of ^{134}Cs and ^{137}Cs activities in surface water between Hasaki and FNPP1 was 500 times or more. After the sudden changes, the difference of ^{134}Cs and ^{137}Cs activities in surface water between both stations decreased to only 30 times indicating that increased southward transport reaching at Hasaki made less activity difference between these two stations. We also can see small decrease of ^{134}Cs and ^{137}Cs activities in surface water at FNPP1 and FNPP2 between the end of May and the beginning of June 2011 as shown in Fig. 5.35. These small decreases occurring at the similar period with increases of Hasaki and Onahama may also indicate enhanced southward transport as discussed above, if we can assume less change on ^{134}Cs and ^{137}Cs fluxes at FNPP1.

5.1.8 Radionuclides in Coastal Biota

5.1.8.1 Radionuclides in Fish

The Japanese MAFF has continuously been monitoring radionuclides in fish and other seafood products in coastal waters. The monitoring data in 2000–2010 (MEXT, 2012a,b,c) reveals that concentrations of ^{137}Cs in fish off Fukushima and adjacent prefectures were about 0.1 Bq/kg ww. For comparison, ^{137}Cs concentrations in fish from the Brazilian coast were reported to be 0.1–0.3 Bq/kg ww (Cunha et al., 1993). The higher ^{137}Cs concentrations were measured in muscle of the Baltic Sea fish (5.1 Bq/kg ww in herring, 5.0 Bq/kg ww in plaice, and 7.1 Bq/kg ww in cod) (Suplinska and Adamczyk, 2009). In the south Adriatic Sea fish mullet, Antovic and Antovic (2011) measured values between 1.0 and 1.5 Bq/kg ww.

Since 23 March 2011, MAFF has been carried out emergency monitoring of radionuclides in seafood products off Fukushima and adjacent prefectures (Chiba, Ibaraki, Miyagi, Iwate and Aomori). On April 2011, high radiocesium (^{134}Cs + ^{137}Cs) and ^{131}I concentrations were measured in fish samples collected offshore the Fukushima prefecture. The highest concentration of radiocesium, which was 14.4 kBq/kg ww, was detected in Japanese Sandlance. The Fisheries Agency of Japan established the "Basic Policy for Inspections on Radioactive Materials in Fishery Products" on 6 May 2011, in which decided sea areas, frequency of sampling, and marine food species to enforce radioactivity monitoring (MAFF, 2011). MAFF have been releasing these data on a regular basis, in a single annual compilation of more than 8500 samples of fish, shellfish, and seaweeds collected at major landing ports and inland freshwater sites, particularly in the most affected coastal areas near Fukushima, which were analyzed by the Marine Ecological Research Institute, JCAC, National Research Institute of Fisheries Science, the prefectural governments and other institutions.

In June 2011, international expedition for the marine radioactivity measurement (Buesseler et al., 2012; Buesseler, 2012) was conducted and reported radionuclide concentrations in biota samples; ^{137}Cs and ^{134}Cs concentrations in mesopelagic fish ranged from 4.9 to 15.8 Bq/kg dw and from 3.3 to 14.8 Bq/kg dw, respectively.

Figure 5.36 shows the temporal variations of radiocesium (^{134}Cs + ^{137}Cs) concentrations in

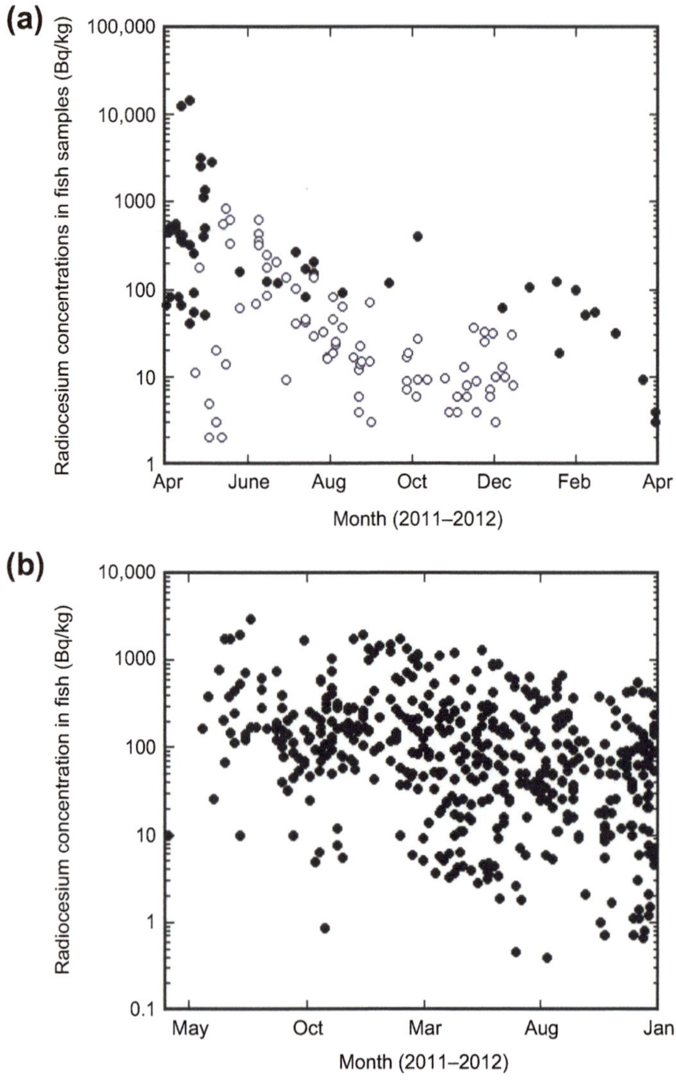

Figure 5.36 Temporal variations of radiocesium ($^{134}Cs + ^{137}Cs$) concentrations in surface-dwelling and bottom-dwelling fish. (a) Surface-dwelling fish (closed circles: Japanese Sandlance, open circles: whitebait). (b) Bottom-dwelling fish Fat greenling (*Hexagrammos otakii*). (Data from MAFF (2012a,b))

surface-dwelling and bottom-dwelling fishes. The radiocesium concentrations in surface-dwelling fish such as whitebait showed high in April–May 2011 and rapid decrease during the period of April–July 2011. The radiocesium concentrations in whitebait were <100 Bq/kg ww after August 2011, which is the new regulatory limit for food consumption. On the other hand, the high radiocesium concentrations in bottom-dwelling fish appeared in May 2011. Although its concentrations in Fat Greenling sampled off Fukushima gradually decreased, they still exceeded the new regulatory limit in the end of 2012. The MAFF results show that total radiocesium levels in demersal (bottom-dwelling) fish, including many important commercial species, are highest off Fukushima and lower in four prefectures to the north and south (Buesseler et al., 2012). Fishing for these species is banned off Fukushima, where 40% of fish are above the new regulatory limit (MAFF, 2012a).

Demersal fish have higher cesium levels than other marine fish types, grouped here as epipelagic (near-surface), pelagic (open ocean), and neuston (surface-dwelling) fish. Contamination levels of demersal fish are comparable only to those of freshwater fish. Radiocesium levels have not decreased 1 year after the accident, except perhaps in neuston, and as of August 2012, fish are still being found with radiocesium levels above 100 Bq/kg ww. The highest total radiocesium levels found to date, more than 25 kBq/kg ww, were from two greenling caught in August 2012, closer to the shore off Fukushima (TEPCO, 2012a,b,c).

In order to understand transfer of radiocesium from seawater to marine biota, CFs have been introduced, which are defined as the ratio of a radionuclide concentration in biota (becquerel per kilogram wet sample) to corresponding dissolved concentration in seawater (becquerel per kilogram). To determine a CF value, the concentrations of ^{137}Cs and stable Cs in some marine fish have previously been studied (Suzuki et al., 1973). Generally, ^{137}Cs concentrations in fish muscle were higher than in viscera, and marine fishes have CF values of 43 ± 12 (arithmetic mean and standard deviation). The CF values for ^{137}Cs were determined in Japanese coastal fishes collected from 1984 to 1990 (Tateda and Koyanagi, 1996), in which $^{137}Cs/Cs$ (stable) atom ratios were also examined to clarify the distribution equilibrium of ^{137}Cs between marine coastal fish and seawater. The geometric mean of the CF in Japanese coastal fishes was 52 ± 4 (standard error of the mean), with values ranging from 14 to 133. The $^{137}Cs/Cs$ atom ratios in both marine fish and seawater, which were $(5.2 \pm 1.4) \times 10^{-9}$ and $(4.5 \pm 1.1) \times 10^{-9}$ atom ratios, respectively, indicate that the distribution of ^{137}Cs was in equilibrium between fish muscle and seawater. The ^{137}Cs CF values for the South Adriatic Sea mullet were reported to be 38–46 in muscle (Antovic and Antovic, 2011). The IAEA (2004) recommended a value of 100 as a generalized CF for ^{137}Cs in fish. This recommended value was confirmed by Steel (1991), who reported the CF of 92 ± 43 (cod), 58 ± 17 (haddock), 39 ± 16 (plaice) and 150 ± 82 (whiting) (arithmetic mean and standard deviation) for North Sea fish collected from 1978 to 1985. These findings suggest that radiocesium accumulates in fish muscle tissues with relatively modest CFs; the ^{137}Cs concentration in fish is typically 100 times that in the surrounding seawater. The CFs increase only slightly as one moves up the food chain (Doi et al., 2012). Bioaccumulation is much higher in general in freshwater fish because of lower salinities, in which the CF of freshwater fish such as pike was 6000 (Saxèn et al., 2010). Uptake of cesium is balanced by loss back to the ocean, which increases with body size and metabolic rate (Doi et al., 2012). The loss rate is a few percentage per day on average and has been shown to be faster if the cesium supply is pulsed rather than steady (Rowan and Rasmussen, 1995). The average ^{137}Cs CF between fish (South Adriatic Sea mullet) and sediment was calculated to be 1.4 ± 0.4 (Antovic and Antovic, 2011). In case when the sediment was seriously contaminated by radiocesium, this CF value suggests that there is a possibility of occurrence of highly radiocesium-contaminated demersal fish even if the radiocesium concentrations in seawater decreased <1 Bq/l.

5.1.8.2 Seaweeds

The background levels of ^{137}Cs concentrations in blown algae were summarized by Morita et al. (2010a). The ^{137}Cs concentrations in blown algae offshore Japan ranged from 0.02 to 0.09 Bq/kg ww during the period of 1998–2008. After the Fukushima accident, MAFF has reported radionuclide levels in seaweed samples. High radiocesium concentrations were recorded in seaweed samples collected offshore Fukushima prefecture; the radiocesium concentration in Wakame seaweed (*Undaria pinnatifida*) reported on 18 May 2011 was 1200 Bq/kg ww. The radiocesium and ^{131}I concentrations in Hiziki seaweed (*Hizikia fusiformis*) reported on 26 May 2011 were 1100 and 2200 Bq/kg ww, respectively (MAFF, 2012a). After June 2011, the concentrations of radiocesium in seaweed including ones collected off Fukushima decreased to less than the new regulation value, although sampling of Wakame and Hiziki seaweeds offshore Fukushima was very limited. The temporal variation of the radiocesium concentration in Arame seaweed (*Eienia bicyclis*) is shown in Fig. 5.37. The radiocesium concentration in Arame seaweed decreased exponentially during the period from May 2011 to January 2012. The apparent biological half-life of radiocesium in Arame seaweed was 50 days (MAFF, 2012b).

The CFs for ^{137}Cs in Japanese coastal algae were studied during 1984–1990 (Tateda and Koyanagi, 1994). $^{137}Cs/Cs$ (stable) atom ratios were also

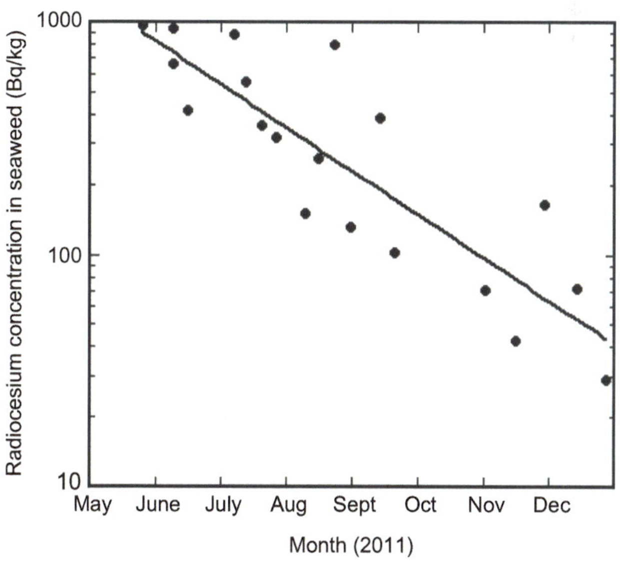

Figure 5.37 Temporal variations of radiocesium ($^{134}Cs + ^{137}Cs$) concentrations in Arame (*Eienia bicyclis*) seaweed samples. (Data from MAFF (2012a,b))

examined to clarify the distribution equilibrium of ^{137}Cs in marine algae and seawater. The CF values in marine algae were within the range from 5.4 to 92, and the geometric mean of CF was 28 ± 2 (standard error) in Japanese coastal species. The CF values in edible species were within the range from 5.4 to 67, and the geometric mean of CF was 26 ± 4 (standard error). The values of $^{137}Cs/Cs$ atom ratios in marine algae and seawater, being $(4.6 \pm 0.6) \times 10^{-9}$ (edible algae) and $(4.5 \pm 0.4) \times 10^{-9}$ atom ratios, respectively, indicated that ^{137}Cs reached an equilibrium state in partition between algae and seawater. Therefore, the CF value obtained in this study can be regarded as an equilibrated value. The temporal variation of the CF value for ^{137}Cs at the Japan Sea coastal area were examined during 1975–2007 (Morita et al., 2010a). The CF values, ranging from 3.6 to 45 for ^{137}Cs, showed gradual increase in this period, although its level was similar to that of the previous reports (Tateda and Koyanagi, 1994). The results revealed that the CF values for ^{137}Cs in Japanese coastal algae were consistent with the Japanese guideline CFs, although they were smaller than the recommended value of 50 by IAEA (2004).

5.1.8.3 Plankton

MAFF has been analyzing radiocesium in zooplankton off Fukushima and Miyagi prefectures. The radiocesium ($^{134}Cs + {}^{137}Cs$) concentrations in zooplankton sampled in July 2011–February 2012 ranged from 0.3 to 8.4 Bq/kg ww (MAFF, 2012b). The radiocesium in zooplankton collected at station K2 (47° N, 160° E) and S1 (30° N, 145° E) in April–May 2011 was determined by Honda et al. (2012). The ^{137}Cs concentration in zooplankton from the surface mixed layer (upper 80 m) was 13.5 Bq/kg dw, and that in zooplankton from the subsurface layer (80–200 m) was 59.2 Bq/kg dw. The corresponding values at S1 were 71.5 Bq/kg dw in the surface mixed layer (upper 50 m) and 57.7 Bq/kg dw in the subsurface layer (50–200 m). ^{134}Cs was also detected in all zooplankton samples, with $^{134}Cs/^{137}Cs$ ratios of nearly 1. The ^{137}Cs concentrations in zooplankton on a wet-weight basis were estimated using the water contents of respective zooplankton samples (about 90%), which corresponded to 1.7 (4.0) and 3.2 (4.3) Bq/kg ww in surface and subsurface from K2 (S1), respectively.

According to the result of the international expedition for the marine radioactivity measurement (Buesseler et al., 2012), conducted in June 2011, ^{137}Cs and ^{134}Cs concentrations in zooplankton ranged from 0.3 to 56.4, and from not detected to 45.5 Bq/kg dw, respectively. Pre-Fukushima ^{137}Cs level in zooplankton around Japan was <0.1 Bq/kg ww (~0.08 Bq/kg ww) (Kasamatsu and Ishikawa, 1997) and 0.01–0.02 Bq/kg ww (Kaeriyama et al., 2008). These findings revealed that the radiocesium levels in post-Fukushima zooplankton were more than one order of magnitude greater than that of pre-Fukushima. These concentrations are comparable to those observed just after the Chernobyl accident, in which the ^{137}Cs concentration in zooplankton in the southern Baltic Sea and the northern Adriatic Sea were 0.4–2.1 Bq/kg ww (Knapinska-Skiba et al., 2003) and about 5 Bq/kg ww (Marzano and Triulzi, 2003), respectively.

The CF was also determined for ^{137}Cs in zooplankton. The reported CF values for zooplankton off Japan are 10–100 (Kasamatsu and Ishikawa, 1997; Kaeriyama et al., 2008). For Fukushima-derived radiocesium, the median values of CF (Buesseler et al., 2012) were 44 for ^{137}Cs and 36 for ^{134}Cs, comparable to the recommended International Atomic Energy Agency (IAEA) value of 40 for zooplankton (IAEA, 2004).

5.1.8.4 Other Radionuclides in Biota

Significant amounts of radiostrontium have also been discharged in marine environment due to the Fukushima accident (Povinec et al., 2012c). MAFF has measured radiostrontium in biota samples (MAFF, 2012c). Although the level of radiostrontium in biota was generally low (<0.1 Bq/kg ww), in which the IAEA recommended CF for fish (IAEA, 2004) is 3. The ^{89}Sr and ^{90}Sr concentrations of 0.45 and 1.2 Bq/kg ww, respectively, were detected in a Rockfish (*Sebastes cheni*) sample collected on 12 December 2011.

Silver is known to be one of the highly biological active elements, which is enriched in squid organs (Oikawa et al., 2003) and in crabs (Morita et al., 2010b). Buesseler et al. (2012) documented that ^{110m}Ag was detected in zooplankton samples with concentrations ranging from 0.5 to 23.6 Bq/kg dw, whereas it was undetectable in all fish samples. Although no direct releases of ^{110m}Ag to the coastal ocean have been reported, the Fukushima-derived ^{110m}Ag was measured in deposition samples as well as in soil samples. Pathway to zooplankton is unclear because dissolved ^{110m}Ag concentrations in seawater could not be measured.

5.2 World Atmosphere

Large radionuclide releases during March–April 2011 from the tsunami-damaged Fukushima Dai-ichi NPP caused measurable radionuclide contamination of the global environment. As the prevailing wind direction in Japan is from the west to the east, the radioactive plume was transported over the Pacific Ocean to be registered first by the CTBTO radionuclide monitoring stations in Hawaii, and later at the west coast of USA (Diaz Leon et al., 2011), in Europe at Island, and later by many national laboratories (Masson et al., 2011; Baeza et al., 2012; Barsanti et al., 2012; Bossev et al., 2012; Carvalho et al., 2012; Clemenza et al., 2012; Cosma et al., 2012; Ioannidou et al., 2012; Kritidis et al., 2012; Lujanienė et al., 2012a,b, 2013; Manolopoulou et al., 2011; Piñero García and Ferro García, 2012; Povinec et al., 2012a,c, 2013a).

The most often detected radionuclides were ^{137}Cs ($T_{1/2}$ = 30.17 years), ^{134}Cs ($T_{1/2}$ = 2.06 years) and ^{131}I ($T_{1/2}$ = 8.02 days). They represent anthropogenic radionuclides without natural contribution, which have been released from nuclear reprocessing facilities and nuclear reactors (Livingston and Povinec, 2000; Mizutani et al., 2009). Under normal conditions only ^{137}Cs, released during atmospheric tests of nuclear weapons and during the Chernobyl accident, is still present in the environment due to its long half-life (Livingston and Povinec, 2002). The ^{137}Cs has been of great radiological concern due to its relatively long physical half-life and its bioavailability. Because of high abundance and accumulation in human tissue, it has been important from the point of view of delivering radiation doses to the public (Aarkrog et al., 1997). The ^{137}Cs is present in atmospheric aerosols, soil and vegetation. It has been used worldwide as a tracer of environmental processes, specifically for studying transport process in atmospheric, marine and terrestrial ecosystems (Livingston and Povinec, 2002; Povinec et al., 2003a,b; Lee et al., 2002; Igarashi et al., 2005; Pham et al., 2010, 2011; Lujaniene et al., 2006a,b, 2009).

Radionuclides released after the Fukushima accident to the atmosphere were mostly in the gaseous and aerosol forms. The first group has been represented mainly by ^{133}Xe (half-life of 5.24 days), which had also reached the highest released activities from a short-term source ($13-20 \times 10^{18}$ Bq as estimated by Stohl et al. (2011)). Radionuclide release rates from the damaged Fukushima NPP have large uncertainties. Atmospheric releases for ^{137}Cs are in the range of 13–15 PBq, and liquid releases directly to the sea were estimated to be from 1 to 41 PBq (Chino et al., 2011; Kawamura et al., 2011; Tsumune et al., 2012; Dietz and Kriest, 2012; Bailly du Bois et al., 2012).

The ^{131}I represents a radionuclide released to the atmosphere during a nuclear accident in large quantities mostly in gaseous, aerosol, acid and organic forms (Povinec et al., 1988; Masson et al., 2011). While ^{131}I is radioecologically one of the most important anthropogenic radionuclides because of its association with thyroid cancer, ^{129}I on the other hand is because of its long half-life (1.57×10^7 years) a useful isotope tracer for studying transport processes in the environment (Raisbeck and Yiou, 1999; Povinec et al., 2010). The last group of radionuclides, represented mainly by ^{137}Cs, is after release to the atmosphere rapidly associated with aerosol particles. This group of radionuclides represents a major reservoir of radioactive contaminants of the atmosphere.

The radionuclides associated with aerosols have been deposited on the Earth's surface by wet and dry depositions. Before the Chernobyl accident, temporal variations of ^{137}Cs in the surface air were strongly related to global fallout, which was stratospherically controlled. Concentrations of ^{137}Cs in the atmosphere had a decreasing trend after the moratorium on testing nuclear weapons in the atmosphere signed in 1963. Similarly as for other global fallout radionuclides, the ^{137}Cs atmospheric concentration peaked in 1963 (Livingston and Povinec, 2002). The second marked ^{137}Cs peak in the surface air occurred in 1986, just after the Chernobyl accident. For example, the maximum ^{137}Cs activity concentration measured in the Bratislava air in 1986 was 4 Bq/m^3, about 200,000 higher than the pre-Chernobyl value (Povinec et al., 1988). A few years after the Chernobyl accident, the summer maxima characteristic for the transport of ^{137}Cs from the stratosphere to the troposphere were still visible, later, during the 1990s, the main source of atmospheric ^{137}Cs became the soil resuspension (Pham et al., 2011, 2013; Povinec et al., 2012a). Therefore, most of the pre-Fukushima ^{137}Cs found in the ground-level air had been from resuspension of the deposited ^{137}Cs (of global fallout and the Chernobyl origin) in soil. The ^{137}Cs levels measured during the pre-Fukushima period ranged in Bratislava from 0.04 to 1.6 $\mu Bq/m^3$, with the mean value of 0.49 $\mu Bq/m^3$ (Povinec et al., 2012a). Short-lived radionuclides such as ^{131}I and ^{134}Cs have already decayed after the Chernobyl accident. Their maximum levels observed during April–May 1986 in the Bratislava air were 14 Bq/m^3 and 2 Bq/m^3, respectively (Povinec et al., 1988).

In this subchapter, we shall discuss global atmospheric dispersion of particles from the Fukushima NPP using a Lagrangian stochastic dispersion model, and compare them with the radionuclide measurements in aerosols. We are focusing on ^{131}I, ^{134}Cs and ^{137}Cs, as they are considered to be the most important for the long-term radiological impact because of large releases, relatively long half-lives and their contributions to radiation doses of the public. Such investigations are important prerequisite of radiological assessment of the Fukushima accident in the terrestrial and marine environments, as well as in future possible accidental releases of radionuclides from nuclear installations.

5.2.1 Radionuclide Source Terms Used in the Modeling Exercises

The discharged radioactive materials from the Fukushima NPP consisted of many types of radionuclides; however, we list in Table 5.3 only the most important ones. Radionuclide release rates from the Fukushima NPP (Chino et al., 2011; Kawamura et al., 2011) were estimated on the basis of TEPCO, Nuclear and Industrial Safety Agency and IAEA press releases to the public (TEPCO, 2011a; NISA, 2011a,b; IAEA, 2011). Following pressure venting of the reactor containments, hydrogen explosions and fires in the Fukushima Dai-ichi NPP, the main atmospheric radionuclide releases occurred between 12 and 16 March 2011, with smaller contributions up to 24 March. The total atmospheric releases of ^{131}I and ^{137}Cs were estimated to be around 160 and 15 PBq, respectively (Table 5.3). The contribution of ^{134}Cs was similar to ^{137}Cs as the $^{134}Cs/^{137}Cs$ activity ratio was close to one. The estimated release rates in the cases of ^{131}I and ^{137}Cs are well above 1 PBq which require specific terrestrial and marine radioactivity assessment studies (Livingston and Povinec, 2000), necessary for the radiation protection of the total environment, i.e. the man, fauna and flora. However, as there are still large uncertainties in the estimation of the radionuclide source terms, we shall use in our simulation experiments as an input a ^{137}Cs release into the atmosphere of 1 PBq. This unit input value will enable to adjust the obtained modeling results by multiplying them by a factor value which may be better known at the later time when more precise radionuclide source-term functions will be available.

5.2.2 Data on Radioactivity of Aerosols over Europe

In order to compare modeling and measurement approaches in the spatial and temporal variations of the distribution of Fukushima-derived radionuclides over the Europe, an integrated data set presented in Fig. 5.38 has been used. The ^{131}I data (which have had the highest spatial resolution and smallest experimental uncertainties) reported by the northern European stations (Masson et al., 2011; JER, 2012) have been combined to present a general view on its spatial and temporal

Table 5.3. Radionuclide Inventories and Releases in the Atmosphere and Ocean (in PBq)

Nuclide	Half-life	Natural Inventory[a]	Global Fallout Atmosphere[a]	Global Fallout Inventory in the Ocean[b]		Chernobyl		Fukushima	
				Total Inventory in the Ocean	Inventory in 2010 in the Ocean	Atmosphere[c]	Ocean[b]	Atmosphere	Ocean
^{3}H	12.32 y	2200		113 000	8 000				0.1[p]
^{129}I	$15.7 \cdot 10^{6}$ y	$0.6\ 10^{-3}$		$0.3 \cdot 10^{-3}$	$0.3 \cdot 10^{-3}$	0.013×10^{-3}			$0.0024 \cdot 10^{-3}$[g]
									$0.0046 \cdot 10^{-3}$[h]
^{131}I	8.02 d	-		-	-	1760		153-160[d,e]	62[i]
									120[i]
^{137}Cs	30.17 y	-	950	600	170	85	16	13-15[d,e]	3.5±0.7[j]
								23-50[f]	4[k]
									0.94-3.5[l]
									10-34[m]
									9-18[n]
									5[o]

[a]UNSCEAR, 2008; [b]IAEA, 2005; [c]IAEA, 2003; [d]TEPCO, 2011; [e]Chino et al., 2011; [f]Stohl et al., 2012; [g]Hou et al., 2013; [h]Atmospheric deposition (this work); [i]Morino et al., 2011; [j]Tsumune et al., 2012; [k]Kawamura et al., 2012; [l]Dietze and Kriest, 2012; [m]Bailly do Bois et al., 2012; [n]Rypina et al., 2013; [o]atmospheric deposition (Kawamura et al., 2011); [p]liquid discharges with atmospheric deposition (this work)

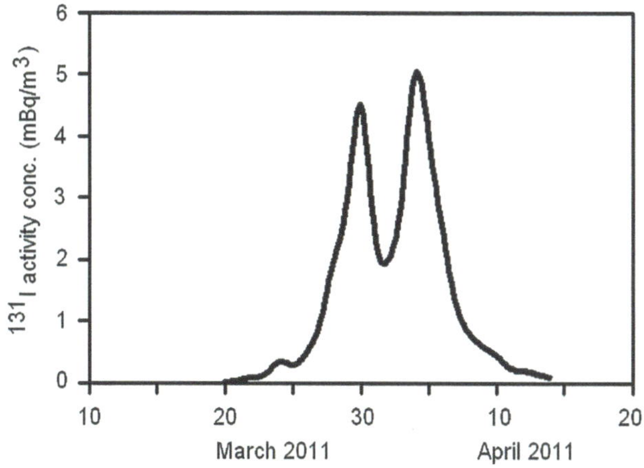

Figure 5.38 Averaged ^{131}I concentrations in aerosols samples collected over the Europe. (Adapted from Povinec et al., 2013a).

distributions over the North Europe. The Fukushima signal was detected first in Europe on 19 March (the CTBTO station in Reykjavik). Two well-distinguished maxima were recorded by the northern European stations: the first one on 30 March (at 4.5 mBq/m^3) and the second one on 3 April 2011 (at 5 mBq/m^3). It is interesting to note that a similar time dependence of ^{131}I distribution was observed in Central and South Europe stations (with differences of a few days only), however, with lower activity concentrations (maxima at 2.5 and 1 mBq/m^3 for Central and South Europe, respectively).

5.2.3 Modeling the Air Mass Transport

5.2.3.1 Modeling of Forward and Backward Trajectories

The simulations of forward air mass trajectories from Fukushima to the Europe were carried out using a numerical model of the European Centre for Medium-Range Weather Forecasts (ECMWF; http://www.ecmwf.int). For modeling the forward trajectories, it was supposed that the particles were released from Fukushima (141.0° E, 37.3° N) at four different pressure levels: 925 hPa (~800 m a.s.l.), 850 hPa (~1500 m a.s.l.), 700 hPa (~3000 m a.s.l.), and 500 hPa (~5000 m a.s.l.). It is expected that the particles during the transport are staying approximately at the same pressure levels. The trajectories were modeled with the time frequency of 6 h. The source of information for simulation of the trajectories was an analysis of the state of the atmosphere during a given time interval, which was available from the ECMWF. The trajectories are presented above a weather background, given as a ground-level pressure at the time of a particle release from the source.

To check the system performance, three-dimensional backward trajectories of air masses at 850 and 700 hPa height between the Central Europe and Fukushima were calculated using the National Oceanic and Atmospheric Administration (NOAA) Hybrid Single-Particle Lagrangian Integrated Trajectory (HYSPLIT) dispersion model (Rolph, 2011; Draxler and Rolph, 2011) with support of the Final Analyses and the Global Data Assimilation System meteorological databases at the NOAA Air Resources Laboratory server.

The zonal westward flow, modified by planetary waves, dominates at 500 hPa level (typically 5000 m a.s.l.). Some synoptic centers from the lower levels can still be recognized at 500 hPa level, for example the Aleut cyclone. At 850 hPa level (around 1500 m a.s.l.), the surface synoptic objects are well expressed, and similar situation is at 700 hPa level (around 3000 m a.s.l.). To see the field, which has driven the particle trajectories, we plotted in Fig. 5.39

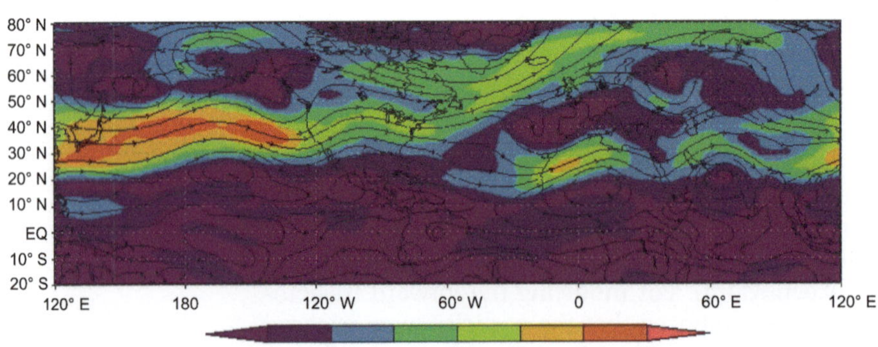

Figure 5.39 Average wind speed (in knots) at 500 hPa (from 12 to 22 March 2011) which driven particle trajectories between Japan and Europe (the lines represent the streamlines). (Adapted from Povinec et al. (2013a)).

the average wind field at 500 hPa (from 12 to 22 March 2011). It can be seen that particles, which were released from Fukushima in different times and different altitudes above the earth surface, had various trajectories. The jet stream has been influencing significantly the particles spreading at upper atmospheric levels. The wind speed reached the values of about 70 km/h.

Forward air mass trajectories for particle releases from the Fukushima NPP are presented in Fig. 5.40. The distance between neighboring points is equivalent to the particle movement during 6 h. The

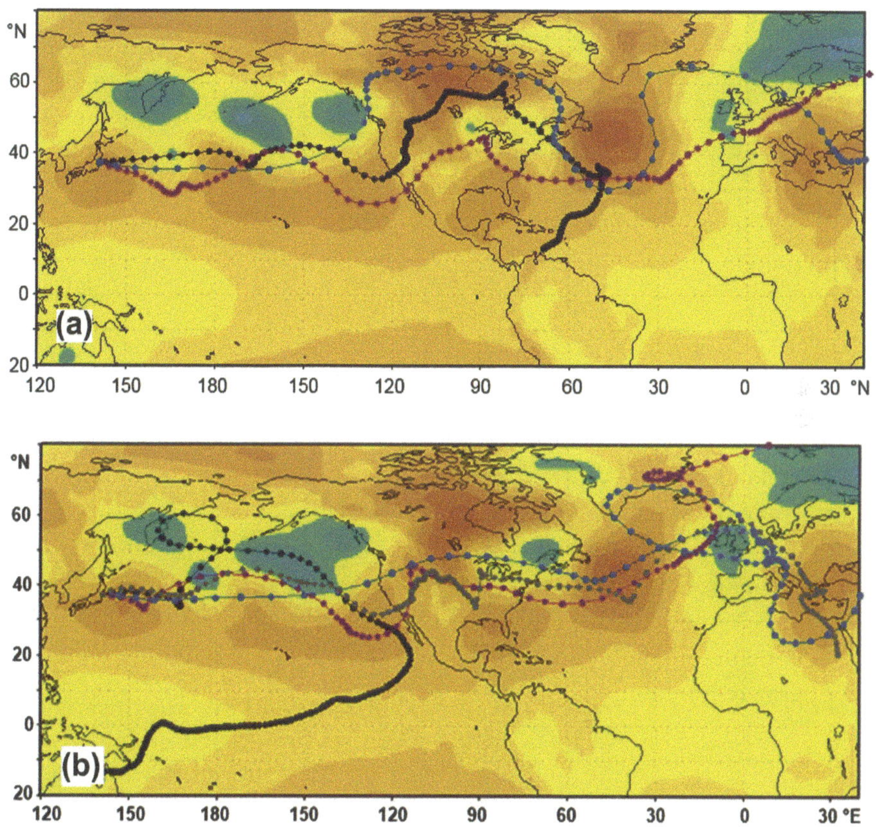

Figure 5.40 Forward air mass trajectories for particle releases from the Fukushima Dai-ichi NPP. Grey points represent 925 hPa pressure level (~800 m a.s.l.), black points 850 hPa (~1500 m a.s.l.), violet points 700 hPa (~3000 m a.s.l.), and blue points 500 hPa (~5000 m a.s.l.). The coloured background represents the sea-level pressure at the start of the simulation. Time of particle release at Fukushima: (a) 12 March at 6:00; (b) 20 March at 6:00; (c) 25 March at 6:00; (d) 30 March at 0:00 (Modified from Povinec et al., 2012b).

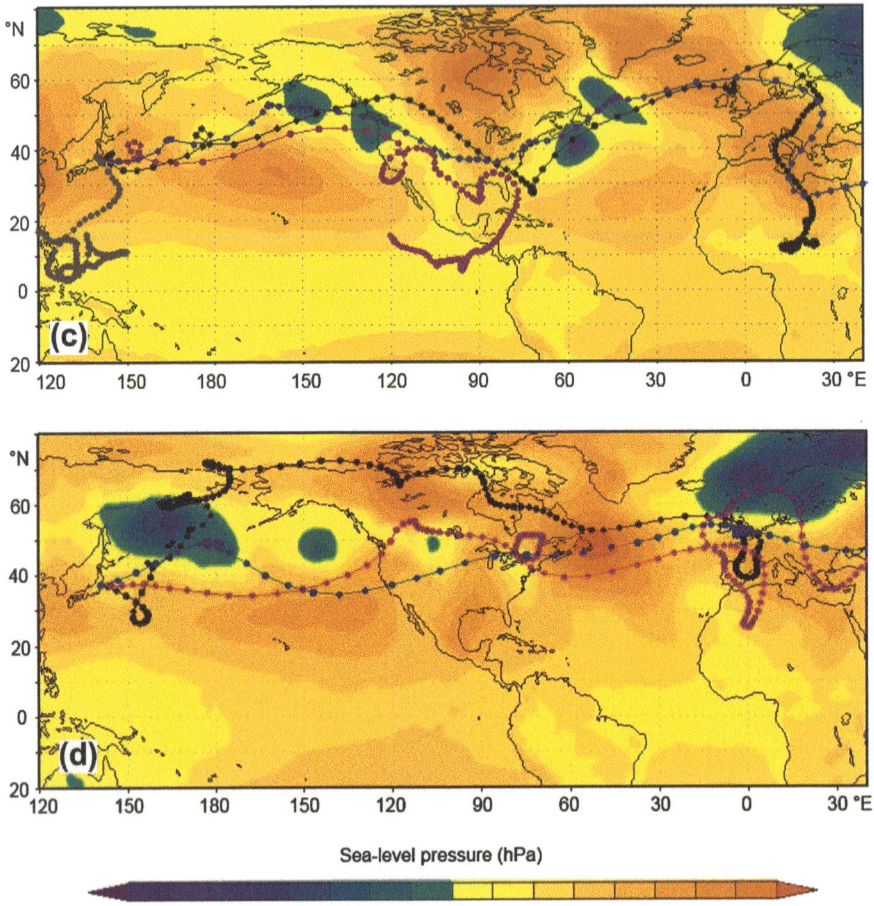

Figure 5.40 Continued

colored background represents the sea level pressure in the time of particle release. It can be seen that particles, which were released in different time and different altitudes above the surface have various trajectories. A typical westerly flow was observed in mid-latitudes. From the climatologic point of view, the main influence on the route of the radioactive cloud to the Europe had the Hawaiian and Aleut cyclones, the Azorean and Greenland anticyclones and the Island cyclone. A subtropical belt of high pressure can be recognized in Fig. 5.40, which was divided into several isolated pressure centers. The isolated centers over the Pacific Ocean are related to Hawaiian anticyclones. In the Atlantic Ocean, the situation differed from the average atmospheric circulation. The Azorean anticyclone center is shifted to northwestern direction, where interaction with the Greenland anticyclone is observed, which is blocking a straight propagation of particles to the east. The low polar pressure belt is broken to a cyclone with center close to the Kamchatka peninsula and the Bering Sea. In the Atlantic Ocean, the center of low pressure is located in the Norwegian and Barents Seas. These pressure objects modified the zonal prevalent flow. Sea level pressures presented in Fig. 5.40 in the background are from initial time releases. During several days of spreading of particles in the air, the pressure fields are naturally modified. This fact has been included in the computation of trajectories.

The simulations revealed that particles released from Fukushima on 11 March arrived to Europe at 850 hPa on 13 April, at 700 hPa on 30 March and at 500 hPa on 20 March. The particles released on 12 March at 00:00 Coordinated Universal Time did not reach the European territory at all. A different situation is observed for particles released just 6 h later—the particles at 700 hPa reached the Europe on 1 April, at 500 hPa on 20 March, and at 850 hPa, they did not reach the European territory.

The origin of the first radionuclide signal observed in the European air should be associated with the first large radionuclide releases, which originated in Fukushima on 11–12 March. The simulations indicate particle trajectories at 500 hPa level arriving to Europe on 20–22 March (Fig. 5.40; as we have limited space available, in all modeling exercises we shall present only a few dispersion maps). The main jet stream of the plume was traveling over the Pacific Ocean, North America, Greenland, the Atlantic Ocean, Island and Scandinavia to the Central Europe.

In other to better understand variations in the air transport at different altitudes and arrival times of the Fukushima radioactive plumes in Europe, we used a backward-trajectory approach based on the ECMWF and NOAA HYSPLIT models. A large number of air mass backward trajectories were calculated over the time of interest. The results obtained using the ECMWF model (Fig. 5.41A) are in a reasonable agreement with results obtained for the forward trajectories, however, as expected, the trajectories in Japan do not fit exactly with the Fukushima site. The trajectories arriving to Bratislava on 21 March at 6:00 started at Fukushima on 12 March in the morning; the trajectories reaching Bratislava on 31 March at noon originated most probably in Fukushima on 21 March; the trajectories reaching Bratislava on 7 April at midnight had origin most probably at Fukushima on March 31, and the trajectories reaching Bratislava on 14 April in the evening originated in Fukushima

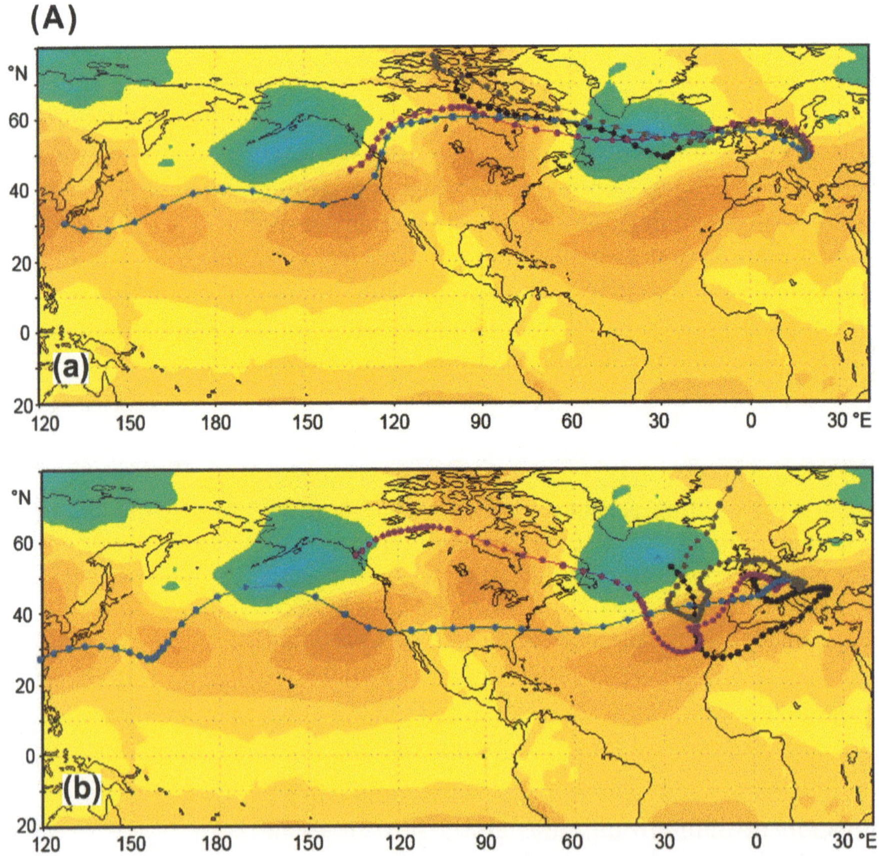

Figure 5.41 (A) Backward air mass trajectories for particle releases from the Fukushima Dai-ichi NPP. Grey points represent 925 hPa pressure level (~800 m a.s.l.), black points 850 hPa (~1500 m a.s.l.), violet points 700 hPa (~3000 m a.s.l.), and blue points 500 hPa (~5000 m a.s.l.). The coloured background represents the sea-level pressure at the start of the simulation. Start of the simulation from Bratislava: (a) 21 March at 6:00; (b) 31 March at 12:00; (c) 7 April at 00:00; (d) 14 April at 18:00. (Modified from Povinec et al., 2012a).

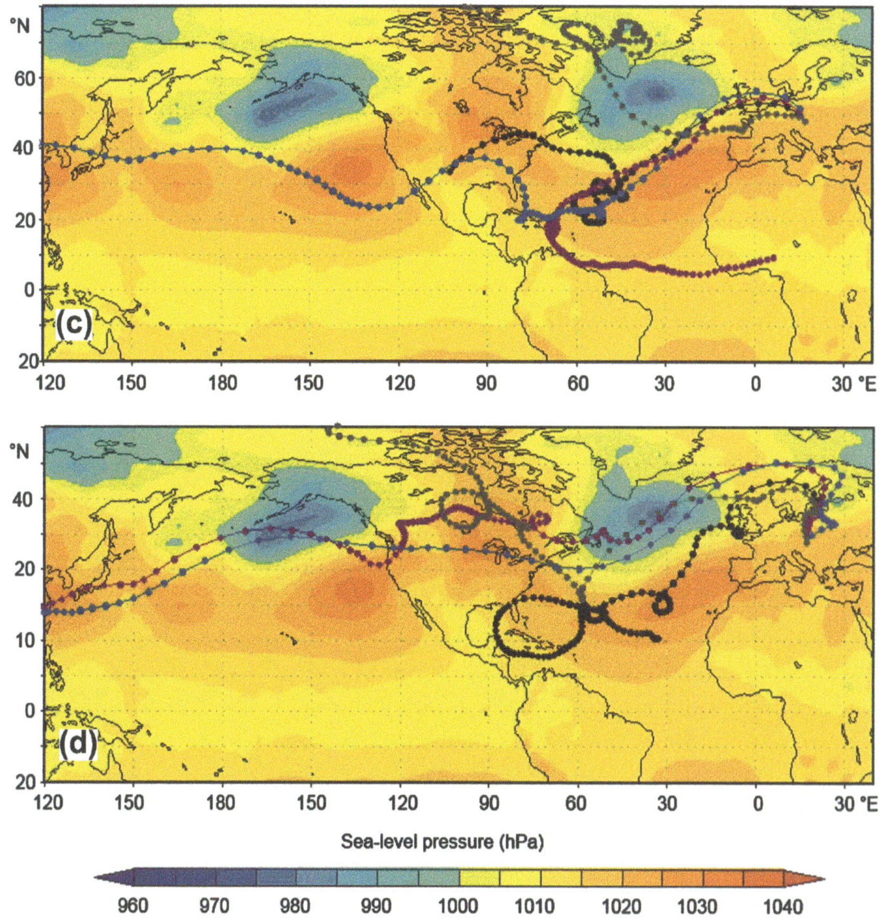

Figure 5.41 Continued

probably at the beginning of April. It can be seen that the dominant air mass transport in all backward simulations was at 500 hPa. The results obtained using the HYSPLIT model (Fig. 5.41B) show backward air mass transport starting at 500 (red triangles), 3000 (blue squares), and 7000 m (green circles) above the ground level. Trajectories are labeled every 24 h by a filled symbol. The vertical projection of the trajectories with time is shown in the panel below the map.

The air masses at 500 m were caught up into a cyclonic system, while air masses at 3000 and 5000 m were lofted and rapidly transported over the North America to Europe. The global view of the

(B)

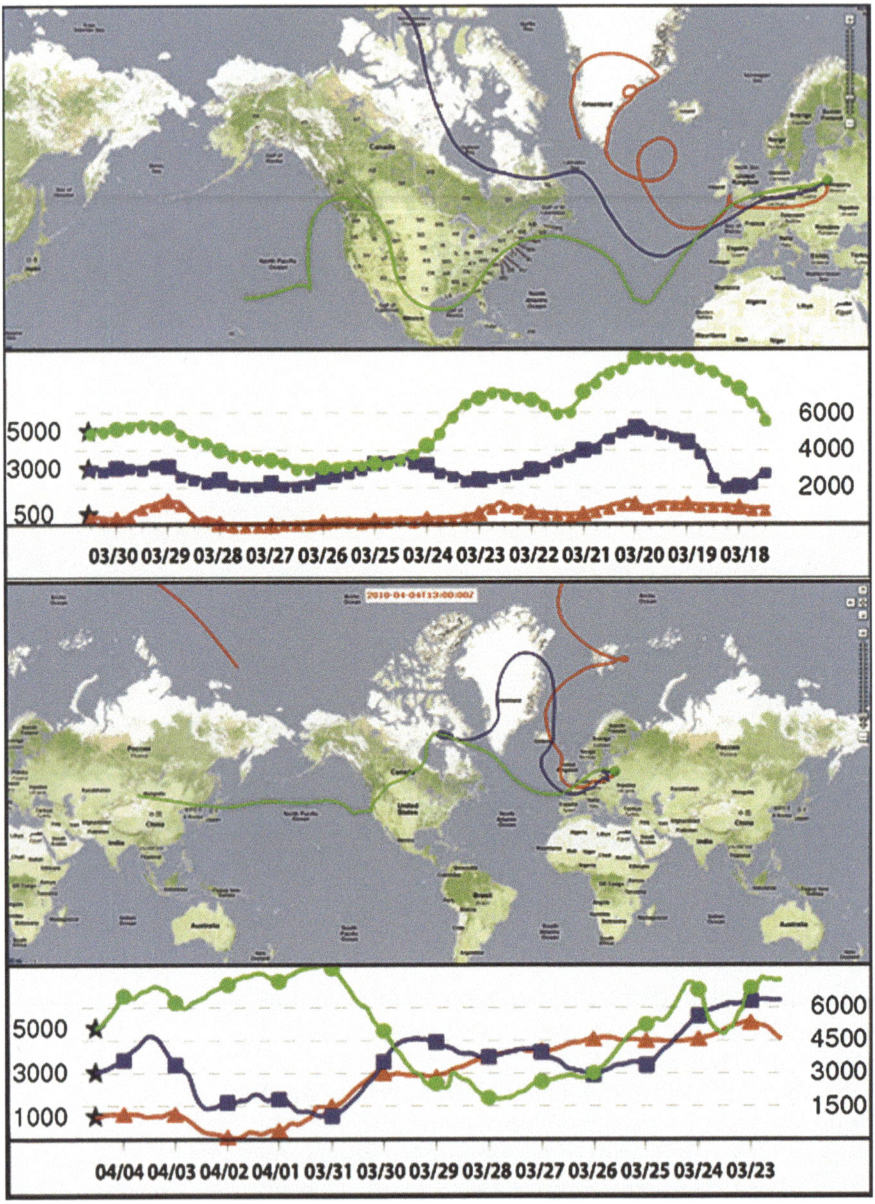

Figure 5.41 (B) Backward trajectories of air mass transport ending on 29 March and 4 April, 2011 in Vilnius (Lithuania) based on the NOAA HYSPLIT model. Backward air mass transport started at 500 m (red triangles), 3000 m (blue squares), and 7000 m (green circles) above the ground. Vertical projection of the trajectories with time is shown in the panel below the map. (Adapted from Lujanienė et al., 2012b).

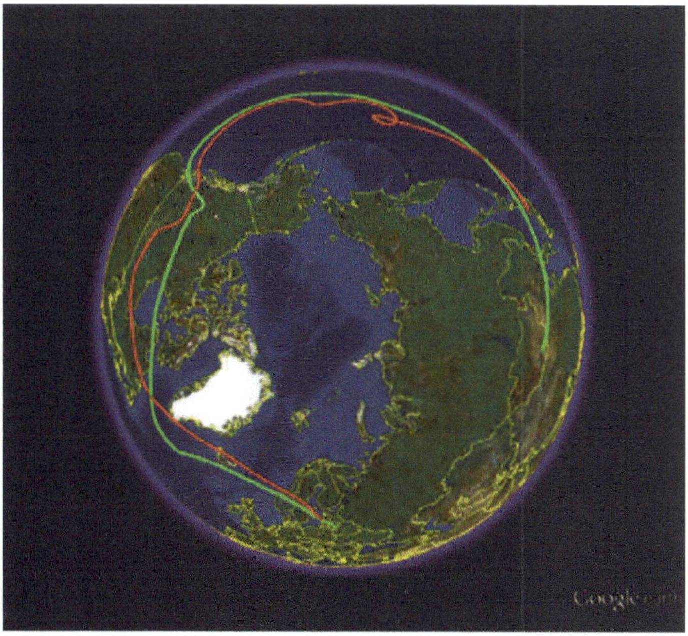

Figure 5.42 The global view of the backward trajectories from Fukushima to Europe. (Adapted from Lujanienė et al. (2012b))

forward/backward trajectories from Fukushima to Europe is presented in Fig. 5.42. The simulations also confirmed that radioactive particles have had a greater chance of being transported to Europe at higher atmospheric levels.

5.2.3.2 Lagrangian Dispersion Model

The dispersion of particles from the Fukushima source was simulated using Lagrangian particle model. In this type of models, the initial plume is broken into fictitious elements/particles, which represent emission portions (Závodský, 2011). The Lagrangian model calculates trajectories of the elements that follow the instantaneous flow in the element position. Under the instantaneous flow, a superposition of mean wind and turbulent wind fluctuations can be expressed as

$$r(t+\Delta t) = r(t) + \Delta t . v(r, t), \text{ where } v = v_{\text{avg}} + v_t.$$

The symbol r denotes the position vector, v is the vector of particle's velocity, v_{avg} is the grid scale mean wind, and v_t represents the turbulent wind fluctuations.

The output particle velocity is a sum of the deterministic velocity and the semirandom stochastic velocity, generated using Monte Carlo techniques. The probability density function for the random component, which simulates the atmospheric turbulence, is dependent on the state of the atmospheric boundary layer. Thus, both basic scales of atmospheric motions are taken into account. The model can also take into account a decay of radioactive particles on their way from the source to the observer, as well as a scavenging by wet and dry depositions.

For the meteorological input, the Integrated Monitoring System Model Suite Lagrangian dispersion model has been used (Gera et al., 2013; MICROSTEP-MIS, 2012). It calculates the spreading of radioactive material with regard to changes in atmospheric conditions, especially changes in the wind direction. The meteorological input for the dispersion model was a time-sequence analysis of atmospheric state in GRIB format (WMO, 2009). The Global Forecast System global weather model was used in simulation time span from 12 March to 27 March 2011. The 3D wind parameters (u, v and vertical velocity) at upper air model levels were needed to simulate dispersion due to large-scale winds. Surface meteorological parameters influencing thermal and mechanical turbulence and surface roughness were needed for simulation of turbulent wind fluctuations. The values of input meteorological parameters in the intergrid locations were bilinear interpolations of the gridded values.

The vertical velocity, particle dissipation and turbulence during the particle transport were considered in the Lagrangian modeling. A single release of 1 PBq of ^{137}Cs, which occurred on 12 March 2011 from the damaged Fukushima nuclear reactors was considered. The initial plume height, as a result of initial vertical velocity and buoyancy, was kept at 2000–3000 m. The meteorological data and simulated trajectories revealed that the arrival times of particles released on different days were not the same. The jet stream affected the transport of emitted particles at upper atmospheric levels as the particles were transported at different altitudes. The simulated trajectories using the Lagrangian dispersion model show that the first signal from the Fukushima accident could be detected in Europe on 19 March 2011 (Island, Fig. 5.43). The color range in the figures should be interpreted as follows: shades of red indicate particles from the bottom up to 3 km, black to dark blue color indicates the middle layer up to 6 km, and light blue represents an upper layer. Particles are organized into two vortices. Figure 5.43 for 20 March shows the predominance of particles in the upper levels above the European territory. There is predominance of dark blue color, later, after 21 March, the red color points are visible as well. On 20 March, the signal could be detected over the northern Europe, and on 21–22 March, the entire Europe was covered with the radioactive plume.

For a comparison of modeling predictions with real radionuclide data, we need to have the following in mind:

1. Particles released from Fukushima at a later stage could arrive to Europe earlier, depending on specific meteorological conditions.
2. Particles traveling in different heights could have different arriving times in Europe.
3. Particles from the upper air levels could be transported to the ground air depending on specific meteorological conditions.

The modeling results of Masson et al. (2011) are in a reasonable agreement with real observations carried out in the North America and in Europe (Island) as the first detection of contaminated air masses occurred on 17 and 19 March, respectively (see also discussion to Fig. 5.43). The northern Scandinavia registered the plume on 20 March, and the other European countries between 21 and 22 March (Masson et al., 2011; Lujanienė et al., 2012a; Povinec et al., 2012a,b, 2013a). Usually two activity concentration maxima of ^{131}I, ^{134}Cs and ^{137}Cs were observed in European aerosols (Fig. 5.38). Depending on the position in the Europe, the ^{131}I activity concentrations varied in the maxima up to 6 mBq/m^3 (~1 mBq/m^3 for ^{137}Cs) for the Northern European countries, up to 3 mBq/m^3 (~0.1 mBq/m^3 for ^{137}Cs) for the Central Europe, and up to 1 mBq/m^3

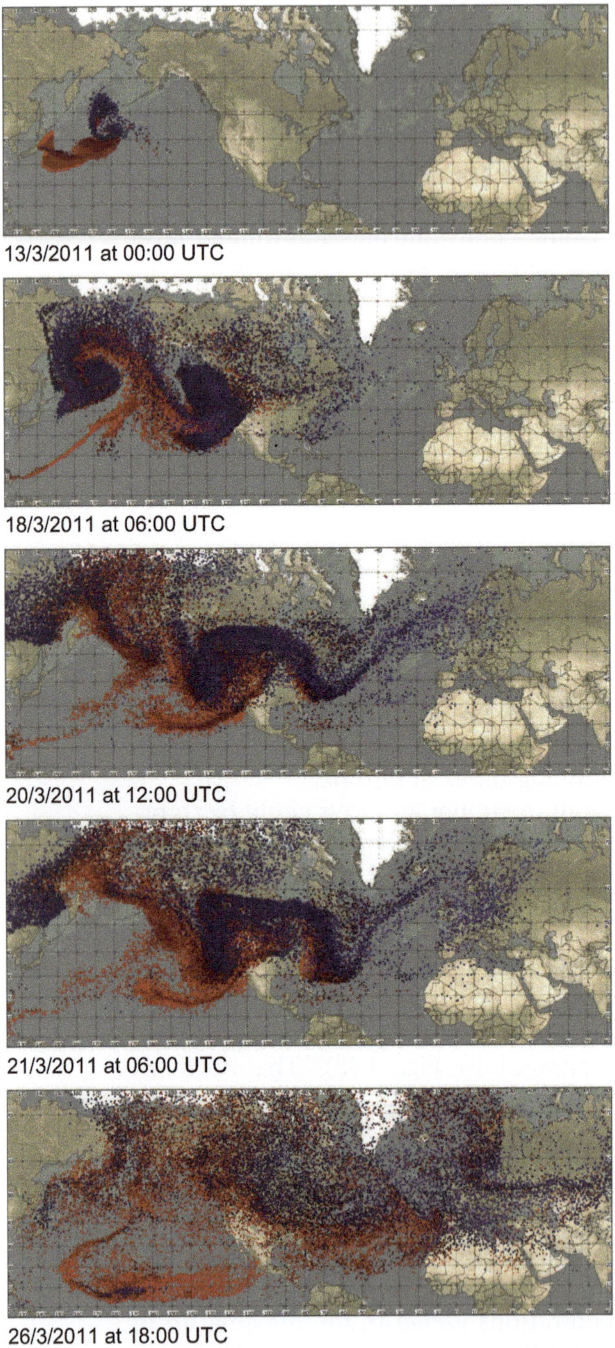

Figure 5.43 Air mass trajectories simulated using the Lagrangian dispersion model. The color range: shades of red indicate particles from the bottom up to 3 km height, black to dark blue color indicates the middle layer up to 6 km height, and light blue represents a layer above 6 km. (Modified from Povinec et al., 2013a).

(~0.05 mBq/m^3 for ^{137}Cs) for the South European countries. While the first maximum was observed in Europe on 30 March, the second one was observed on 3 April 2011 (Fig. 5.38). The origin of the first peak should be associated with first large radionuclide releases, which originated on 12–13 March, and the origin of the second peak could be associated with the largest releases which occurred on 15 March. The decreasing radionuclide levels observed between these peaks may be due to pronounced rainfall events, which occurred over most part of Europe. The increased ^{131}I and $^{134,137}Cs$ concentrations in the European air persisted for approximately 20 days, locally depending on precipitation conditions. The ^{131}I was vertically well mixed, as confirmed by measurements at the high altitudes. The Jungfraujoch station (46.5° N, 8° E; 3580 m a.s.l.) measured ^{131}I levels up to 0.9 mBq/m^3. High-altitude samples taken with airplanes on 30 March above Switzerland (up to 8000 m) and above northern Germany (up to 10,000 m) showed ^{131}I levels around 2 mBq/m^3 (Masson et al., 2011).

Different meteorological conditions at the European sites (e.g. rainfall characteristics, fog formation or growth of aerosol particles and their deposition) could affect the activity concentrations at the ground-level air (Gera et al., in preparation). The amount of precipitation (above 10 mm) could have a certain effect on changes in activity concentrations of studied radionuclides at the ground-level air. However, no correlation was found between the ^{131}I and ^{137}Cs activity concentrations and the amount of precipitation (Lujaniene et al., 2012a,b), indicating that the influence of precipitation on deposition of radionuclides in Europe was generally negligible.

The Fukushima radionuclide levels on aerosols measured in Europe were three to four orders of magnitude lower than the activity levels encountered in Europe after the Chernobyl accident in April 1986. For example, the maximum ^{131}I and ^{137}Cs levels measured on 30 March 2011 in Bratislava (0.5 and 0.07 mBq/m^3, respectively; Povinec et al., 2012a,b) were much lower than the Chernobyl values (14 and 4 Bq/m^3, respectively; Povinec et al., 1988).

Although there have been too many uncertainties (estimation of the source terms, changeable meteorological conditions along the pass from Japan to Europe affecting scavenging efficiency for radionuclides), the dispersion of radionuclides has been reasonably well predicted. It has been found that the air masses mostly traveled from Japan to Europe at 500 hPa (5000 m) height level. The vertical radionuclide distribution was measured only in a few places of the world; therefore, we do not have enough information on the vertical mixing. The horizontal dispersion in the Europe reached about 4000 km, however, as shown by Lagrangian modeling, the entire world has been marked with the Fukushima radionuclides (Fig. 5.43), although at very low levels. If we take an average distance between Japan and Europe over the Pacific Ocean, North America and the Atlantic Ocean (18 000 km), and the travel time of 10–15 days, we can estimate an average speed of the plume to be 50–70 km/h, what is in reasonable agreement with meteorological observations (Lujanienė et al., 2012a,b). It is interesting to notice that the plumes between North America and Europe mostly traveled close to the North Pole (over the Greenland), as is a typical travel journey for commercial airplanes.

We may conclude that the temporal and spatial distributions of Fukushima-derived radionuclides in aerosols revealed the presence of two dominant radionuclide maxima which were observed throughout the Europe with decreasing amplitudes from the North to the South, which were associated with different air masses present in the European air. Modeled forward and backward trajectories indicated a preferential transport of air masses between Fukushima and Europe at 500 hPa (5000 m a. s. l.) air heights. The Lagrangian modeling showed that the horizontal dispersion in the Europe reached about 4000-km-wide belt, however, the entire world has been labeled with the Fukushima radionuclides, although at very low levels. A typical travel time between Fukushima and Europe has been estimated to be of 10–15 days, with an average speed of the plume of 50–70 km/h. An average ^{131}I concentration, which was measured over the Europe ($\sim 1\ mBq/m^3$), would result in the total amount of dispersed ^{131}I of about 1 PBq. Although this represents a high release rate (almost 1% of the total amount of ^{131}I released from the Fukushima NPP), as it was distributed over a huge area, it has not been of any radiological significance for European citizens.

5.3 Pacific Ocean

5.3.1 Radionuclides in Seawater

As for distribution of the radionuclides released from the Fukushima Dai-ichi NPP accident in the ocean, the early distribution is formed mainly by the radionuclides emitted into the atmosphere, then deposited to the ocean, and released directly into the ocean (a direct discharge). And in addition to these two major pathways, radionuclides also may be carried into the ocean by river runoff or groundwater discharge to the ocean. Since larger portion of radionuclides released from the FNPP1 are in dissolved form, they moved with seawater; therefore, the distribution of radionuclides changed with movement of seawater.

The FNPP1 locates on the coastline at Fukushima, Japan, and the strong coastal flow was dominated to the south near the FNPP1 site, therefore it is thought that the radionuclides were transported by the southward flow instead of diffusion. In the open ocean, east of the Honshu Island, the Oyashio Current goes south along the Chishima Islands and the Honshu Island. The Kuroshio Current goes north along Philippines and Okinawa Island. Then both Currents flow to the east as a Kuroshio Extension after they were merged. At the time of FNPP1 accident, the coastal flow went south, and the radionuclides released from FNPP1 were transported to the Kuroshio Extension, south first and then to the east. As for the part which deposited to the ocean via the atmosphere, our observation showed heterogeneous deposition over the wide area in the North Pacific Ocean (Aoyama et al., 2012a) during the radioactive plume traveled over the North Pacific Ocean. We also cannot forget the portion, which interacts with marine sediment, or an interaction with the particulate matter in the ocean, either.

This chapter will describe the research results including the measurements of the radionuclides in the ocean based on an article by Aoyama et al. 2013 and model research results.

On 11 March 2011, an extraordinary earthquake of magnitude 9.0 centered about 130 km off the Pacific coast of Japan's main island, at 38.3° N, 142.4° E, was followed by a huge tsunami with waves reaching up to 40 m height in Iwate region and about 10 m in Fukushima region (The-2011-Tohoku-Earthquake-Tsunami-Joint-Survey-Group, 2011; Mori et al., 2011). These events caused the loss of about 16,000 lives, missing of about 4000 lives and an extensive damage. One of the consequences was a station blackout (total loss of alternating current electric power) at the TEPCO Fukushima Dai-ichi NPP. The station blackout developed into a disaster that left three of the six FNPP1 reactors heavily damaged and caused radionuclides to be discharged into the air and ocean (Chino et al., 2011; Kawamura et al., 2011; Morino et al., 2011; Stohl et al., 2012; Tsumune et al., 2012; Estournel et al., 2012).

^{134}Cs and ^{137}Cs were released to the North Pacific Ocean by two major likely pathways, direct discharge from the FNPP1 accident site and atmospheric deposition off Honshu Islands of Japan, east and northeast of the site. ^{134}Cs and ^{137}Cs activities in the surface water in the North Pacific Ocean were already reported (Aoyama et al., 2012a; Honda et al., 2012) and those ranged a few to 1000 Bq/m^3 in April–May 2011. Distributions of ^{134}Cs and ^{137}Cs activities in the surface water off Honshu and coastal stations around Japan during the period from April 2011 to November 2011 (Inoue et al., 2012a,b; Aoyama et al., 2012a; Miyazawa et al., 2012; Buesseler et al., 2011; Buesseler, 2012) were also reported and discussed.

During the first month of release period, ^{134}Cs and ^{137}Cs activities ratios were very close to one (0.99 ± 0.03 for FNPP1 north and south discharge channels) and extremely uniform (Buesseler et al., 2011). The presence of ^{134}Cs is a unique isotopic signature for tracking these waters and calculating mixing ratios. In the oceans, the behavior of cesium is thought to be conservative, i.e. it is soluble (<1% attached to marine particles) and is carried primarily with ocean waters and as such has been used as a tracer of water mass mixing and transport (Buesseler et al., 2011).

Results of observations of ^{134}Cs and ^{137}Cs activities in surface water from April to December 2011 at Hasaki, a coastal station 180 km south of the FNPP1 site, showed the maximum activity, around 2000 Bq/m^3 in June 2011, representing a delay of 2 months from the corresponding maximum in April 2011 at FNPP1. Directly discharged ^{134}Cs and ^{137}Cs were transported dominantly southward along the coastline of northeastern Honshu. The reasons for the 2-month delay at Hasaki are not yet clear, however, clockwise current associated with a warm water eddy of which center located at 36.5° N, 141.4° E off Iwaki between Onahama and Hasaki in the mid of May 2011 might prevent southward transport of ^{134}Cs and ^{137}Cs released from FNPP1 to Hasaki until the end of May 2011 (Aoyama et al., 2012b).

Fukushima-derived ^{134}Cs and ^{137}Cs were detected throughout waters 30–600 km offshore, with the highest activities associated with near-shore eddies and the Kuroshio Current acting as a southern boundary for transport in June 2011. Buesseler et al. (2012) calculated a total ^{137}Cs inventory of 1.9–2.1 PBq in an ocean area of 150,000 km^2. A similar value of 2.2 ± 0.3 PBq has been obtained by Povinec et al. (2013b) for the same ocean area.

High-density sampling of surface seawater to measure ^{134}Cs and ^{137}Cs activities were carried out by 17 cargo ships as Voluntary Observing Ship (VOS) cruises and several research vessel cruises since March 2011 till March 2012 in the North Pacific Ocean. We also discuss these behaviors of the radioactive plume in the North Pacific Ocean through March 2012.

5.3.1.1 Seawater Sampling and Radionuclide Analyses

We shall discuss in this chapter two types of samples collected in the northwestern Pacific Ocean after the Fukushima accident. The first group of samples was collected at more than 300 stations as shown in Fig. 5.44. A typical sample size was 2 l of surface seawater. The samples were treated by an improved ammonium phosphomolybdate (AMP) procedure developed by M. Aoyama (Hirose et al.,

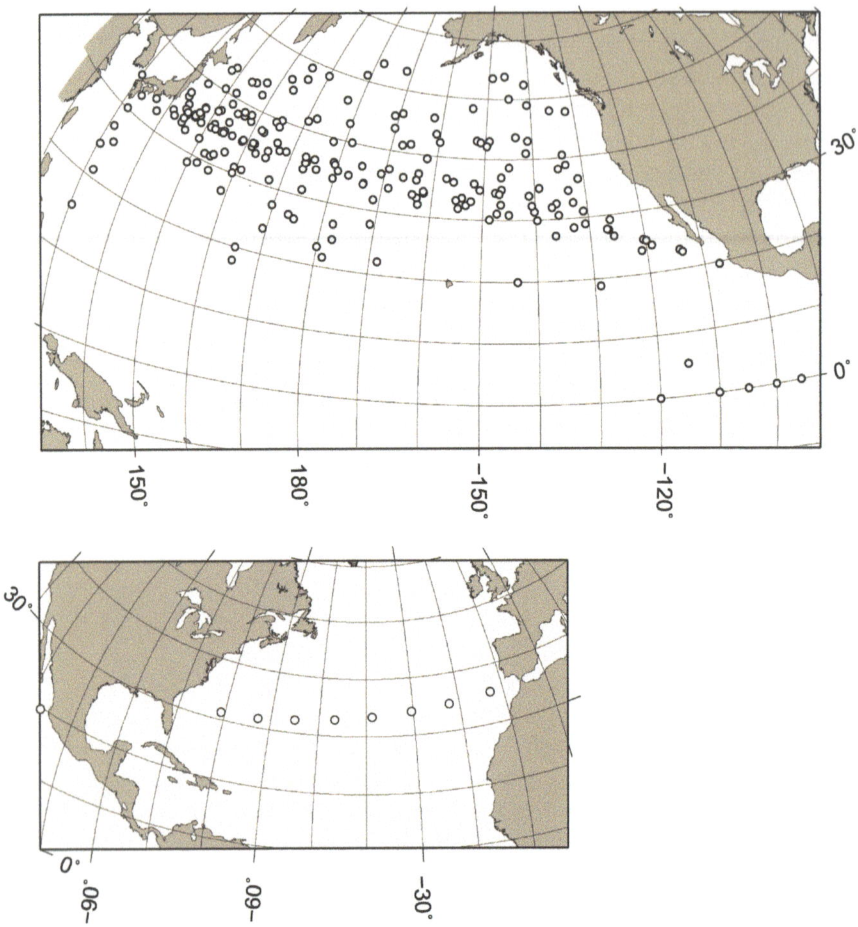

Figure 5.44 Sampling locations in the Pacific Ocean (upper) and the Atlantic Ocean (lower). (Adapted from Aoyama et al., 2013).

2005; Aoyama and Hirose, 2008). This improvement of AMP procedure for 2 l seawater samples increased the weight yield of AMP/Cs compound to over 99%, as well as radiochemical yield of radiocesium. The AMP/Cs compound activities were measured at the Ogoya Underground Facility of the Low-Level Radioactivity Laboratory of Kanazawa University using high-efficiency, well-type ultralow background Ge-detectors (Hamajima and Komura, 2004). One example of the best performance at this underground facility was reported that a detection limit of ^{137}Cs is

0.18 mBq for a counting time of 10,000 min (Hirose et al., 2005). Therefore, this development permits us to use the residue of nuclear weapon tests as useful tracers in oceanography (Aoyama et al., 2011; Povinec et al., 2011a,b; Sanchez-Cabeza et al., 2011) and to measure released ^{134}Cs and ^{137}Cs from the FNPP1 in 2 l samples with activities <1 Bq/m^3, too.

Because reagents can add trace levels of radioactivity, skewing small volume measurements, it is important to know the specific activity of analytes such as ^{137}Cs in the reagents. The ^{137}Cs activity in CsCl was measured to be 0.03 mBq/g by using extremely low background gamma-spectrometry and we neglect this amount of ^{137}Cs because we use only 0.26 g as carrier. The ^{137}Cs activity in AMP we used was 0.024 mBq/g and we subtract the corresponding amount of ^{137}Cs in the AMP used to extract radiocesium from the samples because we use 4–6 g for extraction. There is no serious contamination of ^{137}Cs from other reagents. For ^{134}Cs contaminations, we did not observe any ^{134}Cs contaminations from the reagents.

The second set of seawater samples discussed in this chapter was collected offshore Fukushima during the international research cruise with participation of 13 institutions, which was organized by US scientists on 3–17 June 2011 using the research vessel *Ka'imikai-o-Kanaloa* (KOK' 2011) of the University of Hawaii (Buesseler et al., 2012). The cruise track extended from 34 to 37° N, and from 142 to 147° E (from about 30 to 600 km off Japan), visiting altogether 50 sampling stations (Fig. 5.45). Both surface and water profile samples were collected. Most of the collected seawater samples were analyzed by Buesseler et al. (2012). Eleven depth profiles were analyzed in this work for ^{137}Cs and ^{134}Cs by the International Atomic Energy Agency's Environment Laboratories (IAEA-EL) in Monaco, and four depth profiles were analyzed for ^{129}I and ^{3}H (Hou et al., 2013; Povinec et al., 2013b). The collected seawater samples for tritium and ^{129}I analyses were stored in 1 l glass bottles with air-tight covering so no exchange with the surrounding air was possible. The samples for cesium analysis were stored in plastic containers from 1 l (profiles) to 15 l (surface) volumes.

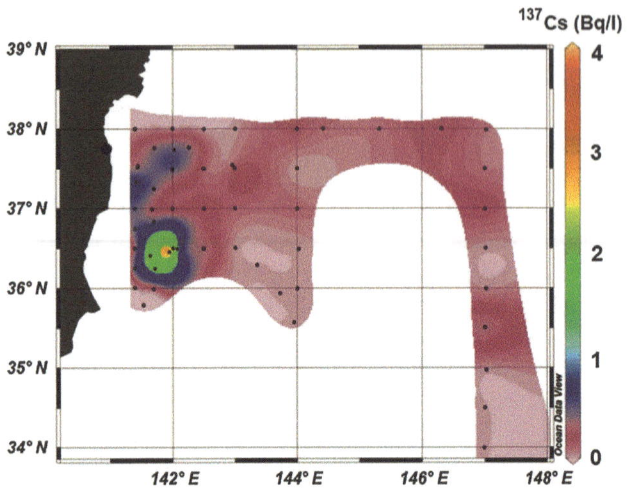

Figure 5.45 Sampling stations and ^{137}Cs levels in surface seawater during the international research cruise carried out on 3–17 June 2011 using the research vessel *Kaìmikai-o-Kanaloa* (KOK' 2011) of the University of Hawaii. (After Buesseler et al., 2012; Povinec et al., 2013b)

Tritium was analyzed in seawater samples using the ^{3}He in-growth method (Palcsu et al., 2010). The method consists of three major steps:

1. The water sample is put into a stainless steel vessel, and the dissolved gases including helium are then removed from the water by vacuum pumping.
2. The samples are stored for several months so that ^{3}He atoms are produced by tritium decay.
3. The helium fraction is admitted to a dual collector noble gas mass spectrometer, the abundance of the tritiogenic ^{3}He is then measured, from which ^{3}H activity is calculated. The sensitivity of the method is 0.01 TU. The measurements were carried out in the Institute for Nuclear Research (ATOMKI), Debrecen, Hungary.

Iodine-129 was analyzed using AMS. The method for preparation of AgI targets for AMS measurements was described by Hou et al. (2013). About 200 ml of seawater samples was spiked with ^{125}I and transferred to a separation funnel. After addition of 0.5 mg ^{127}I carrier (Woodward Inc., with a $^{129}I/^{127}I$ ratio lower than 1×10^{-13}), 3 ml of 1 M $NaHSO_3$ solution and 3 M HNO_3 were added to pH 1–2 to convert all inorganic iodine to iodide. Iodine was then extracted into $CHCl_3$ after addition of $NaNO_2$ to oxidize iodide to I_2. Iodine in $CHCl_3$ phase was then back-extracted to the water phase using 10 ml of 0.1 mM $NaHSO_3$ solution. Half milliliter of 1.0 mol/l $AgNO_3$ solution was added to the back-extracted aqueous phases to precipitate iodide as AgI, which was separated by centrifuge. The obtained AgI precipitate was dried at 70 °C and used for AMS measurement of ^{129}I. ^{125}I in the precipitate was counted using a NaI(Tl) gamma-detector to monitor the chemical yield of iodine in the separation (Hou et al., 2007). An ICP-MS system (X Series II, Thermo Fisher Scientific, Waltham, USA) equipped with an Xs-skimmer cone and standard concentric nebulizer was used for measurement of ^{127}I. Cs^+ (to 2.0 ppb) as internal standard and 1% NH_3 media were applied for measurements Hou et al. (2007, 2009a,b).

The $^{129}I/^{127}I$ ratios in total iodine samples were determined by AMS at the Vienna Environmental Research Accelerator of the University of Vienna using a 3 MV Pelletron (Wallner et al., 2007), and at the 3 MV Pelletron AMS facility of the University of Arizona (Biddulph et al., 2000). The machine $^{129}I/^{127}I$ background ratio was $\sim 3 \times 10^{-14}$. The blanks prepared using the same procedure as the samples gave for the $^{129}I/^{127}I$ ratios values below 1×10^{-13}, which was by about three orders of magnitude lower than the measured $^{129}I/^{127}I$ ratios in seawater samples.

Cesium-137 and ^{134}Cs analysis of seawater samples was carried out using the method described in detail by Levy et al. (2011) and Povinec et al. (2012b). In low-activity samples, radiocesium was preconcentrated by adsorption onto ammonium phosphomolybdate (AMP) at pH 2. The AMP suspension was allowed to settle and the supernatant solution was decanted or siphoned away. The AMP solid particle in the remaining suspension was separated by centrifuge. The separated AMP solid was washed with water. The Cs recovery was obtained by determination of stable ^{133}Cs concentration in water before and after AMP absorption by ICP-MS. The AMP sample was analyzed for the presence of ^{134}Cs and ^{137}Cs in the CAVE (Counting laboratory for environmental radionuclides) underground facility of

IAEA-EL using large efficiency high-purity germanium (HPGe) detectors (Povinec et al., 2005b). High-activity seawater samples were nondestructively counted directly using HPGe spectrometers in the CAVE laboratory.

5.3.1.2 Radionuclides in the North Pacific Ocean

5.3.1.2.1 Cesium Isotopes in Surface Water

Before the FNPP1 accident, ^{137}Cs was already present in the ocean, which originated from the nuclear weapon tests conducted in the late 1950s and in the early 1960s (Aoyama et al., 2011; Aoyama and Hirose, 2004, 2008). In the western North Pacific Ocean, ^{137}Cs activity in surface water was 10–100 Bq/m^3 in the late 1950s and in the early 1960s, then it decreased gradually to around a few Becquerels per cubic meter (Aoyama et al., 2011, 2006). In 1986 Chernobyl accident, however, a contribution of atmospheric deposition of Chernobyl-derived ^{137}Cs in the North Pacific Ocean was around 3% of total deposition derived from global fallout before 1986 (Aoyama et al., 1986). Therefore, the effect of Chernobyl accident in the North Pacific Ocean is negligible. Before the FNPP1 accident, distribution and invention of ^{137}Cs which originated from atmospheric weapons tests had been studied in the Pacific Ocean since the late 1950s. The ^{137}Cs inventory in the North Pacific Ocean was 290 $\pm$ 30 PBq in January 1970 based on 10° by 10° mesh data of the ^{137}Cs deposition (Aoyama et al., 2006). In 2003, ^{137}Cs inventory in the North Pacific Ocean was 86 PBq by the model study (Tsumune et al., 2011) and 85 PBq by the observation (Aoyama et al., 2012a), then it decreased in 2011 due to radioactive decay to 69 PBq (Aoyama et al., 2012a). In 2000s, just before the FNPP1 accident, the ^{137}Cs activity in surface water was a few Becquerels per cubic meter and showed less change compared with the decreasing trend of ^{137}Cs activity we observed before 2000. A horizontal distribution of ^{137}Cs in the 2000s in the surface water showed a very homogeneous distribution, but relatively high ^{137}Cs activity regions in surface water were observed in the western part of the subtropical gyre in both the North Pacific Ocean and the South Pacific Ocean where ^{137}Cs

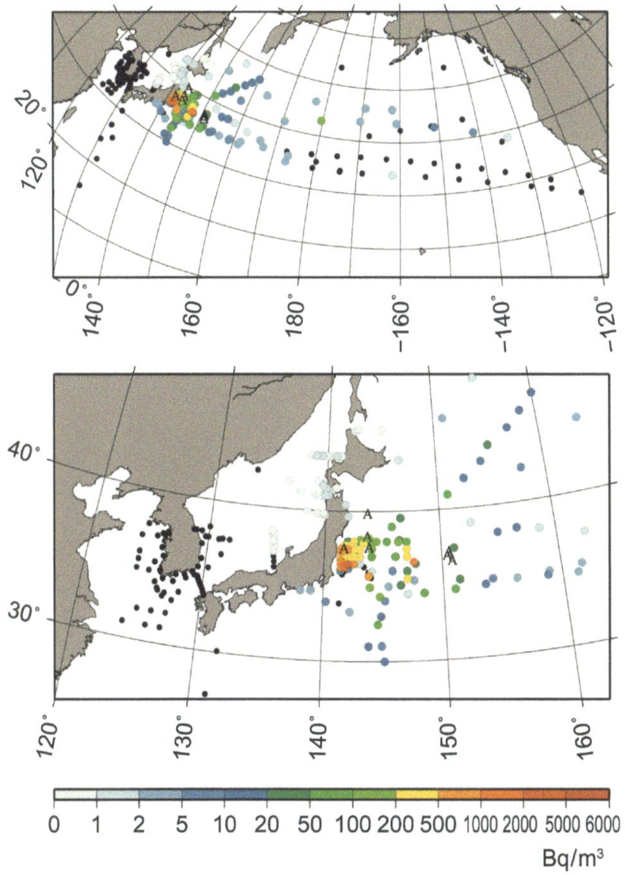

Figure 5.46 ^{134}Cs activity in surface water during the period from April 2011 to June 2011 for the North Pacific Ocean (upper) and close to Japan (lower). Positions of Argo floats on 15 May 2011 are marked "A". (After Aoyama et al. (2013))

activity exceeded 2 and 1.5 Bq/m^3, respectively (Aoyama et al., 2012a).

After the FNPP1 accident, both ^{134}Cs and ^{137}Cs are observed in a wide area in the North Pacific Ocean as shown in Figs 5.46–5.49. The differences between ^{134}Cs and ^{137}Cs activities observed after the FNPP1 accident were consistent with pre-Fukushima ^{137}Cs originated from the nuclear weapons tests as described above. It is also clear that ^{134}Cs and ^{137}Cs activities ratios were close to 1, which is also consistent with observed ^{134}Cs and ^{137}Cs activities ratios of 0.99 $\pm$ 0.03 at very close to the source region of the

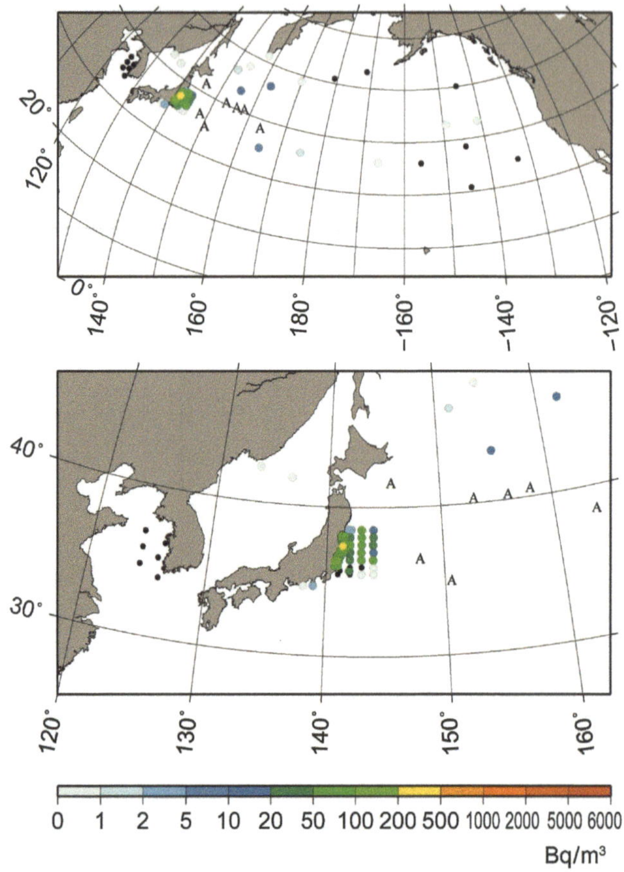

Figure 5.47 ^{134}Cs activity in surface water during the period from July 2011 to September 2011 for the North Pacific Ocean (upper) and close to Japan (lower). Positions of Argo floats on 15 August 2011 are marked "A". (After Aoyama et al., 2013).

FNPP1 (Buesseler et al., 2011). These are clear evidences that observed ^{134}Cs and excess ^{137}Cs originated from the FNPP1 accident. The horizontal distribution of FNPP1-origin ^{134}Cs in the western North Pacific Ocean, except just in front of the FNPP1 site, showed that the high-concentration area located close to the FNPP1 accident site which might have received both atmospheric deposition—showing good consistency with previous atmospheric transport model study (Honda et al., 2012)—and direct discharge (Tsumune et al., 2012) from the FNPP1 site. We see another high-concentration area near the

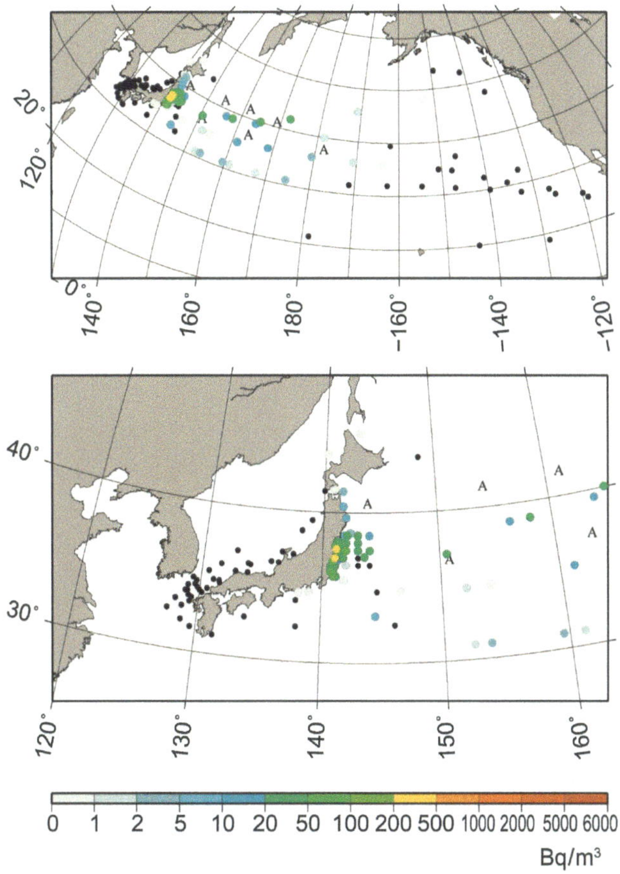

Figure 5.48 ^{134}Cs activity in surface water during the period from October 2011 to December 2011 for the North Pacific Ocean (upper) and close to Japan (lower). Positions of Argo floats on 15 November 2011 are marked "A". (After Aoyama et al., 2013).

International Date Line in April–June 2011 as shown in Fig. 5.46 (upper panel). This high-concentration region may be more likely explained by atmospheric deposition because of the transport distance compared to the surface current. At the sea area east of the International Date Line north of 40° N in the Pacific Ocean, the ^{134}Cs activity in surface water in April 2011 was $<12\ Bq/m^3$.

In July–September 2011, relatively high-concentration area for which ^{134}Cs activity exceed $10\ Bq/m^3$ moved eastward and arrived at 165° E as

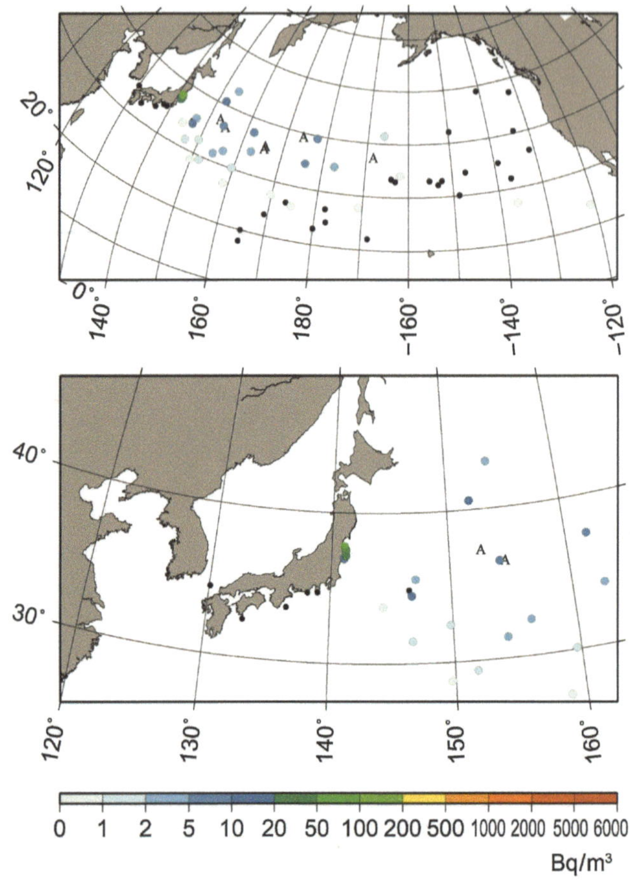

Figure 5.49 ^{134}Cs activity in surface water during the period from January 2012 to March 2012 for the North Pacific Ocean (upper) and close to Japan (lower). Positions of Argo floats on 15 February 2012 are marked "A". (After Aoyama et al., 2013).

shown in Fig. 5.47. In October–December 2011, relatively high-concentration area for which ^{134}Cs activity exceed 10 Bq/m^3 moved more east, and arrived at 172° E, along 40° N as shown in Fig. 5.48. In Janary–March 2012, it arrived at the International Date Line as shown in Fig. 5.49.

The atmospheric deposition occurred mainly in March 2011 (Chino et al., 2011), therefore, ^{134}Cs and ^{137}Cs activities in surface water derived by atmospheric deposition, except close area of the FNPP1 site, should decrease by dispersion with time rapidly, while due to eastward movement of radioactive

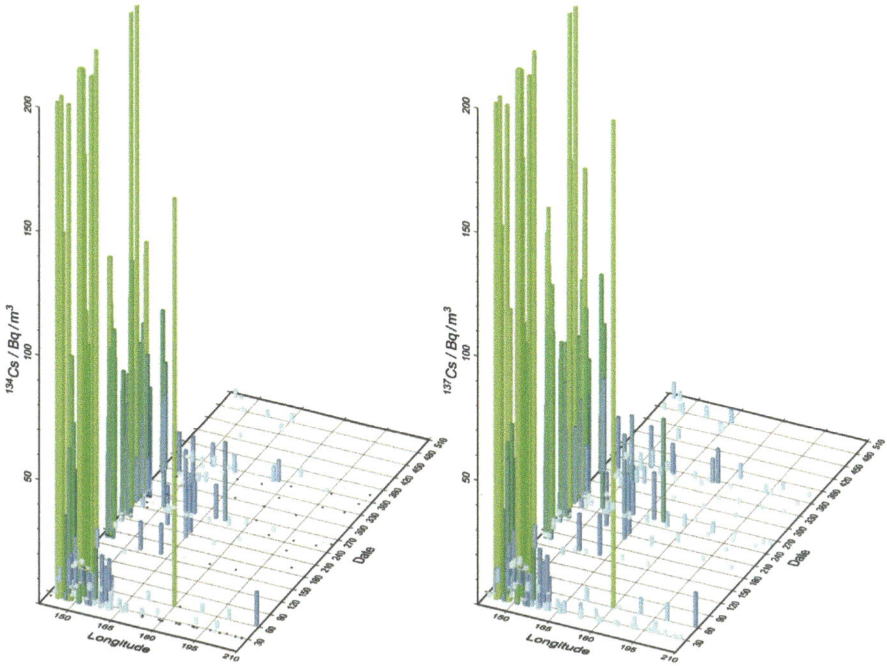

Figure 5.50 ^{134}Cs (left) and ^{137}Cs (right) activity in surface water during the period from 11 March 2011 (day 0) to 31 July 2011 (day 510). (After Aoyama et al., 2013).

plume relatively higher activities exceeding 10 Bq/m^3 were observed as shown in Fig. 5.50 and Figs 5.46–5.49. The radioactive plume was formed by the atmospheric deposition close to the FNPP1 site and direct discharge. It is interesting to estimate a zonal speed of radioactive plume based on our observations. A feature was that the radioactive plume was confined along 40° N when the plume reached International Date Line as stated in Chapter 3.2. The radioactive plume traveled 1800 km (from 160° E to 178° E) for 270 days (9 months) (Fig. 5.51), therefore an average zonal speed (u) of the surface radioactive plume was calculated to be about 8 cm/s which was consistent with a speed of reported surface current of 4–16 cm/s at the region (Maximenko et al., 2009).

Eleven Argo floats were deployed off Fukushima on 31 March–13 April at 37.001° N–37.709° N, 141.250° E–141.399° E after the accident (Argo-Information-Center). Nine of 11 floats were still operational until around January–March 2012,

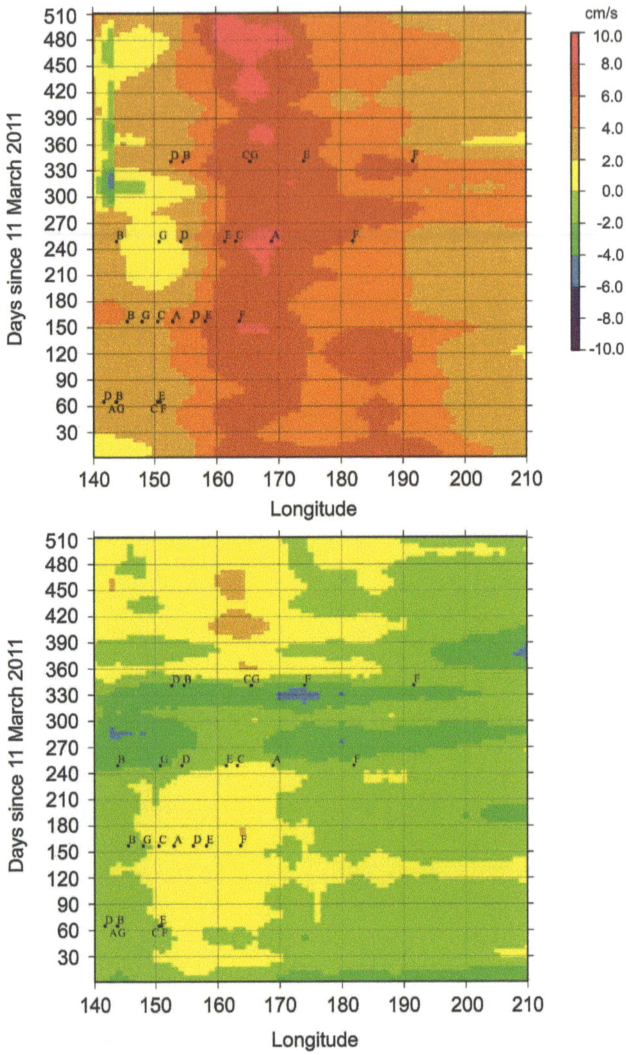

Figure 5.51 Zonal (upper) and meridional (lower) speed by satellite along 40° N (38°–42° N) during the period from 11 March 2011 (day 0) to 3 August 2012 (day 511). Positions of Argo floats are marked from A to G for 15 May 2011, 15 August 2011, 15 November 2011 and 15 February 2012. (After Aoyama et al., 2013).

therefore we can compare our observations and trajectories of nine Argo floats. In Figs 5.46–5.49, positions of Argo floats at mid time of each 3-month period were plotted marked “A”. In April–June 2011, a distribution of Fukushima radioactive plume and positions of Argo floats showed some discrepancy because the distribution of Fukushima radioactive plume was formed by both combination of atmospheric deposition and direct discharge as stated in Chapter 3.2. In July–September 2011, positions of Argo floats were moved more eastward up to 165° E, although observed results were so sparse radioactive plume also moved to east as well as Argo floats as shown in Fig. 5.47. Three months later, both Fukushima radioactive plume and Argo floats moved more east, up to 172° E as shown in Fig. 5.48. Almost 1 year after the accident, again we observed that both Fukushima radioactive plume and Argo floats moved more eastward up to 180° E as shown in Fig. 5.49. A zonal speed, u, based on trajectories of nine Argo floats between May 2011 and August 2011 ranged from 0.1 to 15.6 cm/s with an average of 7.8 cm/s. A zonal speed, u, based on trajectories of nine Argo floats between August 2011 and November 2011 ranged from −1.9 to 20.1 cm/s with an average of 7.7 cm/s. In between November 2011 and February 2012, it ranged from −1.7 to 16.7 cm/s with an average of 8.9 cm/s. These zonal speeds by Argo floats showed excellent agreement with zonal speed of Fukushima radioactive plume (about 8 cm/s) derived by our observations. Therefore, we can say that deploying Argo floats just after nuclear reactor accidents near the coastline might be good to trace radioactive plume, which moves in the surface layer.

We can also assume that Fukushima radioactive plume moved with surface water. Therefore it is also interested to compare between surface current speed by satellite observations and actual movement of Fukushima radiocesium obtained by our observations as stated in Chapter 3.2. We look at surface current observation by satellite (Bonjean and Lagerloef, 2002) along 40° N (38° N–42° N). Zonal speed during the period from April 2011 to March 2012 ranged from 1.5 to 7.4 cm/s. The average speed estimated by advection of Fukushima radioactivity was 8 cm/s in average and this zonal speed by observation showed good agreement with satellite-derived surface current as well as the zonal speed obtained by movement of Argo floats. It should be also noted that there exist larger variabilities in the radioactive plume as shown in Fig. 5.51.

Zonal speed, *u*, and meridian speed, *v*, showed temporal and spatial variations as shown in Fig. 5.51, then resulted positions of Argo floats also showed large variability, which indicates that the movement of radioactive plume varied as well.

We may conclude that ^{134}Cs and ^{137}Cs were released to the North Pacific Ocean by two pathways, direct discharge from the Fukushima NPP1 accident site and atmospheric deposition off Honshu Islands of Japan, east and northeast of the site. High-density observations of ^{134}Cs and ^{137}Cs in the surface water were carried out by 17 VOS cruises and several research-vessel cruises since March 2011 till March 2012. The main body of radioactive surface plume, of which activity was larger than 10 Bq/m^3, had been traveling along 40° N and reached The International Date Line on March 2012 1 year after the accident. A feature was that the radioactive plume was confined along 40° N when the plume reached The International Date Line. A zonal speed of the surface plume was estimated to be about 8 cm/s which was consistent with a zonal speed of surface current at the region observed by Argo floats and satellite.

5.3.1.2.2 Cesium Isotopes in the KOK'2011 Cruise Samples

The distribution of the Fukushima-derived ^{137}Cs in surface waters 30–600 km offshore the Fukushima NPP (Fig. 5.45), based on the results of Buesseler et al. (2012) and Povinec et al. (2013b), shows that the measured ^{137}Cs concentrations in surface waters ranged from 1.8 to 3500 mBq/l, up to 2700 times higher than the global fallout background, although the cruise track did not go closer than 30 km from the coast (Buesseler et al., 2012; Povinec et al., 2013b). The highest activities were observed at sampling sites around 36.5° N, 142° E, southeast of the Fukushima NPP. It is evident that the Kuroshio Current acts as the southern boundary for the transport of the radionuclides in the northwestern Pacific Ocean. The sampling sites around 37.5° N, 141.5° E, east of the Fukushima NPP, showed lower ^{137}Cs levels, only around 1 Bq/l. The elevated ^{137}Cs levels covered an area of around 150,000 km^2 (south of 38° N and west of 147° E). Even at distances around 600 km off

Fukushima, ^{137}Cs activity concentrations of around 0.3 Bq/l were found, i.e. by about a factor of 300 above the global fallout background of 1 mBq/l.

We noticed that a considerable decrease in surface ^{137}Cs levels took place between the measurements, which were carried out earlier and during the KOK'2011 cruise. About 24 Bq/l of ^{137}Cs were measured in seawater 30 km offshore Fukushima on 23 March 2011, which could be compared to 3.5 Bq/l, measured during the KOK'2011 expedition. Figure 5.44 also indicates that an atmospheric deposition of ^{137}Cs occurred within the 600 km zone offshore Fukushima.

The ^{134}Cs and ^{137}Cs water profiles (Fig. 5.52) indicate that mostly surface water maxima were observed, although at some stations (e.g. St. 29, 26, 19, and 40), subsurface maxima (at 20–50 m) were observed as well. The Fukushima-derived ^{134}Cs and ^{137}Cs penetrated up to June 2011 to at least 200 m water depth (St. 4 and 22). This figure also confirms that the $^{134}Cs/^{137}Cs$ activity ratio in the analyzed seawater samples was close to 1, clearly indicating that the source of these radionuclides was the Fukushima accident.

The ^{137}Cs water profiles at St. 31, 22, 14 and 11, which are compared with the pre-Fukushima profiles measured in 1997 during the IAEA 97 expedition in the northwestern Pacific Ocean (Fig. 5.53), show that the Fukushima signal is well visible in all four stations, especially at St. 31, where surface ^{137}Cs levels exceeded the global fallout values by about three orders of magnitude (Povinec et al., 2003a, 2005a).

The ^{137}Cs inventory in the water column of the northwestern Pacific due to the Fukushima accident, estimated using the Buesseler et al. (2012) data and those presented here, is 2.2 ± 0.3 PBq, confirming an essential contribution to its total ^{137}Cs inventory in the northwestern Pacific Ocean (Table 5.1).

Iodine-129 seawater profiles offshore Fukushima presented in Fig. 5.54 show that the highest surface levels were obtained at St. 31 (up to 62×10^7 atoms/l), which is situated about 40 km from the Fukushima NPP (Povinec et al., 2013a). The other three stations (St. 11, 14, and 22) show the ^{129}I concentrations of

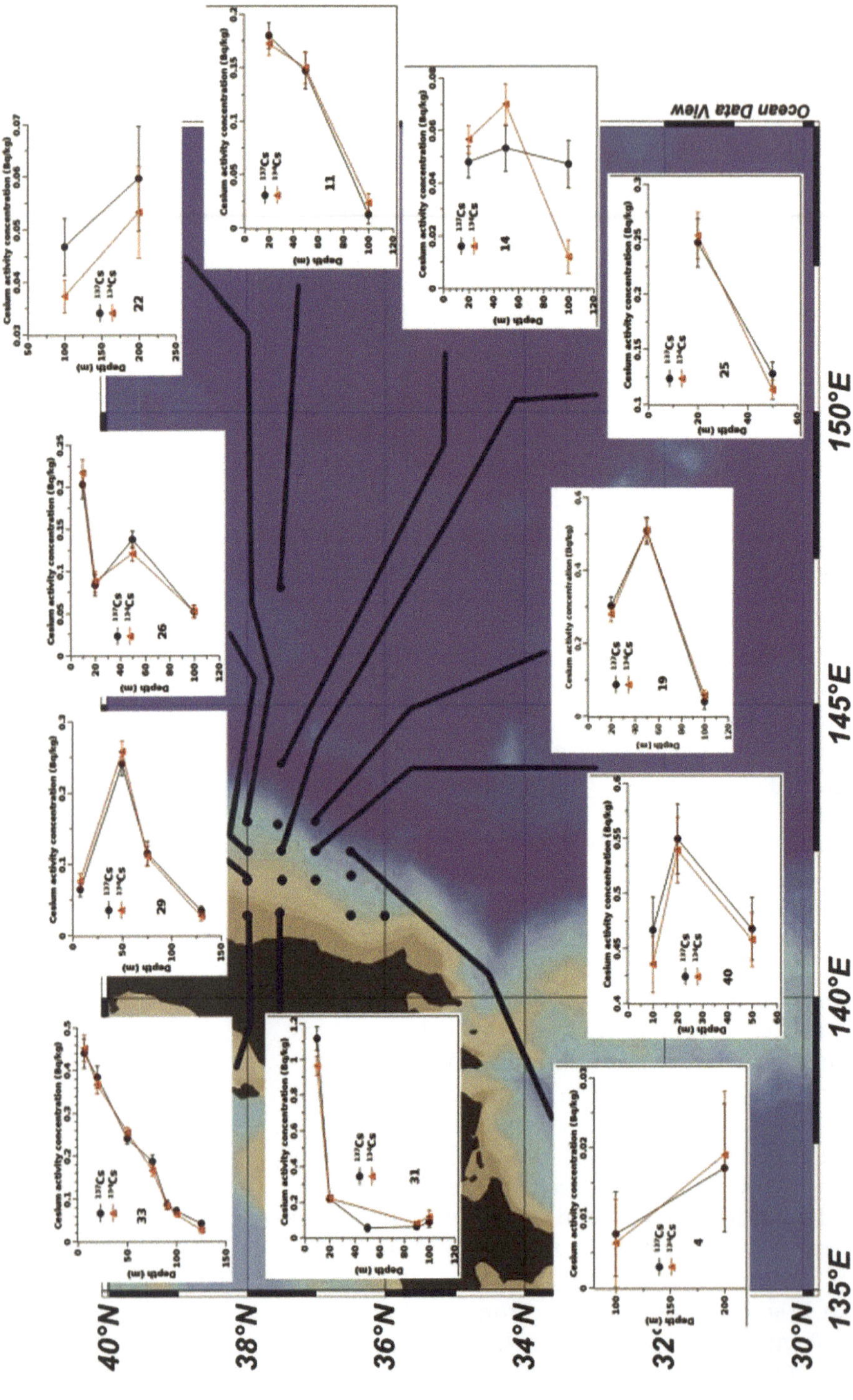

Figure 5.52 The ^{134}Cs and ^{137}Cs water profiles in the northwestern Pacific Ocean measured during the KOK 2011 cruise. (Modified from Povinec et al., 2013b).

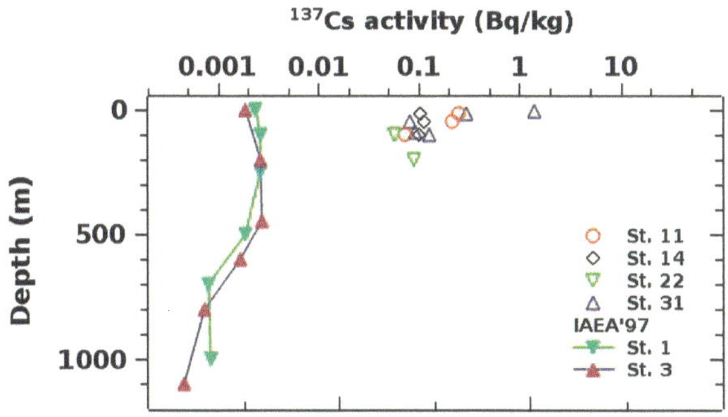

Figure 5.53 The ^{137}Cs water profiles in the northwestern Pacific Ocean measured during the IAEA 97 and KOK 2011 cruises. (Modified from Povinec et al., 2013b).

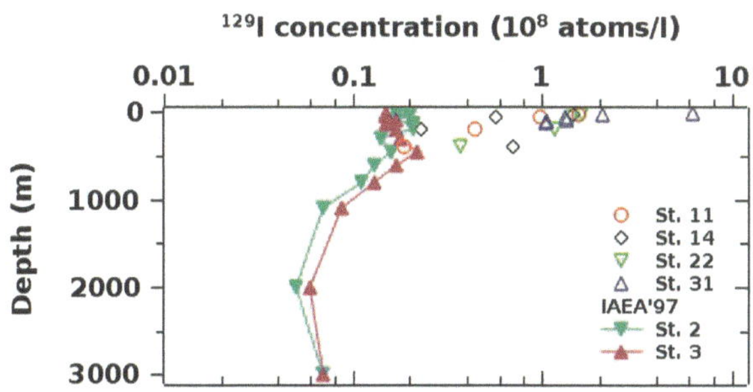

Figure 5.54 The ^{129}I water profiles in the northwestern Pacific Ocean measured during the KOK 2011 cruise. (Modified from Povinec et al., 2013b).

(14–16) × 10^7 atoms/l at 20 m water depth, slightly lower than that at St. 31 (20 × 10^7 atoms/l). It can also be seen that the ^{129}I concentrations were decreasing with depth at all stations. The lowest ^{129}I concentrations (0.77 × 10^7 atoms/l) were observed in 400 m water depths at St. 14, which is about 260 km off Japan, and at St. 11 (1.9 × 10^7 atoms/l), which is the most distant location (530 km off Japan). The distribution of ^{129}I/^{127}I ratio follows the same trend as the ^{129}I concentrations. The highest ^{129}I/^{127}I atom ratio (22 × 10^{-10}) was observed at 10 m depth at St. 31, while the lowest one (0.26 × 10^{-10}) in 400 m at St. 14. For a better comparison with other radionuclides, the observed range of ^{129}I levels in the unit of activity concentration was 0.01–0.8 μBq/l. Comparing the ^{129}I results presented in Fig. 5.54 with published data (Povinec et al., 2000, 2010; Suzuki et al., 2010), it can be concluded that St. 31 and 22 have been heavily influenced by the Fukushima accident, as the observed ^{129}I levels were higher by a factor of 30 and 10, respectively. However, St. 14 and 11, which are the most distant stations from Japan, were also impacted by the Fukushima accident as their surface ^{129}I concentrations were almost by about a factor of 10 higher than the pre-Fukushima levels. Station 31 (measured down to 125 m) and St. 22 (down to 400 m) clearly indicate vertical transport of ^{129}I in the water column, which can also be seen at St. 11, as at 400 m water depth the ^{129}I level is by about a factor of two higher than expected.

The $^{129}I/^{137}Cs$ activity ratios at four stations varied between (0.5–5) $\times$ 10^{-6} indicating that all stations at depths up to 200 m were influenced by the Fukushima accident. A $^{131}I/^{137}Cs$ activity ratio of about 17.8 was measured in the water discharged directly to the sea from the Fukushima NPP. From the estimated atmospheric releases of ^{131}I (~157 PBq) and ^{137}Cs (~14 PBq), we get a value of 11.2 for the $^{131}I/^{137}Cs$ activity ratio, which is not far from a median of 15 estimated by Hirose (2012) in precipitation over the Japan. As ^{131}I and ^{129}I have similar behavior in the environment, the amount of ^{129}I directly discharged to the sea can be estimated from the measured $^{131}I/^{137}Cs$ ratio, and the estimated amount of ^{137}Cs discharged to the sea. Thus, if we take as an example 4 PBq for direct liquid discharges of ^{137}Cs to the sea, the amount of ^{131}I directly discharged to the sea will be 62 PBq. Based on these data, and the measured $^{129}I/^{131}I$ ratio for released radioiodine from the Fukushima NPP, the amount of ^{129}I directly discharged to the sea can be estimated to be 2.4 GBq (Hou et al., 2013). Morino et al. (2011) estimated that about 120 PBq of ^{131}I was deposited over the Pacific Ocean, of which 29 PBq was deposited in the area 34–41° N and 137–145° E. Using these values and the measured $^{129}I/^{131}I$ ratio, we may estimate that about 4.6 GBq of ^{129}I released to the atmosphere was deposited in the Pacific Ocean, and 1.1 GBq in the area 34–41° N and 137–145° E.

The Chernobyl accident released to the atmosphere about 13 GBq of ^{129}I (IAEA, 2003), that is almost by a factor of two higher than the total release of about 7 GBq of ^{129}I from the Fukushima accident. The ^{129}I released from the Chernobyl accident had, however, a negligible impact on the marine

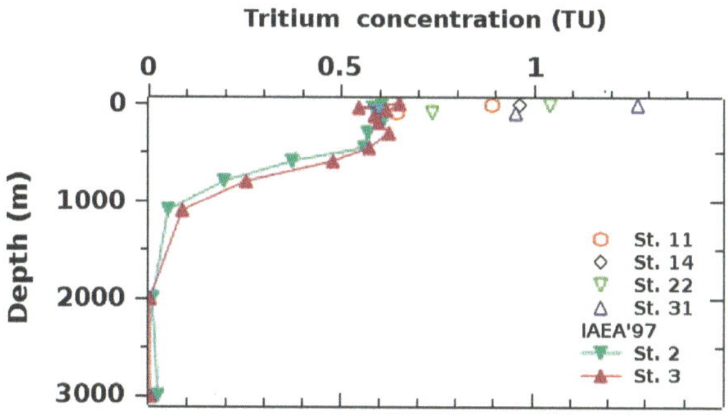

Figure 5.55 The ^{3}H water profiles in the northwestern Pacific Ocean measured during the KOK' 2011 cruise. (Modified from Povinec et al., 2013b).

environment. The European reprocessing plants at La Hague (France) and Sellafield (UK), discharged together much higher ^{129}I amounts to the sea (~35 PBq) and to the atmosphere (~3 PBq) (Hou et al., 2009b). As noted earlier, the liquid releases primarily affected the European seas and the Arctic Ocean; however, the ^{129}I released to the atmospheric was transported over the globe, and partially was also deposited on the Pacific Ocean.

Tritium seawater profiles (Fig. 5.55) generally follow the cesium and ^{129}I water profiles, although surprisingly the ^{3}H levels (in the range of 0.4–1.3 TU, equivalent to 0.05–0.15 Bq/l), are only by about a factor of three above the global fallout background. The lower ^{3}H signal in the seawater offshore Fukushima should be attributed to less release of ^{3}H from the Fukushima Dai-ichi NPP, due to much lower fission yield of ^{3}H compared to ^{137}Cs and ^{129}I. All stations (except St. 14–15, sampling depth of 100 m) were affected by the Fukushima tritium. From the measured $^{3}H/^{129}I$ and $^{3}H/^{137}Cs$ activity ratios in the water column, and the previously estimated ^{129}I and ^{137}Cs releases to the sea we may estimate the total ^{3}H activity released and deposited over the northwestern Pacific Ocean to be of the order of 0.1 PBq.

Generally, we may state that the estimated radionuclide release rates from the damaged reactors of the Fukushima NPP indicate that a relatively smaller amount of radionuclides (in total of about 180 PBq) were released to the environment when compared with the Chernobyl accident (in total of about 1800 PBq). When comparing only the radioecologically the most important ^{137}Cs, about 85 PBq of ^{137}Cs released during the Chernobyl accident to the atmosphere (UNSCEAR, 1993, 2008) may be compared with about 13 PBq released during the Chernobyl accident. When comparing the Fukushima radionuclide release rates with the Chernobyl one, we see that they were lower by a factor of 13 and 7 for ^{131}I and ^{137}Cs, respectively. The highest total radioactivity, released to the atmosphere till now, was from atmospheric nuclear weapons tests carried out during the late 1950s and early 1960s (in total of about 188,000 PBq) (Povinec et al., 2012e).

5.3.2 Radionuclides in Ocean Fish

The Japanese MAFF has continuously been monitoring radionuclides in fish and other seafood products in the open ocean. The monitoring data in 2000–2010 (MEXT, 2012a,b,c) reveal that concentrations of ^{137}Cs in the North Pacific fishes were about 0.1 Bq/kg ww (wet weight based); it ranged from 0.05 to 0.09 Bq/kg ww for chum salmon (*Oncorhynchus keta*) and from 0.15 to 0.27 Bq/kg ww for skipjack tuna (*Katsuwomus pelamis*). The ^{137}Cs concentrations in open ocean fish showed gradual decrease during the period 2000–2010 (Fig. 5.56). The ^{137}Cs concentrations in skipjack tuna were higher than that in chum salmon. An apparent biological half-life of ^{137}Cs is 16 years for both salmon and tuna.

Since 23 March 2011, the MAFF has been carrying emergency monitoring of radionuclides in seafood products in the North Pacific. The radiocesium concentration of 33 Bq/kg ww was measured in skipjack tuna in April 2011. The radiocesium concentrations in blue fin and yellow fin tunas ranged from 15.2 to 41 Bq/kg ww during the period of September–November 2011, and from 0.43 to 10.4 Bq/kg ww during the period of October 2011–March 2012, respectively. The radiocesium concentrations in chum salmon ranged from 0.32 to 8 Bq/kg ww during the period of April 2011–June 2012. The temporal variations of the radiocesium concentrations in skipjack and yellow fin tunas and chum salmon are shown in Fig. 5.57. The

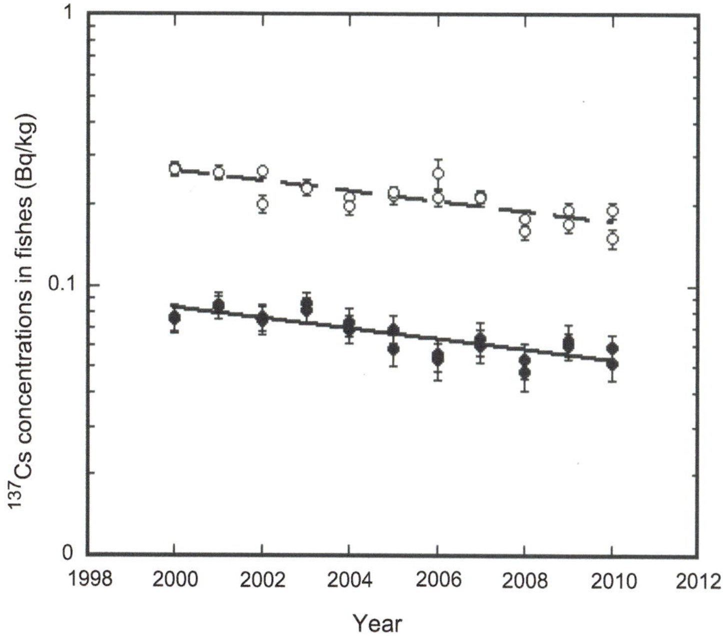

Figure 5.56 The ^{137}Cs concentrations in open ocean fish showing gradual decrease during the period 2000–2010.

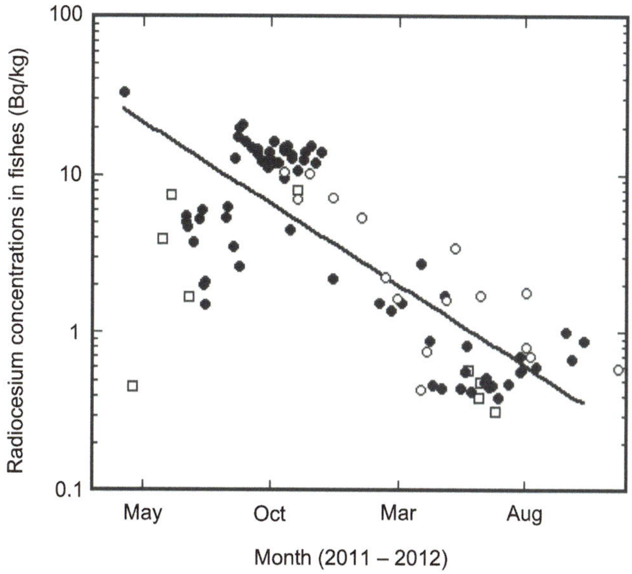

Figure 5.57 Temporal variations of radiocesium concentrations in skipjack and yellow fin tunas, and chum salmon.

radiocesium concentrations in tunas decreased exponentially during the period of April 2011–November 2012, in which an apparent biological half-life of radiocesium in tunas is about 90 days.

The Fukushima-derived radiocesium was detected in Pacific blue fin tuna caught in the remote sea area, California Current region (Madigan et al., 2012). The ^{134}Cs and ^{137}Cs concentrations in 15 Pacific blue fin tunas sampled in August 2011 were measured to be 4.0 and 6.3 Bq/kg dw, respectively. This finding suggests that Pacific blue fin tuna can rapidly transport radionuclides from a point source in Japan coast to distant ecoregions and demonstrate the importance of migratory animals as transport vectors of radionuclides because there were no ^{134}Cs and background levels ($\sim$1 Bq/kg dw) of ^{137}Cs in pre-Fukushima blue fin and post-Fukushima yellow fin tunas in the California.

5.3.3 Modeling of Radionuclide Transport in the Ocean

Several types of circulation models have been developed to simulate radionuclide concentrations in surface and subsurface seawater (e.g. Monte et al., 2009; Tsumune et al., 2011, 2012; Rypina et al., 2013). Tsumune et al. (2011, 2012) recently described a high-resolution Ocean General Circulation Model (OGCM) with vertical mixing, horizontal viscosity and eddy flux parametrization which has been used for tracing global fallout radionuclides in the Southern Hemisphere oceans.

We shall discuss in this chapter simulations of ^{137}Cs dispersion in the world ocean after releases from the Fukushima Dai-ichi NPP using the Long-term Assessment Model of Radionuclides (LAMER) in the oceans code. The model was developed with the aim to assess impacts of multilateral radioactive effluents from nuclear facilities to the ocean during normal or accidental situations (Nakano, 2004, 2006, 2008). The LAMER is based on the OGCM developed for the long-term assessment of ^{137}Cs and other radionuclides with similar behavior in seawater (e.g. ^{90}Sr and ^{129}I) (Nakano and Povinec, 2003a), as well as for particle-reactive radionuclides (e.g. Pu isotopes and ^{241}Am) (Nakano and Povinec, 2003b; Nakano, 2006) in the water column of the world ocean.

5.3.3.1 LAMER Code

The LAMER code is a calculation program simulating radionuclide transport in the ocean via advection, diffusion and scavenging processes. The averaged annual three-dimensional velocity fields, obtained by the OGCM were used in the calculations. The LAMER code represents a medium-resolution OGCM with the grids of 2° for latitudinal and longitudinal directions. In the vertical direction, we used 15 layers from the ocean surface to its bottom for determination of the velocity fields. The model can also deal with scavenging processes, which are especially required when modeling the dispersion of particle-reactive radionuclides such as plutonium and americium (Nakano and Povinec, 2003b). However, for cesium, the distribution coefficient (K_d) is only 2000 (compared to 100,000 for Pu) (IAEA, 2004), therefore scavenging processes in this case are not of great importance. Actually, a part of Cs in seawater would sink to the seafloor, and some would resuspend back to seawater from the sediment. However, this study is focusing only on Cs (or similar radionuclides), therefore, scavenging processes are not considered.

The characteristics of the LAMER code and parameters used in simulations have already been described, therefore we shall not repeat them there (Nakano and Povinec, 2003a, 2003b, 2012; Nakano, 2004). We shall focus on the evaluation of the simulations, and their comparison with experimental data on the ^{137}Cs distribution in the North Pacific Ocean after the Fukushima accident (Povinec et al., 2013a, b).

5.3.3.2 Radionuclide Source Term Used in Ocean Simulations

The radionuclide source term used in marine simulations consisted of two parts:

1. A direct release of radioactive water to the marine environment.
2. A deposition of radionuclides from the atmosphere on the ocean surface.

We shall focus on ^{137}Cs only as this radionuclide is considered to be the most important for the long-term impact due to large releases, relatively long half-life and its contributions to radiation doses of the public (Aarkrog et al., 1997). Following the results presented in previous chapters, we shall do the simulations for the unit of ^{137}Cs release to the ocean equivalent of 1 PBq. The real ^{137}Cs activity concentrations can be then adjusted when more exact values of radionuclide source terms will be available, just by multiplying the simulated results. Using the above information on the ^{137}Cs release rates, the LAMER calculated the three-dimensional distribution of ^{137}Cs in the world ocean up to the year 2041, i.e. 30 years after the accident.

5.3.3.3 Horizontal and Vertical Distribution of ^{137}Cs in the North Pacific Ocean

The ocean global circulation model (with unit release of 1 PBq of ^{137}Cs) shows that the ^{137}Cs plume in surface seawater will move from the Fukushima coast to the eastward with Kuroshio Current and its extension (Fig. 5.58). Then, it will move relatively slowly with the North Pacific Current, as it will take 4–5 years to arrive at the US west coast with ^{137}Cs levels of ~0.3 Bq/m^3. In 2022, the entire North Pacific Ocean will be labeled with Fukushima ^{137}Cs with concentrations bellow 0.1 Bq/m^3. The maximum predicted ^{137}Cs activity concentrations due to the Fukushima accident in 2012 in surface seawater of the open western North Pacific Ocean (at 38° N, 164° E) will be ~2 Bq/m^3.

If we take presently the most probable total ^{137}Cs release rate, composed from direct liquid releases (4 PBq) and the atmospheric deposition over the northwestern Pacific Ocean (5 PBq; Nakano and Povinec, 2012) and multiply the above predicted ^{137}Cs activity concentrations for this total release, then the levels expected at the US coast will be by about 3 Bq/m^3. The model-predicted values of surface ^{137}Cs concentrations in the Pacific Ocean can be compared with the historical observations gathered in the framework of the Worldwide Marine Radioactivity Studies (WOMARS) project (Povinec

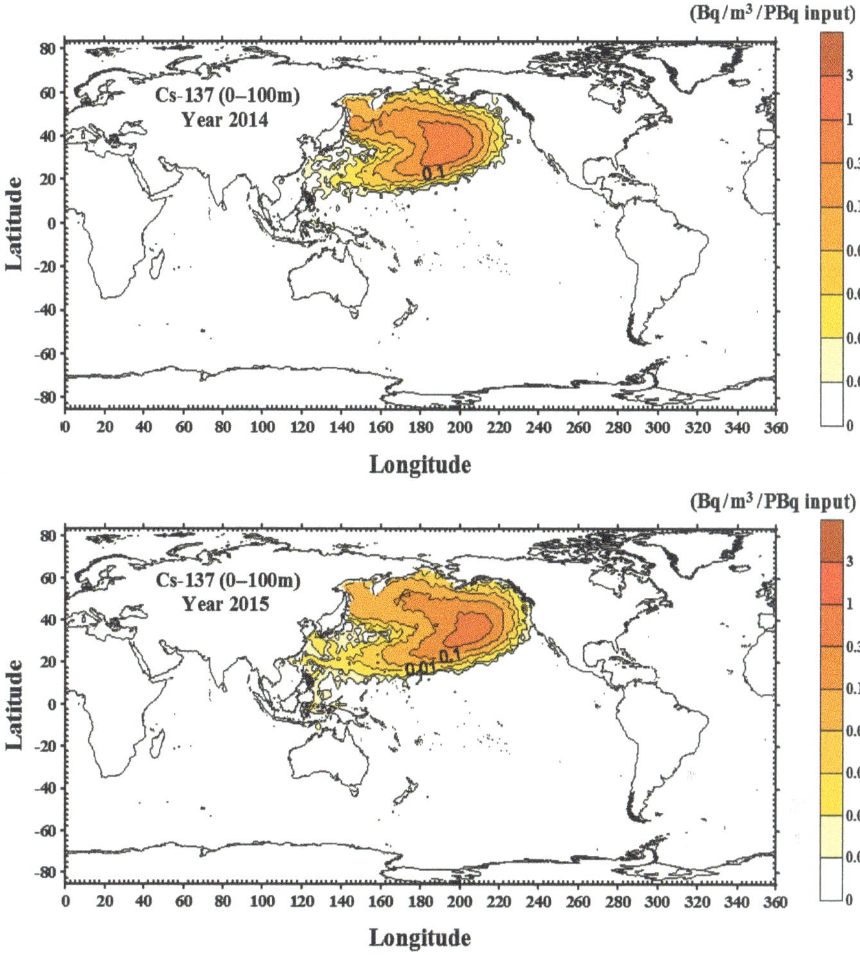

Figure 5.58 Simulations of ^{137}Cs plume (a unit release of 1 PBq) in surface seawater moving from the Fukushima coast to the eastward with Kuroshio Current and its extension. (Modified after Nakano and Povinec, 2012).

et al., 2005) stored in the GLOMARD/MARIS database (Povinec et al., 2004, 2006). If we compare historical ^{137}Cs levels observed in the Pacific Ocean in the latitude belts defined by the WOMARS collaboration (e.g. between 25° and 40° N, either west or east of 180° E), and taking into account the mean effective half-life of ^{137}Cs in surface water of the North Pacific Ocean (13 ± 1 years), the estimated average ^{137}Cs levels for the year 2011 are 1.0 ± 0.1 and 0.9 ± 0.2 Bq/m^3 (Povinec et al., 2013b), for the western and the eastern North Pacific, respectively.

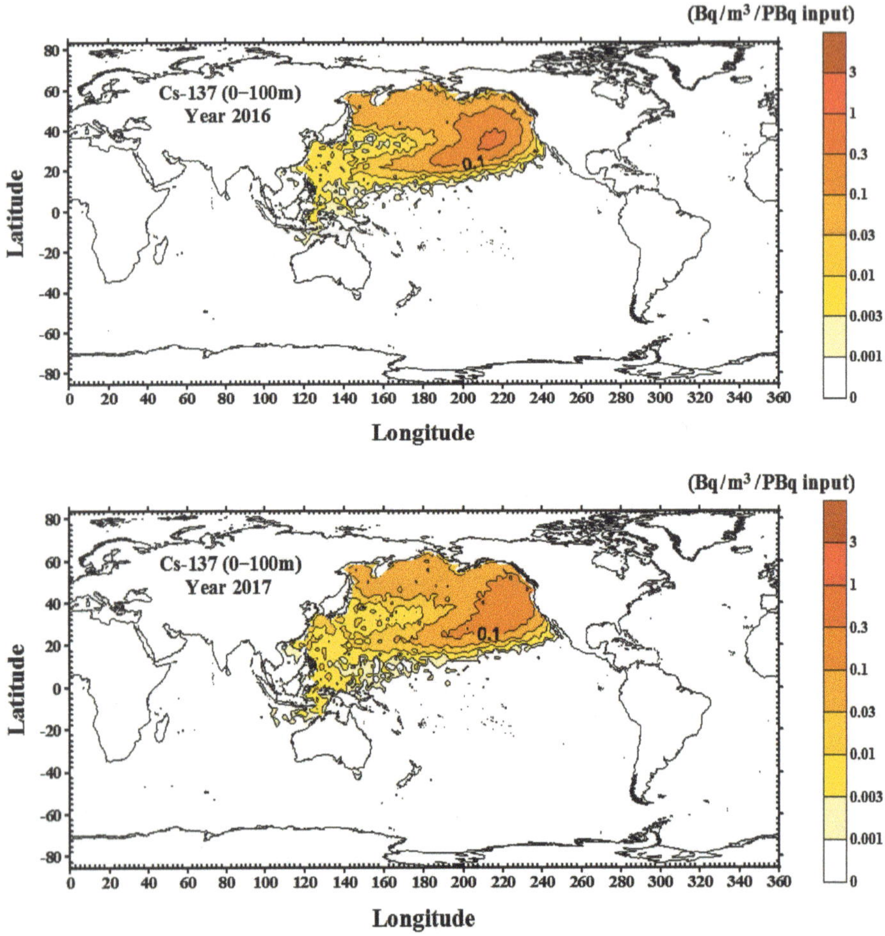

Figure 5.58 Continued

Therefore 10 years after the Fukushima accident, the predicted ^{137}Cs levels (<1 Bq/m^3) will not be detectable over global fallout background in surface waters of the northeastern Pacific Ocean.

The chronological change in the maximum of ^{137}Cs concentration from the Fukushima accident in surface waters of the northwestern Pacific Ocean is shown in Fig. 5.59 (Nakano and Povinec, 2012). The global fallout ^{137}Cs maximum in the world ocean was observed in 1963 after the largest atmospheric nuclear weapons tests carried out during 1961–1962 at Novaya Zemlya in the Kara Sea of the Arctic Ocean

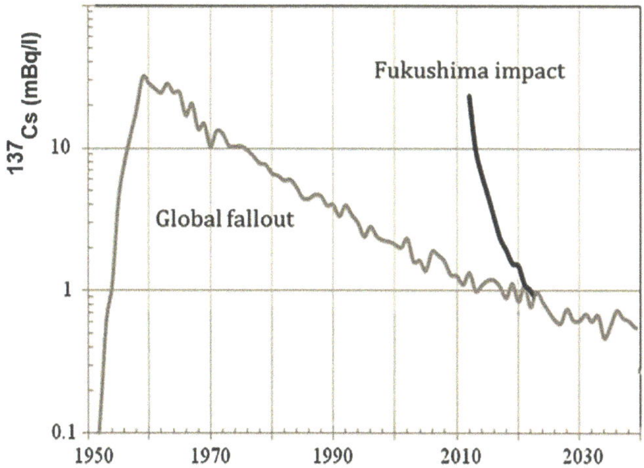

Figure 5.59 Chronological changes in the maximum of ^{137}Cs concentration from the Fukushima accident in surface waters of the northwestern Pacific Ocean. (Modified from Nakano and Povinec (2012))

(Livingston and Povinec, 2002). However, due to nuclear tests carried out at Bikini and Enewetak atolls in the central subtropical Pacific, this maximum in the northwest Pacific Ocean was shifted to the end of the 1950s (Aoyama et al., 2006; Inomata et al., 2009). For comparison, Fig. 5.59 also shows the calculated ^{137}Cs concentrations due to the past atmospheric nuclear weapons tests (global fallout) at a point of 38° N, 164° E in the northwestern Pacific Ocean. In 2012, the maximum ^{137}Cs concentration due to the Fukushima accident will reach in the northwestern Pacific Ocean 23 Bq/m^3, what is comparable with the levels estimated for the early 1960s from global fallout (Nakano and Povinec, 2003a,b; Inomata et al., 2009). The concentration due to the Fukushima accident will be rapidly decreasing with time, reaching <1 Bq/m^3 in 2021.

Because of its low resolution (2 by 2°), the LAMER code is suitable only for simulating the open ocean distribution of ^{137}Cs at the surface and in the water column, about 1 year after the Fukushima accident (Nakano and Povinec, 2012). Unfortunately, there are not many recent ^{137}Cs data available for surface seawaters of the North Pacific Ocean (Kameník et al., 2013). Surface seawater samples were collected by 17 cargo ships and by

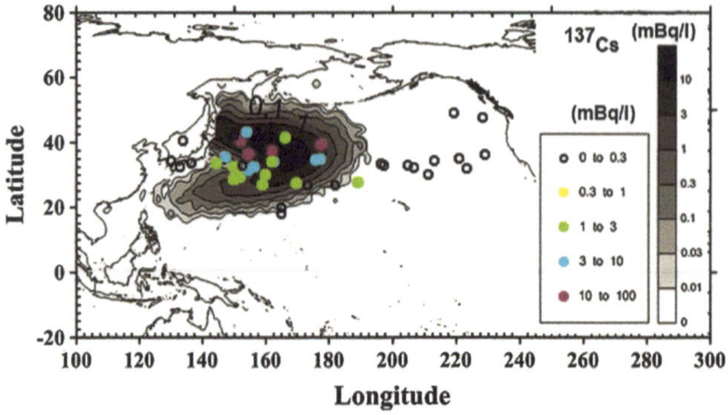

Figure 5.60 Measured and simulated ^{137}Cs concentrations in surface seawater of the north-western Pacific ocean. (After Povinec et al., 2013b).

several research vessels in 2011 and 2012, as described by Aoyama et al. (2013). It can be seen from Fig. 5.60 that a reasonable compromise has been found between the measured and the simulated ^{137}Cs concentrations in surface seawater of the northwestern Pacific Ocean (Povinec et al., 2013a).

The chronological change of the maximum ^{137}Cs concentrations at the surface (0–100 m) and at the 200–300 m water depth presented in Fig. 5.61 indicates that the surface concentration will gradually decrease, and the concentrations at 200–300 m will be

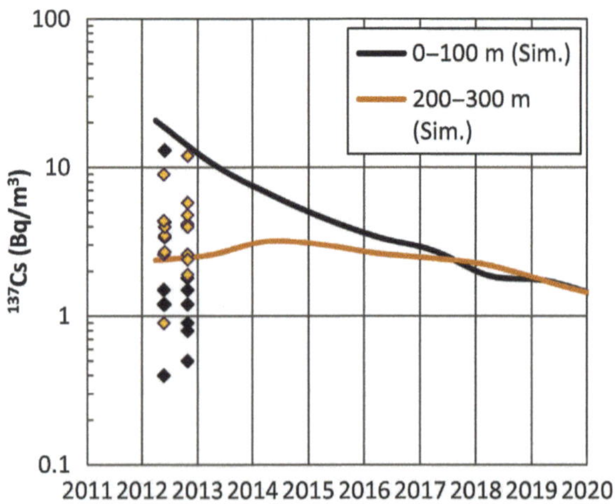

Figure 5.61 Measured and the simulated ^{137}Cs concentrations in the water column of the northwestern Pacific Ocean 1 year after the Fukushima accident. (After Povinec et al., 2013b).

increasing from zero up to the surface level in a few years after the accident (Povinec et al., 2013a). The ^{137}Cs levels in the water column observed almost 1 year after the Fukushima accident are within the range predicted by the LAMER simulations.

References

Aarkrog, A., Baxter, M.S., Bettencourt, A.O., Bojanowski, R., Bologa, A., Charmasson, S., Cunha, I., Delfanti, R., Duran, E., Holm, E., Jeffree, R., Livingston, H.D., Mahapanyawong, S., Nies, H., Osvath, I., Pingyu, Li, Povinec, P.P., Sanchez, A., Smith, J.N., Swift, D.A., 1997. Comparison of doses from ^{137}Cs and ^{210}Po in marine food: a major international study. J. Environ. Radioact. 34, 69–90.

Amachi, S., Kamagata, Y., Kanagawa, T., Muramatsu, Y., 2001. Bacteria mediate methylation of iodine in marine and terrestrial environment. Appl. Environ. Microbiol. 67, 2718–2722.

Amano, H., Akiyama, M., Chunlei, B., Kawamura, T., Kishimoto, T., Kuroda, T., Muroi, T., Odaira, T., Ohota, Y., Takeda, K., Watanabe, Y., Morimoto, T., 2012. Radiation measurements in the Chiba metropolitan area and radiological aspects of fallout from the Fukushima Daiichi nuclear power plants accident. J. Environ. Radioact. 111, 42–52.

Anderson, S.J., Sposito, G., 1991. Cesium-adsorption method for measuring accessible structural surface charge. Soil Sci. Soc. Am. J. 55, 1569–1576.

Antovic, I., Antovic, N.M., 2011. Determination of concentration factors for Cs-137 and Ra-226 in the mullet species Chelon labrosus (Mugilidae) from the South Adriatic Sea. J. Environ. Radioact. 102, 713–717.

Aoyama, M., Hirose, K., Suzuki, Y., Inoue, H., Sugimura, Y., 1986. High level radioactive nuclides in Japan in May. Nature 321, 819–820.

Aoyama, M., 1988. Evidence of stratospheric fallout of caesium isotopes from the Chernobyl accident. Geophys. Res. Lett. 15, 327–330.

Aoyama, M., Hirose, K., Takatani, S., 1992. Particle size dependent dry deposition velocity of the Chernobyl radioactivity. In: Precipitation Scavenging and Atmospheric Exchange Processes; Fifth International Conference, vol. 3. Hemisphere, pp. 1581–1593.

Aoyama, M., Hirose, K., 2004. Artificial radionuclides database in the Pacific Ocean: HAM database. Scientific World J. 4, 200–215.

Aoyama, M., Hirose, K., Igarashi, Y., 2006. Re-construction and updating our understanding on the global weapons tests ^{137}Cs fallout. J. Environ. Monit. 8, 431–438.

Aoyama, M., Hirose, K., 2008a. Radiometric determination of anthropogenic radionuclides in seawater. In: Pavel, P.P. (Ed.), Radioactivity in the Environment. Elsevier, pp. 137–162.

Aoyama, M., Hirose, K., Nemoto, K., Takatsuki, Y., Tsumune, D., 2008b. Water masses labeled with global fallout ^{137}Cs

formed by subduction in the North Pacific. Geophys. Res. Lett. 35, L01604. http://dx.doi.org/10.1029/2007GL031964.

Aoyama, M., Fukasawa, M., Hirose, K., Hamajima, Y., Kawano, T., Povinec, P.P., Sanchez-Cabeza, J.A., 2011. Cross equator transport of ^{137}Cs from North Pacific Ocean to South Pacific Ocean (BEAGLE2003 cruises). Prog. Oceanography 89, 7–16. http://dx.doi.org/10.1016/j.pocean.2010.12.003.

Aoyama, M., Tsumune, D., Hamajima, Y., 2012a. Distribution of ^{137}Cs and ^{134}Cs in the North Pacific Ocean: impacts of the TEPCO Fukushima Dai-chi NPP accident. J. Radioanal. Nucl. Chem. http://dx.doi.org/10.1007/s10967-012-2033-2.

Aoyama, M., Tsumune, D., Uematsu, M., Kondo, F., Hamajima, Y., 2012b. Temporal variation of ^{134}Cs and ^{137}Cs activities in surface water at stations along the coastline near the Fukushima Dai-ichi nuclear power plant accident site, Japan. Geochem. J. 46, 321–325.

Aoyama, M., Uematsu, M., Tsumune, D., Hamajima, Y., 2013. Surface pathway of radioactive plume of TEPCO Fukushima NPP1 released 134Cs and 137Cs, Biogeosciences Discuss. 10, 3067–3078. http://dx.doi.org/10.5194/bgd-10-265-2013.

Arimoto, R., Webb, J.L., Conley, M., 2005. Radioactive contamination of atmospheric dust over southeastern New Mexico. Atmos. Environ. 39, 4745–4754.

Bailly du Bois, P., Laguionie, P., Boust, D., Korsakissok, I., Didier, D., Fievet, B., 2012. Estimation of marine source-term following Fukushima Dai-ichi accident. J. Environ. Radioact. 114, 2–9.

Baeza, A., Corbacho, J.A., Rodríguez, A., Galván, J., García-Tenorio, R., Manjón, G., Mantero, J., Vioque, I., Arnold, D., Grossi, C., Serrano, I., Vallés, I., Vargas, A., 2012. Influence of the Fukushima Dai-ichi nuclear accident on Spanish environmental radioactivity levels. J. Environ. Radioact. 114, 138–145.

Balkanski, Y.J., Jacob, D.J., Gardner, G.M., Graustein, W.C., Turekian, K.K., 1993. Transport and residence time of tropospheric aerosols inferred from a global three-dimensional simulation of ^{210}Pb. J. Geophys. Res. 98, 20573–20586.

Barsanti, M., Conte, F., Delbono, I., Iurlaro, G., Battisti, P., Bartoluzzi, S., Lorenzelli, R., Salvi, S., Zicari, S., Papucci, C., Delfanti, R., 2012. Environmental radioactivity analyses in Italy following the Fukushima Dai-chi nuclear accident. J. Environ. Radioact. 114, 126–130.

Beresford, N.A., Barnett, C.L., Howard, B.J., Howard, D.C., Wells, C., Tyler, A.N., Bradley, S., Copplestone, D., 2012. Observations of Fukushima fallout in Great Britain. J. Environ. Radioact. 114, 48–53.

Biddulph, D.L., Beck, J.W., Burr, G.S., Donahue, D.J., Hatheway, A.L., Jull, A.J.T., 2000. Measurement of the radioisotope ^{129}I at the NSF-Arizona AMS. Nucl. Instrum. Methods Phys. Res. B 172, 693–698.

Biegalski, S.R., Bowyer, T.W., Eslinger, P.W., Friese, J.A., Greenwood, L.R., Haas, D.A., Hayes, J.C., Hoffman, I.,

Keillor, M., Miley, H.S., Moring, M., 2012. Analysis of data from sensitive U.S. monitoring stations for the Fukushima Dai-ichi nuclear reactor accident. J. Environ. Radioact. 114, 15–21.

Bikit, I., Mrda, D., Todorovic, N., Nikolov, J., Krmar, M., Veskovic, M., Slivka, J., Hansman, J., Forkapic, S., Jovancevic, N., 2012. Airborne radioiodine in northern Serbia from Fukushima. J. Environ. Radioact. 114, 89–93.

Bolsunovsky, A., Dementyev, D., 2011. Evidence of the radioactive fallout in the center of Asia (Russia) following the Fukushima nuclear accident. J. Environ. Radioact. 102, 1062–1064.

Bonjean, F., Lagerloef, G.S.E., 2002. Diagnostic model and analysis of the surface currents in the tropical Pacific Ocean. J. Phys. Oceanogr. 32, 2938–2954. http://dx.doi.org/10.1175/1520-0485. 032<2938:DMAAOT>2.0.CO;2.

Bossew, P., Kirchner, G., De Cort, M., De Vries, G., Nishev, A., De felice, L., 2012. Radioactivity from Fukushima Dai-ichi in air over Europe: part 1: spatio-temporal analysis. J. Environ. Radioact. 114, 22–34.

Boulyga, S.F., Heumann, K.G., 2006. Determination of extremely low $^{236}U/^{238}U$ isotope ratios in environmental samples by sector-field inductively coupled plasma mass spectrometry using high-efficiency sample introduction. J. Environ. Radioact. 88, 1–10.

Bowyer, T.W., Biegalski, S.R., Cooper, M., Eslinger, P.W., Haas, D., Hayes, J.C., et al., 2011. Elevated radioxenon detected remotely following the Fukushima nuclear accident. J. Environ. Radioact. 102, 681–687.

Buesseler, K.O., 2012. Fishing for answers off Fukushima. Science 338, 480–482.

Buesseler, K., Aoyama, M., Fukasawa, M., 2011. Impacts of the Fukushima nuclear power plants on marine radioactivity. Environ. Sci. Technol. 45, 9931–9935.

Buesseler, K.O., Jayne, S.R., Fisher, N.S., Rypina, I.I., Baumann, H., Baumann, Z., Breier, C.F., Douglass, E.M., George, J., Macdonald, A.M., Miyamoto, H., Nishikawa, J., Pike, S.M., Yoshida, S., 2012. Fukushima-derived radionuclides in the ocean and biota off Japan. Proc. Natl. Acad. Sci. U S A 109, 5984–5988.

Carvalho, F.P., Reis, M.C., Oliveira, J.M., Malta, M., Silva, L., 2012. Radioactivity from Fukushima nuclear accident detected in Lisbon. Portugal. J. Environ. Radioact. 114, 152–156.

Chamberlain, A.C., 1953. Experiments on the deposition of iodine-131 vapour onto surfaces from an airstream. Phil. Mag. 44, 1145–1153.

Chamberlain, A.C., 1991. Radioactive Aerosols. In: Cambridge Environmental Chemistry Series, vol. 3. University Press, Cambridge.

Chino, M., Nakayama, H., Nagai, H., Terada, H., Katata, G., Yamazawa, H., 2011. Preliminary estimation of release amounts of ^{131}I and ^{137}Cs accidentally discharged from the Fukushima Daiichi nuclear power plant into the atmosphere. J. Nucl. Sci. Technol. 48, 1129–1134.

Choi, H.J., Amistadi, M., Chorover, J., 2005. Clay mineral weathering and contaminant dynamics in a caustic aqueous system I. Wet chemistry and aging effects. Geochim. Cosmochim. Acta 69, 4425–4436.

Clemenza, M., Fiorini, E., Previtali, E., Sala, E., 2012. Measurement of airborne ^{131}I, ^{134}Cs and ^{137}Cs due to the Fukushima reactor incident in Milan (Italy). J. Environ. Radioact. 114, 113–118.

Cosma, C., Iurian, A.R., Nita, D.C., Begy, R., Cindea, C., 2012. Indicators of the Fukushima radioactive release in NW Romania. J. Environ. Radioact. 114, 94–99.

CPSMA (Commission on Physical Sciences, Mathematics and Applications), 1996. Radiochemistry in Nuclear Power Reactors. The National Academy of Sciences, USA. http://www.nap.adu/openbook/NI000156/htm/50.html.

Crecelius, E.A., Robertson, D.E., Abel, K.H., Cochran, D.A., Weimer, W.C., 1978. Atmospheric deposition of 7Be and other elements on the Washington coast, Pacific Northwest Laboratory Annual Report for 1977 to the DOE Assistant Secretary for Environment: Ecological Sciences, PNL-2500 PT-2, Battelle, Pacific Northwest Laboratory, pp. 7.25–7.26.

Cunha, I.L., Minita, C.S., Paiva, R.P., Teixeira, A., 1993. Levels of cesium-137 in seawater and fish from the Brazilian coast. Sci. Total Environ. 140, 431–435.

Diaz Leon, J., Jaffe, D.A., Kaspar, J., Knecht, A., Miller, M.L., Robertson, R.G.H., Schubert, A.G., 2011. Arrival time and magnitude of airborne fission products from the Fukushima, Japan, reactor incident as measured in Seattle, WA, USA. J. Environ. Radioact. http://dx.doi.org/10.1016/j.jenvrad.2011.06.005.

Dietze, H., Kriest, I., 2012. ^{137}Cs off Fukushima Dai-ichi, Japan: model based estimates of dilution and fate. Ocean Sci. 8, 319–332.

Doi, T., Masumoto, K., Toyoda, A., Tanaka, A., Shibata, Y., Hirose, K., 2013. Anthropogenic radionuclides in the atmosphere observed at Tsukuba: characteristics of the radionuclides derived from Fukushima. J. Environ. Radioact. 122, 55–62.

Doi, H., Takahata, T., Tanaka, K., 2012. Trophic position and metabolic rate predict the long-term decay process of radioactive cesium in fish: a meta-analysis. PLOS One 7, e29295.

Draxler, R.R., Rolph, G.D., 2011. HYSPLIT (Hybrid Single-particle Lagrangian Integrated Trajectory). NOAA Air Resources Laboratory. http://ready.arl.noaa.gov/HYSPLIT.php.

Ehhalt, D.H., 1973. Turnover times of ^{137}Cs and HTO in the troposphere and removal rates of natural particles and water vapor. J. Geophys. Res. 78, 7076–7086.

Endo, S., Kajimoto, T., Shizuma, K., 2013. Paddy-field contamination with ^{134}Cs and ^{137}Cs due to Fukushima Dai-ichi nuclear power plant accident and soil-to rice transfer coefficients. J. Environ. Radioact. 116, 59–64.

Endo, S., Kimura, S., Takatsuji, T., Nanasawa, K., Imanaka, T., Shizuma, K., 2012. Measurement of soil contamination by radionuclides due to the Fukushima Daiichi nuclear power plant accident and associated estimated cumulative external dose estimation. J. Environ. Radioact. 111, 18–27.

Environmental Impacts of the Fukushima Accident (PART II), HYPERLINK "http://www.sciencedirect.com/science/journal/0265931X/114" Volume 114, pp. 1-170 (December 2012).

Estournel, C., Bosc, E., Bocquet, M., Ulses, C., Marsaleix, P., Winiarek, V., Osvath, I., Nguyen, C., Duhaut, T., Lyard, F., Michaud, H., Auclair, F., 2012. Assessment of the amount of cesium-137 released into the Pacific Ocean after the Fukushima accident and analysis of its dispersion in Japanese coastal waters. J. Geophys. Res. 117, C11014. http://dx.doi.org/10.1029/2012JC007933.

Evrard, O., Van Beek, P., Gateuille, D., Pont, V., Lefévre, I., Lansard, B., Bonté, P., 2012. Evidence of the radioactive fallout in France due to the Fukushima nuclear accident. J. Environ. Radioact. 114, 54–60.

Fujiwara, H., Fukuyama, T., Shirato, Y., Ohkuro, T., Taniyama, I., Zhang, T.-H., 2007. Deposition of atmospheric ^{137}Cs in Japan associated with the Asian dust event of March 2002. Sci. Total Environ. 384, 306–315.

Fujiwara, T., Saito, T., Muroya, Y., Sawahata, H., Yamashita, Y., Nagasaki, S., Okamoto, K., Takahashi, H., Uesaka, M., Katsumura, Y., Tanaka, S., 2012. Isotopic ratio and vertical distribution of radionuclides in soil affected by the accident of Fukushima Dai-ichi nuclear power plants. J. Environ. Radioact., 13–44.

Furuta, S., Sumiya, S., Watanabe, H., Nakano, M., Imaizumi, K., Takeyasu, M., Nakada, A., Fujita, M., Mizutani, T., Morisawa, M., Kokubun, Y., Kono, T., Nagaoka, M., Yokoyama, H., Hokama, T., Isozaki, T., Nemoto, M., Hiyama, Y., Onuma, T., Kato, C., Kurachi, T., 2011. Results of the environmental radiation monitoring following the accident at the Fukushima Daiichi nuclear power plant – Interim report (ambient radiation dose rate, radioactivity concentration in the air and radioactivity concentration in the fallout). JAEA-Review 2011-035, 1–84 (In Japanese).

Garland, J.A., Cambray, R.S., 1988. Deposition, resuspension and the long-term variation of airborne radioactivity from Chernobyl. In: Proceedings of IVth Symposium International de Radioecologie de Caderache. Centre d'Etudie Nucleaire de Cadarache, France (tome 1), pp. B36–B31.

Gera, M., Povinec, P.P., et al. (in preparation).

Haba, H., Kanaya, J., Mukai, H., Kambara, T., Kase, M., 2012. One-year monitoring of airborne radionuclides in Wako, Japan, after the Fukushima Dai-ichi nuclear power plant accident in 2011. Geochem. J. 46, 271–278.

Hamada, N., Ogino, H., 2012. Food safety regulations: what we learned from the Fukushima nuclear accident. J. Environ. Radioact. 111, 83–99.

Hamajima, Y., Komura, K., 2004. Background components of Ge detectors in ogoya underground laboratory. Appl. Radiat. Isot. 61, 179–183. http://dx.doi.org/10.1016/j.apradiso.2004.03.041.

Harvey, M.J., Matthews, K.M., 1989. ^{7}Be deposition in a high-rainfall area of New Zealand. J. Atmos. Chem. 8, 299–306.
Hernández-Ceballos, M.A., Hong, G.H., Lozano, R.L., Kim, Y.I., Lee, H.M., Kim, S.H., Yeh, S.W., Bolívar, J.P., Baskaran, M., 2012. Tracking the complete revolution of surface westerlies over Northern Hemisphere using radionuclides emitted from Fukushima. Sci. Total Environ. 438, 80–85.
Hirose, K., 1995. Geochemical studies on the Chernobyl radioactivity in environmental samples. J. Radioanal. Nucl. Chem. 197, 315–335 (Articles).
Hirose, K., 2000. Dry and wet deposition behaviors of thorium isotopes. J. Aerosol Res. Japan 15, 256–263.
Hirose, K., 2012. 2011 Fukushima Daiichi nuclear power plant accident: summary of regional radioactivity deposition monitoring results. J. Environ. Radioact. 111, 13–17.
Hirose, K., 2013. Temporal variation of monthly ^{137}Cs deposition observed in Japan: effects of the Fukushima Daiichi nuclear power plant accident, Appl. Radiat. Isot., http://dx.doi.org/10.1016/j.apradiso.2013.03.076
Hirose, K., Aoyama, M., Sugimura, Y., 1990. Plutonium and cesium isotopes in river waters in Japan. J. Radioanal. Nucl. Chem. 141, 191–202.
Hirose, K., Igarashi, Y., Aoyama, M., Kim, C.K., Kim, C.S., Chang, B.W., 2003. Recent trends of plutonium fallout observed in Japan: plutonium as a proxy for desertification. J. Environ. Monit. 5, 1–7.
Hirose, K., Aoyama, M., Igarashi, Y., Komura, K., 2005. Ultra-sensitive mass spectrometric and other methods applied to environmental problems. J. Radioanal. Nucl. Chem. 263, 349–353. http://dx.doi.org/10.1007/s10967-005-0593-0.
Hirose, K., Igarashi, Y., Aoyama, M., 2007. Recent trends of plutonium fallout observed in Japan: Comparison with national lithogenic radionuclides, thorium isotopes. J. Radioanal. Nucl. Chem. 273, 115–118.
Hirose, K., Igarashi, Y., Aoyama, M., 2008. Analysis of 50 years records of atmospheric deposition of long-lived radionuclides in Japan. Appl. Radiat. Isot. 66, 1675–1678.
Hirose, K., Igarashi, Y., Aoyama, M., Inomata, Y., 2010. Depositional behaviors of plutonium and thorium at Tsukuba and Mt. Haruna in Japan indicate the sources of atmospheric dust. J. Environ. Radioact. 101, 106–112.
Hirose, K., Kim, C.K., Kim, C.S., Chang, B.W., Igarashi, Y., Aoyama, M., 2004. Plutonium deposition observed in Daejeon, Korea: wet and dry depositions of plutonium. Sci. Total Environ. 332, 243–252.
Hirose, K., Kikawada, Y., Igarashi, Y., 2012. Temporal variation and provenance of thorium deposition observed at Tsukuba, Japan. J. Environ. Radioact. 108, 24–28.
Hirose, K., Takatani, S., Aoyama, M., 1993. Wet deposition of long-lived radionuclides derived from the Chernobyl accident. J. Atmos. Chem. 17, 61–71.
Honda, M., Aono, T., Aoyama, M., Hamajima, Y., Kawakami, H., Kitamura, M., Masumoto, Y., Miyazawa, Y., Takigawa, M.,

Saino, T., 2012. Dispersion of artificial caesium-134 and -137 in the western North Pacific one month after the Fukushima accident. Geochem. J. 46, 1–9.

Hou, X.L., Aldahan, A., Nislen, S., Possnert, G., Nies, H., Hedfords, J., 2007. Speciation of ^{129}I and ^{127}I in seawater and implications for sources and transport pathways in North Sea. Environ. Sci. Technol. 41, 5993–5999.

Hou, X.L., Aldahan, A., Nielsen, S.P., Possnert, G., 2009a. Time series of ^{129}I and ^{127}I speciation in precipitation from Denmark. Environ. Sci. Technol. 43, 6522–6528.

Hou, X.L., Hansen, V., Aldahan, A., Possnert, G., Lind, O.C., Lujaniene, G., 2009b. A review on speciation of iodine-129 in the environmental and biological samples. Anal. Chim. Acta 632, 181–196.

Hou, X.L., Povinec, P.P., Zhang, L.Y., Biddulph, D., Chang, C.-C., Fan, Y.K., Golser, R., Ješkovský, M., Jull, A.J.T., Liu, Q., Shi, K.L., Steier, P., Zhou, W.J., 2013. Iodine-129 in seawater offshore Fukushima: distribution, speciation, sources, and budget. Environ. Sci. Technol. 47, 3091–3098.

Huh, C.A., Hsu, S.-C., Lin, C.-Y., 2012. Fukushima-derived fission nuclides monitored around Taiwan: free tropospheric versus boundary layer transport. Earth Planet. Sci. Lett. 319–320, 9–14.

IAEA, International Atomic Energy Agency, 1986. Summary Report on the Post-Accident Review Meeting on the Chernobyl Accident, Safety Series No. 75-INSAG-1. IAEA, Vienna.

IAEA, International Atomic Energy Agency, 2003. Chernobyl's Legacy: Health, Environmental and Socio-economic Impacts. IAEA, Vienna.

IAEA, International Atomic Energy Agency, 2004. Sediment Distribution Coefficients and Concentration Factors for Biota in the Marine Environment. Technical Reports Series no. 422. IAEA, Vienna.

IAEA, International Atomic Energy Agency, 2005. Worldwide Marine Radioactivity Studies (WOMARS). Radionuclide Levels in Oceans and Sea. IAEA-TECDOC-1429. IAEA, Vienna.

IAEA, International Atomic Energy Agency, 2011. Briefings on Fukushima nuclear accident. www.iaea.org/.

Igarashi, Y., Aoyama, M., Hirose, K., Miyao, T., Yabuki, S., 2001. Is it possible to use ^{90}Sr and ^{137}Cs as tracers for the aeolian transport? Water Air Soil Pollut. 130, 349–354.

Igarashi, Y., Aoyama, M., Hirose, K., Miyao, T., Nemoto, K., Tomita, M., Fujikawa, T., 2003. Resuspension: decadal monitoring time series of the anthropogenic radioactivity deposition in Japan. J. Radiat. Res. 44, 319–328.

Igarashi, Y., Aoyama, M., Hirose, K., Povinec, P.P., Yabuki, S., 2005. What anthropogenic radionuclides (^{90}Sr and ^{137}Cs) in atmospheric deposition, surface soils and Aeolian dusts suggest for dust transport over Japan. Water Air Soil Pollut. Focus 5, 51–69.

Igarashi, Y., Inomata, Y., Aoyama, M., Hirose, K., Takahashi, H., Shinoda, Y., Sugimoto, N., Shimizu, A., Chiba, M., 2009. Possible change in Asian dust source suggested by atmospheric anthropogenic radionuclides during the 2000s. Atmos. Environ. 43, 2971–2980.

Inomata, Y., Aoyama, M., Hirose, K., 2009. Analysis of 50-y record of surface ^{137}Cs concentrations in the global ocean using the HAM-global database. J. Environ. Monit. 11, 116–125.

Inoue, M., Kofuji, H., Hamajima, Y., Nagao, S., Yoshida, K., Yamamoto, M., 2012a. ^{134}Cs and ^{137}Cs activities in coastal seawater along Northern Sanriku and Tsugaru Strait, northeastern Japan, after Fukushima Dai-ichi nuclear power plant accident. J. Environ. Radioact. 111, 116–119. http://dx.doi.org/10.1016/j.jenvrad.2011.09.012.

Inoue, M., Kofuji, H.Y., Hamajima, Y., Nagao, S., Yoshida, K., Yamamoto, M., 2012b. ^{134}Cs and ^{137}Cs activities in coastal seawater along Northern Sanriku and Tsugaru Strait, northeastern Japan, after Fukushima Dai-ichi nuclear power plant accident. J. Environ. Radioact. 111, 116–119.

Inoue, M., Kofuji, H., Nagao, S., Yamamoto, M., Hamajima, Y., Fujimoto, K., Yoshida, K., Suzuki, A., Takashiro, H., Hayakawa, K., 2012c. Low levels of ^{134}Cs and ^{137}Cs in surface seawaters around the Japanese archipelago after the Fukushima Dai-ichi nuclear power plant accident in 2011. Geochem. J. 46, 311–320.

Ioannidou, A., Manenti, S., Gini, L., Groppi, F., 2012. Fukushima fallout at Milano, Italy. J. Environ. Radioact. 114, 119–125.

JER, Journal of Environmental Radioactivity, 2012. Special issues: Environmental Impacts of the Fukushima Accident (Part I), "http://www.sciencedirect.com/science/journal/0265931X/111" Volume 111, pp. 1–126 (September 2012); Environmental Impacts of the Fukushima Accident (PART II), "http://www.sciencedirect.com/science/journal/0265931X/114" Volume 114, pp. 1–170 (December 2012).

JG, Japanese Government Report. http://www.kantei.go.jp/jp/Topics/2011/iaea_houkokusho.html (accessed 10.07.12.).

Kaeriyama, H., Watabe, T., Kusakabe, M., 2008. ^{137}Cs concentration in zooplankton and its relation to taxonomic composition in the western North Pacific Ocean. J. Environ. Radioact. 99, 1838–1845.

Kameník, J., Dulaiova, H., Buesseler, K.O., Pike, S.M., Šťastná, K., 2013. Cesium-134 and 137 activities in the central North Pacific Ocean after the Fukushima Dai-ichi nuclear power plant accident. Biogeosciences Discuss.

Kanai, Y., 2012. Monitoring of aerosols in Tsukuba after Fukushima nuclar power plant incident in 2011. J. Environ. Radioact. 111, 33–37.

Kaneyasu, N., Ohashi, H., Suzuki, F., Okuda, T., Ikemori, F., 2012. Sulfate aerosol as a potential transport medium of radiocesium from the Fukushima nuclear accident. Environ. Sci. Technol. 46, 5720–5726.

Karlsson, L., Hernandez, F., Rodríguez, S., López-Pérez, M., Hernandez-Armas, J., Alonso-Pérez, S., Cuevas, E., 2008. Using ^{137}Cs and ^{40}K to identify natural Saharan dust contributions to PM10 concentrations and air quality impairment in the Canary Islands. Atmos. Environ. 42 7034–7032.

Kasamatsu, F., Ishikawa, Y., 1997. Natural variation of radionuclide ^{137}Cs concentration in marine organisms with special reference to the effect of food habits and trophic level. Mar. Ecol. Prog. Ser. 160, 109–120.

Katayama, E., Kawakami, I., Okumura, T., 2012. Contribution of Plant-opal to Radioactive-Cesium Contamination of Forested-Terrain. 49th Conference of Radioisotope and Radiation Study, Tokyo, Abstract Book, pp. 59. (In Japanese).

Kato, H., Onda, Y., Teramage, M., 2012. Depth distribution of ^{137}Cs, ^{134}Cs, and ^{131}I in soil profile after the Fukushima Daiichi nuclear plant accident. J. Environ. Radioact. 111, 59–64.

Katsuragi, Y., 1983. A study of ^{90}Sr fallout in Japan. Pap. Meteor. Geophys. 33, 277–291.

Kawamura, H., Kobayashi, T., Furuno, A., In, T., Ishikawa, Y., Nakayama, T., Shima, S., Awaji, T., 2011. Preliminary numerical experiments on oceanic dispersion ^{131}I and ^{137}Cs discharged into the ocean because of the Fukushima Dai-ichi nuclear power plant disaster. J. Nucl. Sci. Technol. 48, 1349–1356.

Kelley, J.M., Bond, L.A., Beasley, T.M., 1999. Global distribution of Pu isotope and Np. Sci. Total Environ. 237/238, 483–500.

Kim, Y., Cygan, R.T., Kirkpatrick, R.J., 1996. ^{133}Cs NMR and XPS investigation of cesium adsorbed on clay minerals and related phases. Geochim. Cosmochim. Acta 60, 1041–1052.

Kim, Y., Kirkpatrick, R.J., 1997. ^{23}Na and ^{133}Cs NMR study of cation adsorption on mineral surfaces: local environments, dynamics and effects of mixed cations. Geochim. Cosmochim. Acta 61, 5199–5208.

Kim, C.-K., Byun, J.-I., Chae, J.-S., Choi, H.-Y., Choi, S.-W., Kim, D.-J., et al., 2012. Radiological impact in Korea following the Fukushima nuclear accident. J. Environ. Radioact. 111, 70–82.

Kinoshita, N., Sueki, K., et al., 2011. Assessment of individual radionuclide distributions from the Fukushima nuclear accident covering central-east Japan. Proc. Natl. Acad. Sci. 108 (49), 19526–19529.

Knapinska-Skiba, D., Bojanowski, R., Piekos, R., 2003. Activity concentration of caesium-137 in seawater and plankton of the Pomeranian Bay (The southern Baltic Sea) before and after flood in 1997. Mar. Pollut. Bull. 46, 1558–1562.

Kritidis, P., Florou, H., Eleftheriadis, K., Evangeliou, N., Gini, M., Sotiropoulou, M., Diapouli, E., Vratolis, 2012. Radioactive pollution of Athens, Greece due to the Fukushima nuclear accident. J. Environ. Radioact. 114, 100–104.

Lambert, G., Polian, G., Sanak, J., Ardouin, B., Buisson, A., Jegou, A., Leroulley, J.C., 1982. Cycle du radon et de ses descendants: application a l'étude des èchanges troposphère-stratosphère. Ann. Geophys. 38, 497–531.

Lee, S.H., Pham, M.K., Povinec, P.P., 2002. Radionuclide variations in the air over Monaco. J. Radioanal. Nucl. Chem. 254, 445–453.

Lee, S.-H., Heo, D.-H., Kang, H.-B., Oh, P.-J., Lee, J.-M., Park, T.-S., Lee, K.B., Oh, J.S., Suh, J.-K., 2012. Distribution of ^{131}I, ^{134}Cs,

^{137}Cs and $^{239,240}Pu$ concentrations in Korean rainwater after the Fukushima nuclear power plant accident. J. Radioanal. Nucl. Chem.. http://dx.doi.org/10.1007/s10967-012-2030-5.

Levy, I., Povinec, P.P., Aoyama, M., Hirose, K., Sanchez-Cabeza, J.A., Comanducci, J.-F., Gastaud, J., et al., 2011. Marine anthropogenic radiotracers in the Southern Hemisphere: new sampling and analytical strategies. Prog. Oceanogr. 89, 120–133.

Livingston, H.D., Povinec, P.P., 2000. Anthropogenic marine radioactivity. Ocean Coastal Manage. 43, 689–712.

Livingston, H.D., Povinec, P.P., 2002. A millennium perspective on the contribution of global fallout radionuclides to ocean science. Health Phys. 82, 656–668.

Loaiza, P., Brudanin, V., Piquemal, F., Reyss, J.L., Stekl, I., Warot, G., Zampaolo, M., 2012. Air radioactivity levels following the Fukushima reactor accident measured at the LaboratoireSouterrain de Modane, France. J. Environ. Radioact. 114, 66–70.

Lozano, R.L., Hernández-Ceballos, M.A., Adame, J.A., Casas-Ruíz, M., Sorribas, M., San Miguel, E.G., Bolívar, J.P., 2011. Radioactive impact of Fukushima accident on the Iberian Peninsula: evolution and plume previous pathway. Environ. Int. 37, 1259–1264.

Lujanienė, G., Jokšas, K., Šilobritienė, B., Morkūnienė, R., 2006a. Physical and chemical characteristics of ^{137}Cs in the Baltic Sea. Radioactivity Environ. 8, 165–179.

Lujanienė, G., Šapolaite, J., Remeikis, V., Lujanas, V., Jermolajev, A., 2006b. Caesium, americium and plutonium isotopes in ground level air of Vilnius. Czechoslovak J. Phys. 56, D55–D61.

Lujanienė, G., Aninkevicius, A., Lujanas, V., 2009. Artificial radionuclides in the atmosphere over Lithuania. J. Environ. Radioact. 100, 108–119.

Lujanienė, G., Byčenkienė, S., Povinec, P.P., Gera, M., 2012a. Radionuclides from the Fukushima accident in the air over Lithuania: measurement and modeling approaches. J. Environ. Radioact. 114, 71–80.

Lujanienė, G., Valiulis, D., Byčenkienė, S., Šakalys, J., Povinec, P.P., 2012b. Plutonium isotopes and ^{241}Am in the atmosphere of Lithuania: a comparison of different source terms. Atmos. Environ. 61, 419–427.

Lujanienė, G., Byčenkienė, S., Povinec, P.P., 2013. Radionuclides in aerosol samples from the Fukushima accident in Vilnius/Lithuania: modeling of aerosol transport. Appl. Rad. Isot. http://dx.doi.org/10.1016/j.apradiso.2013.03.072.

Lyons, C., Colton, D., 2012. Aerial measuring system in Japan. Health Phys. 102, 509–515.

MAFF, 2011. Basic policy for inspection on radioactive materials in fishery products. http://www.jfa.maff.go.jp/j/press/sigen/110506.html.

MAFF, 2012a. Results of the inspection on radioactivity materials in fisheries products (press release March 2011 to December 2012). www.jfa.maff.go.jp/e/inspection/index.html.

MAFF, 2012b. Report on investigation to assess the effects of radioactive materials. http://www.jfa.maff.go.jp/j/sigen/housyaseibussitutyousakekka/pdf/h23report_final_1.pdf.

MAFF, 2012c. Results of the inspection on strontium in fisheries products (press release August 2012). www.jfa.maff.go.jp/j/press/kenkyu/110830html.

Madigan, D.J., Baumann, Z., Fisher, N.S., 2012. Pacific Blue fin tuna transport Fukushima-derived radionuclides from Japan to California. Proc. Natl. Acad. Sci. U S A 109, 9483–9486.

Manolopoulou, M., Vagena, E., Stoulos, A., Ioannidou, A., Papastefanou, C., 2011. Radioidine and radiocesium in Thessaloniki, Northern Greece due to the Fukushima nuclear accident. J. Environ. Radioact. 102, 796–797.

Marzano, F.N., Triulzi, C., 2003. A radiological survey of northern and middle Adriatic Sea before and after the Chernobyl event. Mar. Pollut. Bull. 28, 244–253.

Masson, O., Piga, D., Gurriaran, R., D'Amico, D., 2010. Impact of an exceptional Saharan dust outbreak in France: PM_{10} and artificial radionuclides concentrations in air and in dust deposit. Atmos. Environ. 44, 2478–2486.

Masson, O., Baeza, A., Bieringer, J., Brudecki, K., Bucci, S., Cappai, M., Carvalho, F.P., Connan, O., Cosma, C., Dalheimer, A., Didier, D., Depuydt, G., De Geer, L.E., De Vismes, A., Gini, L., Groppi, F., Gudnason, K., Gurriaran, R., Hainz, D., Halldorsson, O., Hammond, D., Hanley, O., Holey, K., Homoki, Zs., Ioannidou, A., Isajenko, K., Jankovick, M., Katzlberger, C., Kettunen, M., Kierepko, R., Kontro, R., Kwakman, P.J.M., Lecomte, M., Leon Vintro, L., Leppänen, A.-P., Lind, B., Lujaniene, G., Mc Ginnity, P., Mc Mahon, C., Mala, H., Manenti, S., Manolopoulou, M., Mattila, A., Mauring, A., Mietelski, J.W., Møller, B.S., Nielsen, P., Nikolick, J., Overwater, R.M.W., Palsson, S.E., Papastefanou, C., Penev, I., Pham, M.K., Povinec, P.P., Ramebäck, H., Reis, M.C., Ringer, W., Rodriguez, A., Rulík, P., Saey, P.R.J., Samsonov, V., Schlosser, C., Sgorbati, G., Silobritiene, B.V., Söderström, C., Sogni, R., Solier, L., Sonck, M., Steinhauser, G., Steinkopff, T., Steinmann, P., Stoulos, S., Sykora, I., Todorovic, D., Tooloutalaie, N., Tositti, L., Tschiersch, J., Ugron, A., Vagena, E., Vargas, A., Wershofen, H., Zhukova, O., 2011. Tracking of airborne radionuclides from the damaged Fukushima Dai-Ichi nuclear reactors by European networks. Environ. Sci. Technol. 45, 7670–7677.

Matsunaga, T., Amano, H., Yanase, N., 1991. Discharge of dissolved and particulate ^{137}Cs in the Kuji River, Japan. Appl. Geochem. 6, 159–167.

Matsuzaki, H., Muramatsu, Y., Kato, K., Yasumoto, M., Nakano, C., 2007. Development of ^{129}I-AMS system at MALT and measurements of ^{129}I concentrations in several Japanese soils. Nucl. Instr. Methods Phys. Res. B259, 721–726.

Maximenko, N., Niiler, P., Centurioni, L., Rio, M.-H., Melnichenko, O., Chambers, D., Zlotnicki, V., Galperin, B., 2009. Mean dynamic topography of the ocean derived from satellite and drifting buoy data using three different

techniques*. J. Atmos. Oceanic Technol. 26, 1910–1919. http://dx.doi.org/10.1175/2009jtecho672.1.
McNeary, D., Baskaran, M., 2003. Depositional characteristics of ^{7}Be and ^{210}Pb in southeastern Michigan. J. Geophys. Res. 108 (D7), 4201. http://dx.doi.org/10.1029/2002JD003021.
MEXT, Ministry of Education, Culture, Sports, Science and Technology, 2012a. http://www.mext.go.jp/amenu/saigaijyouhou/index.htm.
MEXT, Ministry of Education, Culture, Sports, Science and Technology, 2012b. Environmental radioactivity database. http://search.kankyo-hoshano.go.jp/servlet/search.top.
MEXT, Ministry of Education, Culture, Sports, Science and Technology, 2012c. Monitoring information of environmental radioactivity levels. http://radioactivity.mext.go.jp/en/ (accessed 10.10.12.).
MHLW, Ministry of Health, Labour and Welfare, 2010. National health and nutrition examination survey in 2008 (in Japanese). Available at: http://www.mhlw.go.jp/bunya/kenkou/eiyou/dl/h20-houkoku-01b.pdf (accessed 20.06.11).
MICROSTEP-MIS, 2012. MICROSTEP-MIS Report, Bratislava.
Mizutani, T., Koarashi, J., Takeishi, M., 2009. Monitoring of low-level radioactive liquid effluent in Tokai reprocessing plant. J. Nucl. Sci. Technol. 46, 665–672.
Miyake, Y., Matsuzaki, H., Fujiwara, T., Saito, T., Yamagata, T., Honda, M., Muramatsu, Y., 2012. Isotopic ratio of radioactive iodine ($^{129}I/^{131}I$) released from Fukushima Daiichi NPP accident. Geochem. J. 46 (4), 327–333.
Miyazawa, Y., Masumoto, Y., Varlamov, S.M., Miyama, T., 2012. Transport simulation of the radionuclide from the shelf to open ocean around Fukushima. Continental Shelf Res. 16–29, 50–51.
Momoshima, N., Sugihara, S., Ichikawa, R., Yokoyama, H., 2012. Atmospheric radionuclides transported to Fukuoka, Japan remote from the Fukushima Dai-ichi nuclear power complex following the nuclear accident. J. Environ. Radioact. 111, 28–32.
Monte, L., Periañez, R., Boyer, P., Smith, J.T., Brittain, J.E., 2009. The role of physical processes controlling the behaviour of radionuclide contaminants in the aquatic environment: a review of state-of-the-art modelling approaches. J. Environ. Radioact. 100, 779–784.
Moore, H.E., Poet, S.E., Martell, E.A., 1973. ^{222}Rn, ^{210}Pb, ^{210}Bi, and ^{210}Po profiles and aerosol residence times versus altitude. J. Geophys. Res. 78, 7065–7075.
Mori, N., Takahashi, T., Yasuda, T., Yanagisawa, H., 2011. Survey of 2011 Tohoku earthquake tsunami inundation and run-up. Geophys. Res. Lett. 38. http://dx.doi.org/10.1029/2011gl049210.
Morino, Y., Ohara, T., Nishizawa, M., 2011. Atmospheric behavior, deposition, and budget of radioactive materials from the Fukushima Daiichi nuclear power plant in March 2011. Geophys. Res. Lett. 38, L00G11.
Morita, T., Fujimoto, K., Kasai, H., Yamada, H., Nishiuchi, K., 2010a. Temporal variations of ^{90}Sr and ^{137}Cs concentrations

and the $^{90}Sr/^{137}Cs$ activity ratio in marine brown algae, *Undaria pinnatifida* and *Laminaria longissima*, collected in coastal areas of Japan. J. Environ. Monit. 12, 1179–1186.
Morita, T., Ohtsuka, Y., Fujimoto, K., Minamisako, Y., Iida, R., Namamura, M., Kayama, T., 2010b. Concentrations of ^{137}Cs, ^{90}Sr, ^{108m}Ag, $^{239+240}Pu$ and atom ratio of $^{240}Pu/^{239}Pu$ in tanner crabs, Chionoecetes japonicus and Chionoecetes opilio collected around Japan. Mar. Poll. Bull. 60, 2311–2322.
Muramatsu, Y., Yoshida, S., 1995. Volatilization of methyl iodine from soil-plant system. Atmos. Environ. 29, 21–25.
Muramatsu, Y., Takada, Y., Matsuzaki, H., Yoshida, S., 2008. AMS analysis of ^{129}I in Japanese soil samples collected from background areas far from nuclear facilities. Quarter. Geochronol. 3, 291–297.
Muramatsu, Y., Yoshida, S., Tanaka, A., 2003. Determination of Pu concentration and its isotope ratio in Japanese soils by HR-ICP-MS. J. Radioanal. Nucl. Chem. 255, 477–480.
Nakano, M., 2004. Long-term assessment model of radionuclides in the oceans (LAMER) – Development and validation of the diffusion model in global oceans. JNC Tech. Rev. 22, 67–76 (in Japanese with English abstracts) JNC, Tokai.
Nakano, M., 2006. Simulation of the advection-diffusion-scavenging processes for ^{137}Cs and $^{239,240}Pu$ in the Japan Sea. In: Povinec, P.P., Sanchez-Cabeza, J.A. (Eds.), Radionuclides in the Environment. Elsevier, Amsterdam, pp. 433–448.
Nakano, M., 2008. LAMER: Long-term Assessment Model of Radionuclides in the Oceans. JAEA-Data/Code 2007-024 (in Japanese). JAEA, Tokai.
Nakano, M., Povinec, P.P., 2003a. Oceanic general circulation model for the assessment of the distribution of ^{137}Cs in the world ocean. Deep-sea Res. II 50, 2803–2816.
Nakano, M., Povinec, P.P., 2003b. Modelling the distribution of plutonium in the Pacific Ocean. J. Environ. Radioact. 69, 85–106.
Nakano, M., Povinec, P.P., 2012. Long-term simulations of the ^{137}Cs dispersion from the Fukushima accident in the world ocean. J. Environ. Radioact. 111, 109–115.
NERH, Nuclear Emergency Response Headquarters, Government of Japan: report of the Japanese government to the IAEA ministerial conference on nuclear safety – the accident at TEPCO's Fukushima nuclear power stations, 7 June, Available at: http://www.iaea.org/newscenter/focus/fukushima/japan-report/ (accessed 10.10.12.).
Nicholson, K.W., 1988. A review of particle resuspension. Atmos. Environ. 22, 2639–2651.
NISA, Nuclear and Industrial Safety Agency, 2011a. Regarding the Evaluation of the Conditions on Reactor Cores of Unit 1, 2 and 3 Related to the Accident at Fukushima Daiichi Nuclear Power Station. Available at:. Tokyo Electric Power Co. Inc. (accessed 20.06.11.) http://www.nisa.meti.go.jp/english/press/2011/06/en20110615-5.pdf.
NISA, Nuclear and Industrial Safety Agency, 2011b. Regarding the discharge of the waste water, of which the concentration of

radioactive materials exceeds the concentration limit by the notification, to the sea. http://www.nisa.meti.go.jp/english/files/en20110416-10.pdf (accessed on 20.06.11.).

NSC, Nuclear Safety Commission, 2011. The equivalent dose at thyroid (trial calculation by SPEEDI from 6 o'clock in 12th March to 0 o'clock in 24th March). Available at: http://www.nsc.go.jp/mext_speedi/0312-0324_in.pdf#page=1 (accessed 20.06.11.).

NSCJ: Nuclear Safety Commission of Japan, 2011. Trial estimation of emission of radioactive materials (I-131, Cs-137) into the atmosphere from Fukushima Dai-ichi nuclear power station, Tokyo. Available at: http://www.nsc.go.jp/NSCenglish/geje/2011%200412%20press.pdf.

Ohno, T., Muramatsu, Y., Miura, Y., Oda, K., Inagawa, N., Ogawa, H., Yamazaki, A., Toyama, C., Sato, M., 2012. Depth profiles of radioactive cesium and iodine released from the Fukushima Daiichi nuclear power plant in different agricultural fields and forests. Geochem. J. 46 (4), 287–295.

Oikawa, S., Song, S.J., Maeyama, T., Kishimoto, T., Tomura, K., Higuchi, H., 2003. Determination of trace metals in squid organs by inductively coupled plasma mass spectrometry and neutron activation analysis. Bunseki Kagaku 52, 551–557.

ORNL, Oak Ridge National Laboratory, 2004. Origine-ARP-2, ORNL.

Oura, Y., Ebihara, M., 2012. Radioactivity concentrations of ^{131}I, ^{134}Cs and ^{137}Cs in river water in the greater Tokyo metropolitan area after the Fukushima Daiichi nuclear power plant accident. Geochem. J. 46 (4), 303–309.

Paatero, J., Vira, J., Siitari-Kauppi, M., Hatakka, J., Holmen, K., Viisanen, Y., 2012. Airborne fission products in the high Arctic after the Fukushima nuclear accident. J. Environ. Radioact. 114, 41–47.

Palcsu, L., Major, Z., Köllő, Z., Papp, L., 2010. Using an ultrapure ^{4}He spike in tritium measurements of environmental water samples by the 3He-ingrowth method. Rapid Commun. Mass Spectrom. 24, 698–704.

Papastefanou, C., Ioannidou, A., Stoulos, S., Manolopoulou, M., 1995. Atmospheric deposition of cosmogenic ^{7}Be and ^{137}Cs from fallout of the Chernobyl accident. Sci. Total Environ. 170, 151–156.

Peirson, D.H., Cawse, P.A., Salmon, L., Cambray, R.S., 1973. Trace elements in the atmospheric environment. Nature 241, 252–256.

Perrot, F., Hubert, Ph., Marquet, Ch., Pravikoff, M.S., Bourquin, P., Chiron, H., Guernion, P.-Y., Nachab, A., 2012. Evidence of ^{131}I and ^{134,137}Cs activities in Bordeaux, France due to the Fukushima nuclear accident. J. Environ. Radioact. 114, 61–65.

Pham, M.K., Betti, M., Povinec, P.P., Alfimov, V., Biddulph, D., Gastaud, J., Kieser, W.E., Lopez Gutierez, J.M., Possnert, G., Sanchez-Cabeza, J.A., Suzuki, T., 2010. Certified reference material IAEA-418: ^{129}I in Mediterranean Sea water. J. Radioanal. Nucl. Chem. 286, 121–127.

Pham, M.K., Betti, M., Nies, H., Povinec, P.P., 2011. Temporal changes of ^{7}Be, ^{137}Cs and ^{210}Pb activity concentrations in surface air at Monaco and their correlation with meteorological parameters. J. Environ. Radioact. 102, 1045–1054.

Pham, M.K., Eriksson, M., Levi, I., Nies, H., Osvath, I., Betti, M., 2012. Detection of Fukushima Dai-ichi nuclear power plant accident radioactive traces in Monaco. J. Environ. Radioact. 114, 131–137.

Pham, K.K., Povinec, P.P., Nies, H., Betti, M., 2013. Dry and wet deposition of ^{7}Be, ^{210}Pb and ^{137}Cs in Monaco air during 1998–2010: seasonal variations of deposition fluxes. J. Environ. Radioact. 120, 45–57.

Piňero García, F., Ferro García, M.A., 2012. Traces of fission products in southeast Spain after the Fukushima nuclear accident. J. Environ. Radioact. 114, 146–151.

Pittauerová, D., Hettwig, B., Fischer, H.W., 2011. Fukushima fallout in Northwest German environmental media. J. Environ. Radioact. 102, 877–880.

PMJC (Prime Minister of Japan and his Cabinet), 2011. Report of Japanese government to the IAEA ministerial conference on nuclear safety-the accident at TEPCO's Fukushima nuclear power stations. http://www.kantei.go.jp/jp/Topics/2011/iaea_houkokusho.html.

Povinec, P., Chudý, M., Sýkora, I., Szarka, J., Pikna, M., Holý, K., 1988. Aerosol radioactivity monitoring in Bratislava following the Chernobyl accident. J. Radioanal. Nucl. Chem. Lett. 126, 467–478.

Povinec, P.P., Oregioni, B., Jull, A.J.T., Kieser, W.E., Zhao, X.-L., 2000. AMS measurements of ^{14}C and ^{129}I in seawater around radioactive waste dump sites. Nucl. Instrum. Methods Phys. Res. B 172, 672–678.

Povinec, P.P., Livingston, H.D., Shima, S., Aoyama, M., Gastaud, J., Goroncy, I., Hirose, K., Huyhn-Ngoc, L., Ikeuchi, Y., Ito, To., La Rosa, J., Liong Wee Kwong, L., Lee, S.-H., Moriya, H., Mulsow, S., Oregioni, B., Pettersson, H., Togawa, O., 2003a. IAEA'97 expedition to the NW Pacific Ocean – results of oceanographic and radionuclide investigations of the water column. Deep-Sea Res. II 50, 2607–2638.

Povinec, P.P., Bailly Du Bois, P., Kershaw, P.J., Nies, H., Scotto, P., 2003b. Temporal and spatial trends in the distribution of ^{137}Cs in surface waters of Northern European Seas – a record of 40 years of investigations. Deep-Sea Res. Part II 50, 2785–2801.

Povinec, P.P., Hirose, K., Honda, T., Ito, T., Scott, E.M., Togawa, O., 2004. Spatial distribution of ^{3}H, ^{90}Sr, ^{137}Cs and $^{239,240}Pu$ in surface waters of the Pacific and Indian Oceans – GLOMARD database. J. Environ. Radioact. 76, 113–137.

Povinec, P.P., Aarkrog, A., Buesseler, K.O., Delfanti, R., Hirose, K., Hong, G.-H., Ito, T., Livingston, H.D., Nies, H., Noshkin, V.E., Shima, S., Togawa, O., 2005a. ^{90}Sr, ^{137}Cs and $^{239,240}Pu$ concentration surface water time series in the Pacific and Indian Oceans – WOMARS results. J. Environ. Radioact. 81, 63–87.

Povinec, P.P., Comanducci, J.F., Levy-Palomo, I., 2005b. IAEA-MEL's underground counting laboratory (CAVE) for the analysis of radionuclides in the environment at very low-levels. J. Radioanal. Nucl. Chem. 263, 441–445.

Povinec, P.P., Scotto, Ph., Osvath, I., Ramadan, H., 2006. The Marine Information System (MARIS). Isotopes in Environmental Studies. Aquatic Forum 2004. IAEA, Vienna, pp. 68–69.

Povinec, P.P., Lee, S.H., Liong Wee Kwong, L., Oregioni, B., Jull, A.J.T., Kieser, W.E., Morgenstern, U., Top, Z., 2010. Tritium, radiocarbon, ^{90}Sr and ^{129}I in the Pacific and Indian Oceans. Nucl. Instrum. Methods Phys. Res. B 268, 1214–1218.

Povinec, P.P., Breier, R., Coppola, L., Groening, M., Jeandel, C., Jull, A.J.T., Kieser, W.E., Top, Z., 2011a. Tracing of water masses using a multi-isotope approach in the southern Indian Ocean. Earth Planetary Sci. Lett. 302, 14–26.

Povinec, P.P., Aoyama, M., Fukasawa, M., Hirose, K., Komura, K., Sanchez-Cabeza, J.A., Gastaud, J., Ješkovský, M., Levy, I., Sýkora, I., 2011b. ^{137}Cs water profiles in the South Indian Ocean – an evidence for accumulation of pollutants in the subtropical gyre. Prog. Oceanogr. 89, 17–30. http://dx.doi.org/10.1016/j.pocean.2010.12.004.

Povinec, P.P., Sýkora, I., Holý, K., Gera, M., Kováčik, A., Brest'áková, L., 2012a. Aerosol radioactivity record in Bratislava/Slovakia following the Fukushima accident – a comparison with global fallout and the Chernobyl accident. J. Environ. Radioact. 114, 81–88.

Povinec, P.P., Sýkora, I., Gera, M., Holý, K., Brest'áková, L., Kováčik, A., 2012b. Fukushima-derived radionuclides in ground-level air of Central Europe: a comparison with simulated forward and backward trajectories. J. Radioanal. Nucl. Chem. http://dx.doi.org/10.1007/s10967-012-1943-3.

Povinec, P.P., Hirose, K., Aoyama, M., 2012c. Radiostronium in the western North Pacific: characteristics, behavior, and the Fukushima impact. Environ. Sci. Technol. 46, 10356–10363.

Povinec, P.P., Holý, K., Chudý, M., Šivo, A., Sýkora, I., Ješkovský, M., Richtáriková, M., 2012d. Long-term variations of ^{14}C and ^{137}Cs in the Bratislava air – implications of different atmospheric transport processes. J. Environ. Radioact. 108, 33–40.

Povinec, P.P., Eriksson, M., Scholten, J., Betti, M., 2012e. Marine radioactivity analysis. In: L'Annunziata, M.F. (Ed.), Handbook on Radioactivity Analysis. Academic Press, Amsterdam, pp. 770–832.

Povinec, P.P., Gera, M., Hirose, K., Lujanienė, G., Nakano, M., Plastino, W., Bartok, J., 2013a. Dispersion of Fukushima radionuclides in the global atmosphere and the ocean. Appl. Rad. Isot. http://dx.doi.org/10.1016/j.apradiso.2013.03.058.

Povinec, P.P., Aoyama, M., Biddulph, D., Breier, R., Buesseler, K., Chang, C.C., Golser, R., Hou, X.L., Ješkovský, M., Jull, A.J.T., Kaizer, J., Nakano, M., Nies, H., Palcsu, L., Papp, L., Pham, M.K., Steier, P., Zhang, L.Y., 2013b. Cesium, iodine and tritium in NW Pacific waters – a comparison of the Fukushima

impact with global fallout. Biogeosciences Discuss 10, 6377–6416. http://dx.doi.org/10.5194/bgd-10-6377-2013.
Qin, H., Yokoyama, Y., Fan, Q., Iwatani, H., Tanaka, K., Sakaguchi, A., Kanai, Y., Zhu, J., Onda, Y., Takahashi, Y., 2012. Investigation of cesium adsorption on soil and sediment samples from Fukushima prefecture by sequential extraction and EXAFS technique. Geochem. J. 46 (4), 297–302.
Raisbeck, G.M., Yiou, F., 1999. ^{129}I in the oceans: origins and applications. Sci. Total Environ. 237/238, 31–41.
Rolph, G.D., 2011. Real-time Environmental Applications and Display system (READY) (http://ready.arl.noaa.gov). NOAA Air Resources Laboratory, Silver Spring, USA.
Rosner, G., Winkler, R., 2001. Long-term variation (1986–1998) of post-Chernobyl ^{90}Sr, ^{137}Cs, ^{238}Pu and ^{239,240}Pu concentrations in air, depositions to ground, resuspension factors and resuspension rates in south Germany. Sci. Total Environ. 273, 11–25.
Rosner, G., Hotzl, H., Winkler, R., 1997. Long-term behaviour of plutonium in air, and deposition and the role of resuspension in semi-rural environment in Germany. Sci. Total Environ. 196, 255–261.
Rowan, D.J., Rasmussen, J.B., 1995. The elimination of radiocesium from fish. J. Appl. Ecol. 32, 739–744.
Rypina, I.I., Jayne, S.R., Yoshida, S., Macdonald, A.M., Douglass, E., Buesseler, K., 2013. Short-term dispersal of Fukushima-derived radionuclides off Japan: modeling efforts and model-data intercomparison. Biogeosciences Discuss. http://dx.doi.org/10.5194/bgd-10-1517-2013.
Sakaguchi, A., Kadokura, A., Steier, P., Tanaka, K., Takahashi, Y., Chiga, H., Matsushima, A., Nakashima, S., Onda, Y., 2012. Isotopic determination of U, Pu and Cs in environmental waters following the Fukushima Daiichi nuclear power plant accident. Geochem. J. 46 (4), 355–360.
Sanchez-Cabeza, J.A., Levy, I., Gastaud, J., Eriksson, M., Osvath, I., Aoyama, M., Povinec, P.P., Komura, K., 2011. Transport of North Pacific ^{137}Cs labeled waters to the south-eastern Atlantic Ocean. Prog. Oceanogr. 89, 31–37. http://dx.doi.org/10.1016/j.pocean.2010.12.005.
Saxén, R., Heinävaara, S., Rask, M., Ruuhijärvi, J., Rand, H., 2010. Transfer of ^{137}Cs into fish in small forest lakes. J. Environ. Radioact. 101, 647–653.
Schmel, G.A., 1980. Particle and gas dry deposition: a review. Atmos. Environ. 14, 983–1011.
Shinn, J.H., Homan, D.N., Gray, D.D., 1983. Plutonium aerosol fluxes and pulmonary exposure rates during re-suspension from bare soils near a chemical separation facility. In: Pruppacher, H.R., Semonin, R.G., Slinn, W.G.N. (Eds.), Precipitation Scavenging, Dry Deposition and Re-suspension. Elsevier, Amsterdam, pp. 1131–1143.
Slinn, W.G.N., 1978. Parameterizations for resuspension and for wet and dry deposition of particles and gases for use in radiation dose calculations. Nucl. Saf. 19, 205–219.

Small, S.H., 1960. Wet and dry deposition of fallout materials at Kjeller. Tellus 12, 308–314.

Stewart, K., 1966. The resuspension of particulate material from surface. In: Fish, B.R. (Ed.), Surface Contamination. Pergamon Press, Oxford, pp. 63–74.

Steel, A.K., 1991. Derived concentration factors for caesium-137 in edible species of North Sea fish. J. Rad. Res. 14, 382–391.

Stohl, A., Seibert, P., Wotawa, G., Arnold, D., Burkhart, J.F., Eckhardt, S., Tapia, C., Vargas, A., Yasunari, T.J., 2011. Xenon-133 and caesium-137 releases into the atmosphere from the Fukushima Dai-ichi nuclear power plant: determination of the source term, atmospheric dispersion, and deposition. Atmos. Chem. Phys. Discuss. 11, 28319–28394.

Stohl, A., Seibert, P., Wotawa, G., Arnold, D., Burkhart, J.F., Eckhardt, S., Tapia, C., Vargas, A., Yasunari, T.J., 2012. Xenon-133 and caesium-137 releases into the atmosphere from the Fukushima Dai-ichi nuclear power plant: determination of the source term, atmospheric dispersion, and deposition. Atmos. Chem. Phys. 12, 2313–2343.

Suplinska, M., Adamczyk, A., 2009. Cs-137, Ra-226 and K-40 in the southern Baltic Sea fish flesh, Helsinki Commission, HELCOM MORS-PRO 14/2009.

Suzuki, Y., Nakahara, R., Ueda, T., 1973. Caesium-137 contamination of marine fishes from the coasts of Japan. J. Radiat. Res. 14, 382–391.

Suzuki, T., Minakawa, M., Amano, H., Togawa, O., 2010. The vertical profiles of iodine-129 in the Pacific Ocean and the Japan Sea before the routine operation of a new nuclear fuel reprocessing plant. Nucl. Instr. Methods Phys. Res. B 268, 1229–1231.

Sýkora, I., Povinec, P.P., Brest'áková, L., Florek, M., Holý, K., Masarik, J., 2012. Resuspension processes control variations of ^{137}Cs activity concentrations in the ground-level air. J. Radioanal. Nucl. Chem. 293, 595–599.

Tagami, K., Uchida, S., 2010. Can elemental composition data of crop leaves be used to estimate radionuclide transfer to tree leaves. Radiat. Environ. Biophys. 49, 583–590.

Tagami, K., Uchida, S., Uchihori, Y., Ishii, N., Kitamura, H., Shirakawa, Y., 2011. Specific activity and activity ratios of radionuclides in soil collected about 20 km from the Fukushima Daiichi nuclear power plant: radionuclide release to the south and southwest. Sci. Total Environ. 409 (22), 4885–4888.

Tagami, K., Uchida, S., Ishii, N., Kagiya, S., 2012. Translocation of radiocesium from stems and leaves of plants and the effect on radiocesium concentrations in newly emerged plant tissues. J. Environ. Radioact. 111, 65–69.

Takemura, T., Nakamura, H., Takigawa, M., Kondo, H., Satomura, T., Miyasaka, T., Nakajima, T., 2011. A numerical simulation of global transport of atmospheric particles emitted from the Fukushima Daiichi nuclear power plant. Sora 7, 101–104.

Tanaka, K., Takahashi, Y., Sakaguchi, A., Umeo, M., Hayakawa, S., Tanida, H., Saito, T., Kanai, Y., 2012. Vertical profiles of iodine-131 and cesium-137 in soils in Fukushima prefecture related to the Fukushima Daiichi nuclear power station accident. Geochem. J. 46 (1), 73–76.
Tateda, Y., Koyanagi, T., 1994. Concentration factors for Cs-137 in marine algae from Japanese coastal waters. J. Rad. Res. 35, 213–221.
Tateda, Y., Koyanagi, T., 1996. Concentration factors for ^{137}Cs in Japanese coastal fish (1984–1990). J. Radiat. Res. 37, 71–79.
TEPCO: Tokyo Electric Power Company, 2011a. http://www.tepco.co.jp/en/press/corp-com/release/11042103-e.html (accessed 20.06.11.).
TEPCO, Tokyo Electric Power Company, 2011b. Countermeasures to outflow of radioactive water off the site near water intake of unit 3 at Fukushima Daiichi nuclear power station. http://www.tepco.co.jp/en/press/corp-com/release/11052102-e.html (accessed 20.06.11.).
TEPCO, Tokyo Electric Power Company, 2012a. Present status of Fukushima Daiichi NPS and implemented countermeasure. http://www.tepco.co.jp/en/nu/fukushima-np/review/review2_1-e.html (accessed 30.09.12.).
TEPCO, Tokyo Electric Power Company, 2012b. Fukushima Daiichi nuclear power station unit 2: countermeasures to stop the outflow of contaminated water and the water amount flowed out into the sea. http://www.tepco.co.jp/en/press/corp-com/release/11042103-e.html (accessed 10.10.12.).
TEPCO, Tokyo Electric Power Company, 2012c. Nuclide analysis results of fish and shellfish (August 2012). www.tepco.co.jp/en/nu/fukushima-np/images/handouts_120821_01-e.pdf.
The-2011-Tohoku-Earthquake-Tsunami-Joint-Survey-Group, 2011. Nationwide field survey of the 2011 off the Pacific coast of Tohoku earthquake tsunami. J. Jpn Soc. Civil Eng. Ser. B 67, 63–66.
Todd, J.F., Wong, G.T.F., Olsen, C.R., Larsen, I.L., 1989. Atmospheric depositional characteristics of beryllium-7 and lead-210 along southeastern Virginia coast. J. Geophys. Res. 94, 11106–11116.
Topcuoglu, S., Güngör, N., Köse, A., Varinlioglu, A., 1997. Translocation and depuration of ^{137}Cs in tea plants. J. Radioanal. Nucl. Chem. 218, 263–266.
Tsukada, H., Hasegawa, H., Hisamatsu, S., Yamasaki, S., 2002. Transfer of ^{137}Cs and stable Cs from paddy soil to polished rice in Aomori, Japan. J. Environ. Radioact. 59, 351–363.
Tsumune, D., Aoyama, M., Hirose, K., Bryan, F.O., Lindsay, K., Danabasoglu, G., 2011. Transport of ^{137}Cs in the Southern Hemisphere in an ocean general circulation model. Prog. Oceanogr. 89, 38–48.
Tsumune, D., Tsubono, T., Aoyama, M., Hirose, K., 2012. Distribution of oceanic ^{137}Cs from the Fukushima Dai-ichi nuclear power plant simulated numerically by a regional ocean model. J. Environ. Radioact. 111, 100–108.

Turekian, K.K., Benninger, L.K., Dion, E.P., 1983. ^{7}Be and ^{210}Pb total deposition fluxes at New Haven, Connecticut and at Bermuda. J. Geophys. Res. 88, 5411–5415.

Underwood, B.Y., 2001. Review of Deposition Velocity and Washout Coefficient. AEA Technology, Harwell.

UNSCEAR, United Nations Scientific Committee on the Effects of Atomic Radiation, 1993. Sources and Effects of Ionizing Radiation. Report to the General Assembly. United Nations, New York, p. 922.

UNSCEAR, United Nations Scientific Committee on the Effects of Atomic Radiation, 2000. Sources and Effects of Ionizing Radiation, vol. 1: Sources, United Nations, New York, p. 654.

UNSCEAR, United Nations Scientific Committee on the Effects of Atomic Radiation, 2008. Sources and Effects of Ionizing Radiation. Report to the General Assembly. United Nations, New York, USA.

Wallner, G., Steier, P., Brandl, T., Friesacher, M.E., Hille, P., Kutschera, W., Tatzber, M., Ayromlou, S., 2007. Developments toward the measurement of I-129 in lignite. Nucl. Instr. Meth. B 259, 714–720.

Watanabe, T., Tsuchiya, N., Oura, Y., Ebihara, M., Inoue, C., Hirano, N., Yamada, R., Yamasaki, S., Okamoto, A., Watanabe-Nara, F., Nunohara, K., 2012. Distribution of artificial radionuclides (^{110m}Ag, ^{129m}Te, ^{134}Cs, ^{137}Cs) in surface soils from Miyagi prefecture, northeast Japan, following the 2011 Fukushima Dai-ichi nuclear power plant accident. Geochem. J. 46 (4), 279–285.

WMO, World Meteorological Organization, 2009. Manual on Codes. International Codes. vol. I.2. Part B – Binary Codes. Part C – Common Features to Binary and Alphanumeric Codes. WMO-No. 306. WMO, Geneva, p. 740.

Yamamoto, M., Takada, T., Nagao, S., Koike, T., Shimada, K., Hoshi, M., Zhumadilov, K., Shima, T., Fukuoka, M., Imanaka, T., Endo, S., Sakaguchi, A., Kimura, S., 2012. An early survey of the radioactive contamination of soil due to the Fukushima Dai-ichi nuclear power plant accident, with emphasis on plutonium analysis. Geochem. J. 46 (4), 341–353.

Yonezawa, C., Yamamoto, Y., 2011. Measurements of artificial radionuclides in surface air by CTBTO. Bunseki, 451–458 (in Japanese).

Yoshida, N., Takahashi, Y., 2012. Land-surface contamination by radionuclides from the Fukushima Daiichi nuclear power plant accident. Elements 8, 201–206.

Young, J.A., Silker, W.B., 1980. Aerosol deposition velocities on the Pacific and Atlantic Oceans calculated from ^{7}Be measurements. Earth Planet. Sci. Lett. 50, 92–104.

Závodský, D., 2011. Atmospheric Chemistry and Air Pollution Modelling. Leonardo da Vinci Programme. Matej Bel University, Banská Bystrica, p. 127.

Zhang, W., Bean, M., Benotto, M., Cheung, J., Ungar, K., Ahier, B., 2011. Development of a new aerosol monitoring system and its application in Fukushima nuclear accident related aerosol radioactivity measurement at the CTBT radionuclide station in Sidney of Canada. J. Environ. Radioact. 102, 1065–1069.

Zheng, J., Tagami, K., Watanabe, Y., Uchida, S., Aono, T., Ishii, N., Yoshida, S., Kubota, Y., Fuma, S., Ihara, S., 2012. Isotopic evidence of plutonium release into the environment from the Fukushima DNPP accident. Scientific Rep. 2, 304. http://dx.doi.org/10.1038/step00304.

6

PRE-FUKUSHIMA RADIOACTIVITY OF THE ENVIRONMENT

CHAPTER OUTLINE

6.1 World Atmosphere

6.1.1 Atmospheric Tests of Nuclear Weapons

Anthropogenic radionuclides were initially injected into the atmosphere by atmospheric nuclear explosions at Alamogordo (New Mexico, USA), Hiroshima and Nagasaki (Japan) in 1945. During the period of 1945–1980, 541 atmospheric nuclear weapons tests had been conducted by the United States, the former USSR, China, France and the United Kingdom (UNSCEAR, 1982). The time-series

Fukushima Accident. http://dx.doi.org/10.1016/B978-0-12-408132-1.00006-1

record of annual atmospheric nuclear explosion yields is shown in Fig. 6.1. The major atmospheric nuclear explosions were conducted during 1951–1958 and 1961–1962. Within the atmospheric nuclear testing, the thermonuclear test in 1954 (Bravo test: 15 Mt (millions of tons) of TNT equivalent), conducted in Bikini by USA, was seriously suffered in fishermen (Lucky Dragon) and marine environments by radioactive fallout. The largest thermonuclear teat of 50 Mt was conducted in Novaya Zemlya in 1962 by the former Soviet Union. The cumulative amount of the total yields of atmospheric nuclear explosions reached 150 Mt until 1963 as shown in Fig. 6.2. Large-scale atmospheric nuclear testing carried out by the United States and the former USSR was stopped due to signing the Limited Test Ban Treaty in 1963. However, the atmospheric nuclear tests were continued by China and France, which were conducted in the Lop Nor (China) and Mururoa (French Polynesia), respectively. The last atmospheric nuclear test was conducted in 1980 by China. The total explosion yield due to the atmospheric nuclear testing until 1980 is 161 Mt.

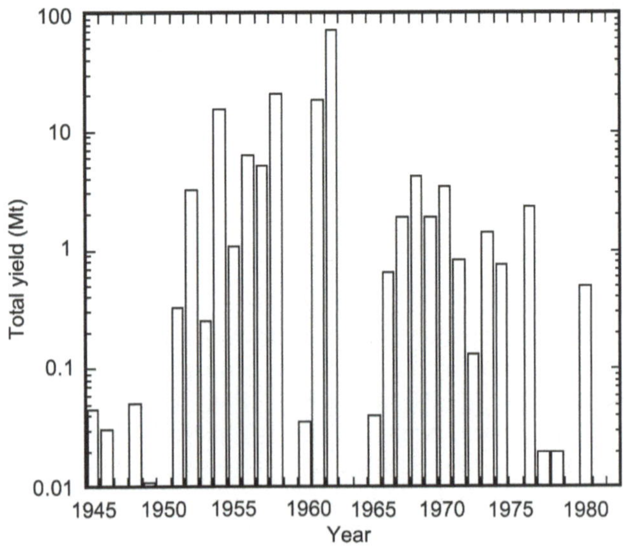

Figure 6.1 Annual total yields of atmospheric nuclear explosions. (Data from UNSCEAR (2000))

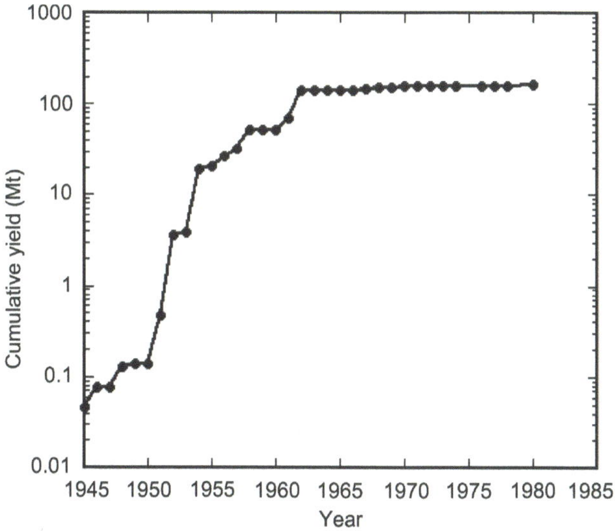

Figure 6.2 Cumulative amount of atmospheric nuclear explosion yield. (Data from UNSCEAR (2000))

Anthropogenic radionuclides released from atmospheric nuclear tests are spread over the globe and are almost globally detectable in the environment (UNSCEAR, 1993). The typical radionuclides produced by atmospheric nuclear testing, which involved fission products (^{137}Cs and ^{90}Sr), fissile materials (plutonium) and neutron activation nuclides, are summarized in Table 6.1. The dominant source of anthropogenic radionuclides in the current environment is the atmospheric nuclear weapons testing. The monitoring of fallout radioactivity started during 1954–1957 at several worldwide monitoring networks. The radionuclide measurements in the deposition samples were performed in USA (HASL, 1976), the UK (Crooks et al., 1959), Japan (Katsuragi, 1983) and Denmark (Aarkrog and Lippert, 1959).

The temporal variation of the annual ^{137}Cs and $^{239,240}Pu$ depositions observed in Japan is shown in Fig. 6.3. A peak of annual depositions of anthropogenic radionuclides occurred in 1963 in the northern hemisphere stations, just after the 1961–1962 large-scale atmospheric nuclear testing conducted by the US and the former USSR, whereas maximum

Table 6.1. Total Amounts of Globally Dispersed Radionuclides from Atmospheric Nuclear Testing (after UNSCEAR, 2000)

Radionuclide	Fission Yield (%)	Normalized Production (PBq/Mt)	Global Release (EBq)
^{3}H		740	186
^{14}C		0.85	0.213
^{89}Sr	3.17	730	117
^{90}Sr	3.50	3.88	0.622
^{91}Y	3.76	748	120
^{95}Zr	5.07	921	148
^{103}Ru	5.20	1540	247
^{106}Ru	2.44	76.0	12.2
^{125}Sb	0.40	4.62	0.741
^{131}I	2.90	4210	675
^{140}Ba	5.18	4730	759
^{141}Ce	4.58	1640	263
^{144}Ce	4.69	191	30.7
^{137}Cs	5.57	5.90	0.948

The total releases of ^{239}Pu, ^{240}Pu and ^{241}Pu were estimated to be 6.52, 4.35 and 1.42 PBq, respectively (from the ratios to ^{90}Sr).

deposition in the southern hemisphere stations appeared in 1964. After the Limited Test Ban Treaty in 1963, the depositions of the anthropogenic radionuclides decreased with a half-life of about 1 year. The ^{137}Cs deposition in Japan was maintained at the level of 10–100 Bq/m^2 year during the 1970s due to the atmospheric nuclear tests by China and France. After the Chinese nuclear test in 1980, a peak of the ^{137}Cs deposition appeared in 1981. After occurrence of the maximum deposition in 1981, the ^{137}Cs deposition deceased with a half-life of about 1 year and reached the lowest level in 1985 since beginning of the radioactive monitoring in 1957. In the 1970s and the early 1980s, the atmospheric levels of

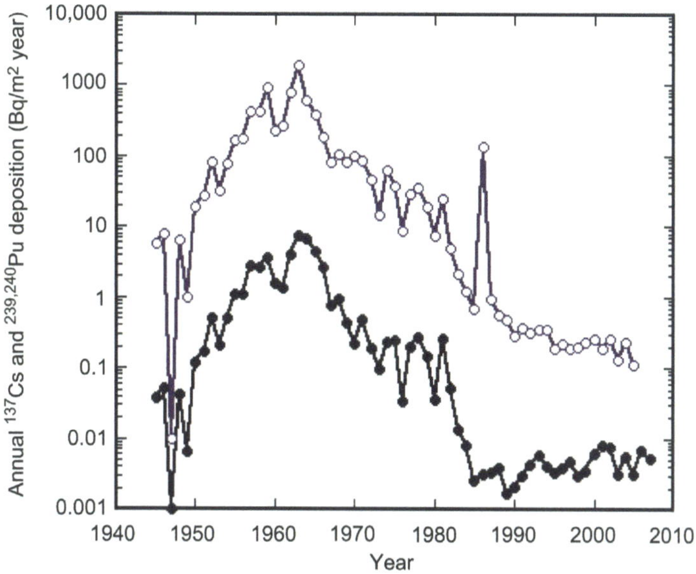

Figure 6.3 Temporal variations of annual ^{137}Cs and ^{239,240}Pu depositions in Japan. 1945–1956: reconstruction from annual explosion yield, ice core profile and soil inventories. 1957 onwards: observed values at Meteorological Research Institute (1957–1980: Koenji, Tokyo, 1980 onwards: Tsukuba).

anthropogenic radionuclides were dominantly supported by China and France nuclear explosions.

The geographical distribution of the radionuclides from global fallout was initially provided as the latitudinal distribution of the cumulative ^{90}Sr deposition by UNSCEAR (2002). However, many questions have arisen during 1970–2010; e.g. the water column inventory of ^{137}Cs in the North Pacific Ocean was two to three times higher than the cumulative decay corrected fallout at the same latitude as stated in the United Nations Scientific Committee on the Effects of Atomic Radiation (UNSCEAR) reports (UNSCEAR, 2002, 2008). Aoyama et al. (2006) showed more precise spatial distribution of global ^{137}Cs fallout primarily on the basis of global measurements in rain, seawater and soil (data in $10° \times 10°$ grids). A typical feature of the geographical distribution is that two high global ^{137}Cs fallout areas exist in the northern hemisphere as shown in Fig. 6.4, where the highest ^{137}Cs fallout was observed in the globe. These high-fallout areas correspond to crossovers of areas where larger precipitation amounts were expected

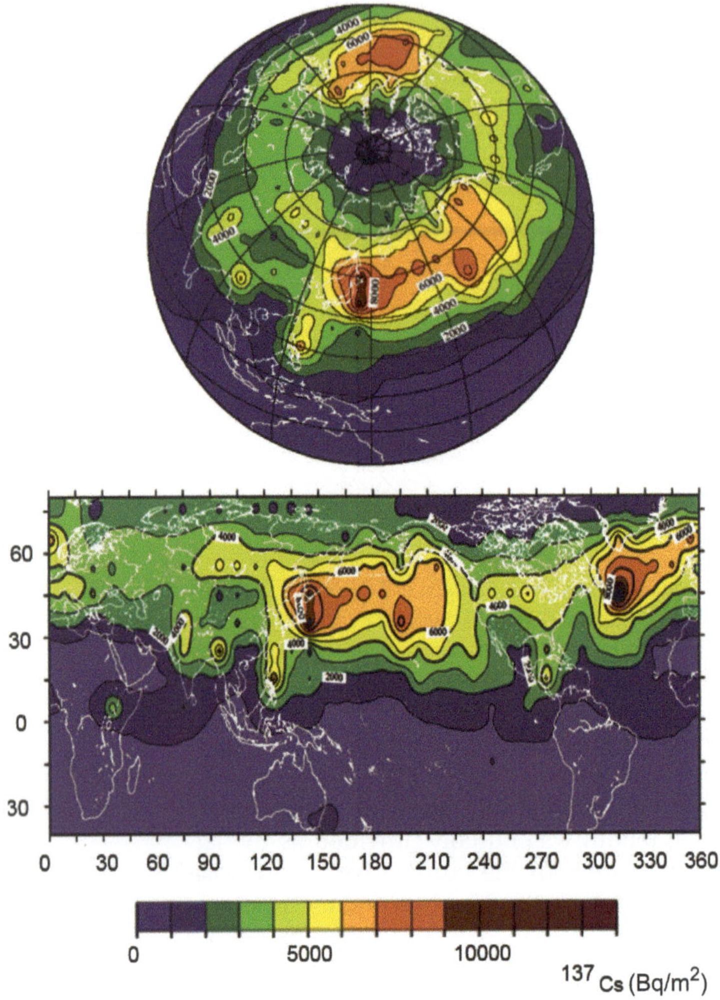

Figure 6.4 Geographical distribution of the cumulative ^{137}Cs deposition in globe in 1970. (Adapted from Aoyama et al. (2006))

and where higher stratosphere–troposphere exchange was expected. A new estimate of 765 ± 79 PBq as global ^{137}Cs fallout for the northern hemisphere is 1.4 times higher than that of 545 PBq in the UNSCEAR reports (UNSCEAR, 2002, 2008).

Anthropogenic radionuclides originating from the atmospheric nuclear explosions were injected into stratosphere and upper troposphere. The injection altitudes were dependent on the scale and height of nuclear explosion (Reiter and Bauer, 1975). Figure 6.5

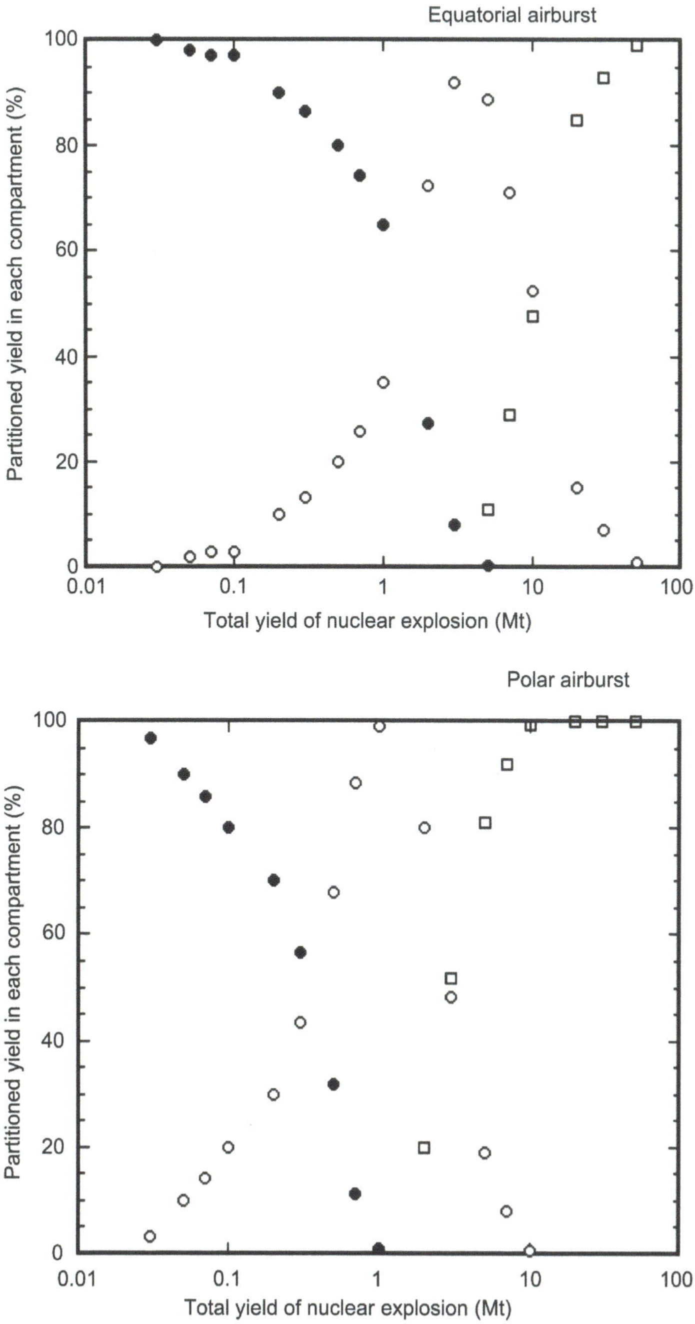

Figure 6.5 The relationship between nuclear explosion yield and injection height of nuclear debris (Data from UNSCEAR, 2000). Equatorial airburst (0°–30° latitude), atmospheric heights: troposphere <17 km, lower stratosphere 17–24 km, and upper stratosphere 24–50 km. Polar airburst (30–90° latitude), atmospheric heights: troposphere <9 km, lower stratosphere 9–17 km, and upper stratosphere 17–50 km

shows the portioning of explosion yield between compartments (troposphere, lower stratosphere and upper stratosphere) for equatorial and polar airbursts. Although the portioning of radionuclides injected in the upper atmosphere depended on the location of airburst, most of the anthropogenic radionuclides were injected into the stratosphere for the thermonuclear explosions of >1 Mt conducted in atmosphere. The stratosphere had been of scientific interest because of major injections of radionuclides due to the nuclear explosions and the possibility of tracing stratospheric transport of bomb-derived radionuclides. The stratosphere aerosol layer was established (Junge et al., 1961).

Anthropogenic radionuclides produced by the nuclear explosion immediately attach to submicrometer aerosols, although several fission products initially are present in gaseous forms (e.g. ^{137}Xe (half-life: 3.82 min) → ^{137}Cs). Stratospheric sampling of anthropogenic radionuclides revealed that anthropogenic radioactivity derived from the atmospheric nuclear testing persists in the stratosphere on a time scale of years and that radionuclides were associated with particles below 0.02 μm radius above 27 km, and with particles nearly 0.1 μm radius between 21 km and tropopause (Martell, 1966). The change of the particle size distribution of radionuclide-bearing particles reflects the change of natural aerosols in stratosphere. The sulfate aerosol layer with submicrometer radius exists at around 21 km altitude (Junge et al., 1961). For troposphere circulation, trade wind circulation in the lower latitude was successfully traced earlier based on the dispersion of radionuclides emitted from the nuclear weapon tests at or near the earth's surface near the equator in 1952 and 1954 (Machta et al., 1956). The westerly circulation in the mid-latitude was traced by using radioactivity measurements and the air trajectory analysis for the Chinese nuclear explosion in 1965 (Kuroda et al., 1965).

To have better and conceptual understanding of natural processes, classical compartment models (Krey and Krajewksi, 1970) are effective rather than computer model simulation; corresponding atmospheric residence times are important values to constraint time scale of aerosols existing in the compartment. Land-based sampling of anthropogenic radionuclides can provide important information on stratospheric motion of aerosols. Long-range monitoring of ^{90}Sr, ^{137}Cs, $^{239,240}Pu$ and ^{238}Pu revealed that their temporal variations reflect global air motion of radionuclide-bearing particles in stratosphere (Hirose et al., 1987). The long-term measurements of anthropogenic radioactivity in surface aerosols and deposition allowed us to have information on transport processes and residence times of aerosols in the upper troposphere and the stratosphere (Ehhalt, 1973; Hirose et al., 1987; Katsuragi, 1983; Reiter and Bauer, 1975). Long-term observation of atmospheric radionuclides in the northern hemisphere mid-latitude region indicated that the annual deposition of radioactive debris from the thermonuclear tests varied with apparent stratospheric residence times of 0.5–1.7 years, which suggested that the three layers with different time scales—that is, upper stratosphere, lower stratosphere below 21 km and active mixing and exchange (AME) layer just above troposphere—exist in stratosphere. The half transport rates from the upper to lower stratosphere compartment, from the lower stratosphere and the AME layer compartment, and from the AME layer to troposphere are 0.5, 0.7 and 0.3 years, respectively (Hirose et al., 1987). The lower stratosphere with the longest half residence time corresponds to the stratospheric sulfate layer.

After the 1990s, the anthropogenic radionuclides observed in surface air and rainwater over the globe are considered to have been derived from the resuspension of radionuclides deposited on the land surface (Arimoto et al., 2005; Hirose et al., 2003, 2007; Karlsson et al., 2008; Rosner and Winkler, 2001). After more than 10 years since the cessation of the atmospheric nuclear testing, the contribution of stratospheric fallout of ^{137}Cs, ^{90}Sr and $^{239,240}Pu$ derived from nuclear tests is negligible. It is likely that the anthropogenic radionuclides and their activity ratios in surface air and rainwater reflect their redistribution processes on the land surface. Especially, regional transport of soil dust from the desert and arid regions, e.g. the Sahara dust in Europe (Lee et al., 2002; Masson et al., 2010; Pham et al.,

2005, 2011, 2013) or the Asian dust in East Asia (Fujiwara et al., 2007; Igarashi et al., 2001, 2003, 2005, 2009; Hirose et al., 2003, 2004, 2010), has been an important factor for resuspension of the anthropogenic radionuclides. Although levels of dominant anthropogenic radionuclides in surface air in the 1990s and 2000s were supported by resuspension, atmospheric behavior differs among the anthropogenic radionuclides. In fact, ^{90}Sr and ^{137}Cs deposition fluxes observed in Japan since 1990 have been decreasing slowly, whereas $^{239,240}Pu$ deposition has been almost constant since 1985, although there is interannual variability of the $^{239,240}Pu$ deposition (Fig. 6.6). The monthly anthropogenic radionuclide depositions in East Asia exhibited seasonal changes, with a maximum in spring (Hirose et al., 2003, 2008; Igarashi et al., 2001, 2003); especially, $^{239,240}Pu$ deposition showed the most typical seasonal change among ^{137}Cs, ^{90}Sr, and $^{239,240}Pu$ (Hirose et al., 2003,

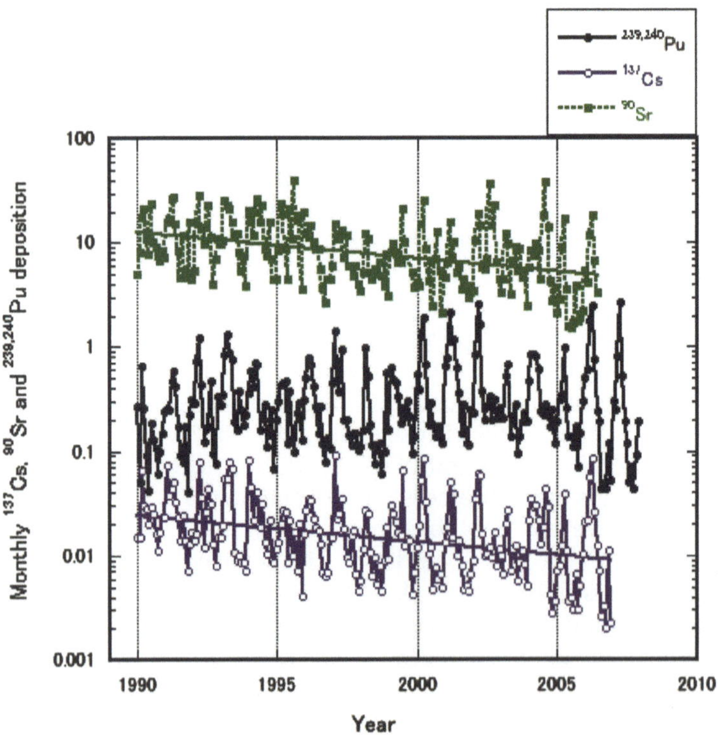

Figure 6.6 Monthly ^{137}Cs, ^{90}Sr and $^{239,240}Pu$ depositions observed at Tsukuba (Japan) since 1990. Unit: Bq/m^2 (^{137}Cs), mBq/m^2 (^{90}Sr), and mBq/m^2 ($^{239,240}Pu$).

2008). The annual and seasonal changes in $^{239,240}Pu$ deposition coincide with the occurrence of the Kosa (Asian dust) event observed in spring (typically, March and April) in Japan (Hirose et al., 2007). These findings suggest that the major origin of resuspension of the anthropogenic radioactive aerosols in Japan in the 1990s and 2000s is aeolian dust produced in the East Asian deserts and arid areas (Hirose et al., 2003; Igarashi et al., 2001).

The situation in Europe has been, however, different. While from time to time the Saharan dust events dominated in the ^{137}Cs activity of aerosols, as is documented in Fig. 6.7 (Masson et al., 2010; Pham et al., 2005, 2011, 2013), during most of the 2000s, a resuspension of ^{137}Cs from soil (which was introduced there by global fallout and during the Chernobyl accident) was the main source of observed ^{137}Cs variations in the atmosphere (as illustrated in Fig. 6.8). While before the 1990s the main source of ^{137}Cs in the atmosphere was global fallout causing summer ^{137}Cs maxima due to the transport of lower stratosphere air masses to the ground level, after the Chernobyl accident the main source became soil resuspension of both global fallout and Chernobyl deposited ^{137}Cs, represented by ^{137}Cs maxima observed in winter months (Fig. 6.9) (Povinec et al., 2012).

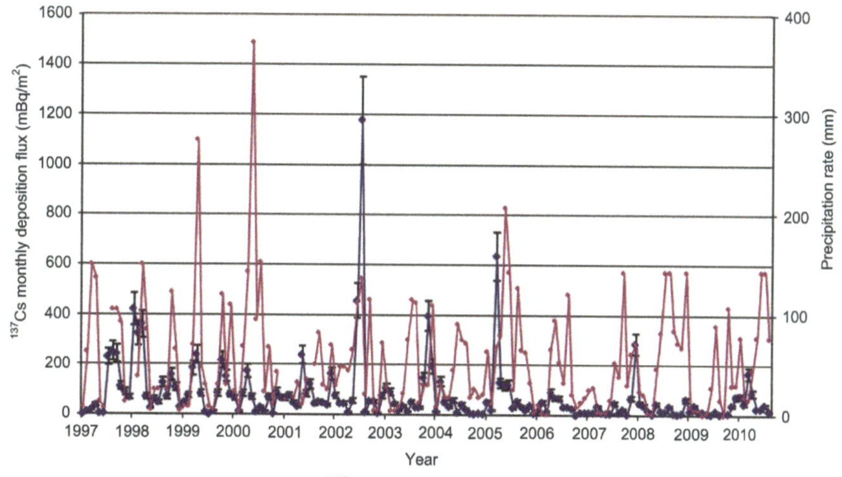

Figure 6.7 Recent measurements of ^{137}Cs monthly deposition fluxes in Monaco. The dominant peak observed in 2002 was due to the Saharan dust event. (Adapted from Pham et al. (2013))

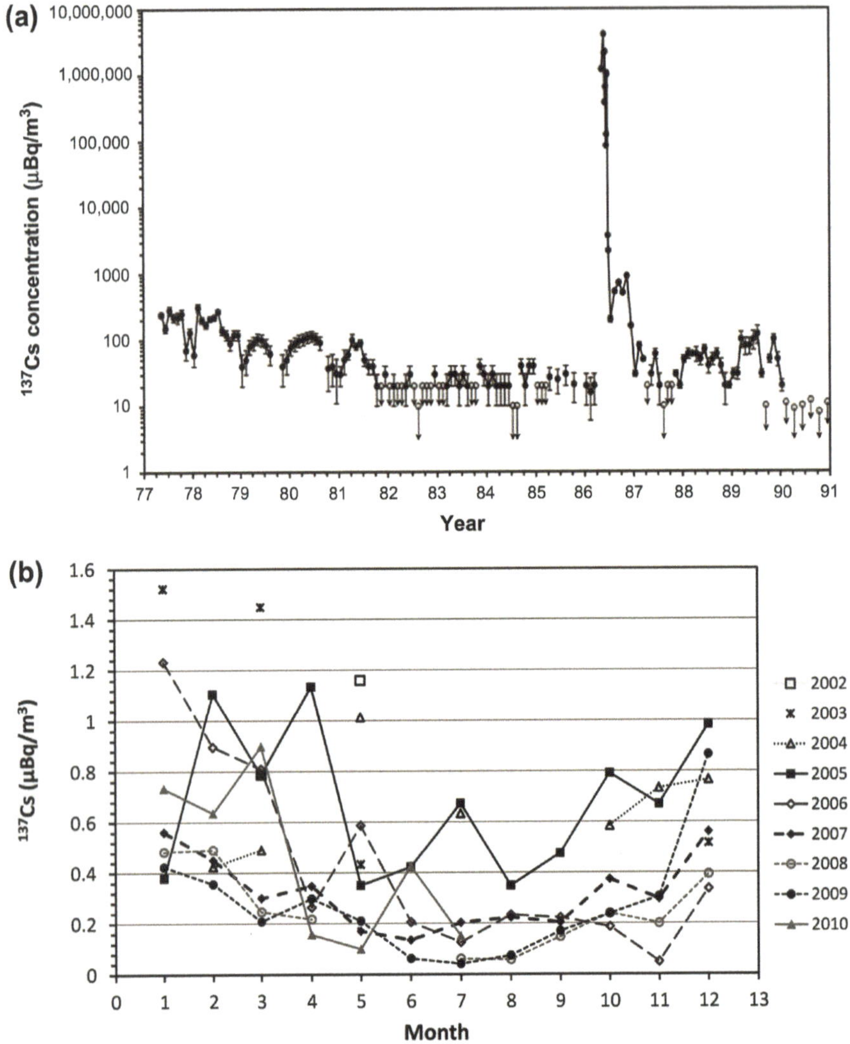

Figure 6.8 (a) Detailed ^{137}Cs monthly records in the Bratislava air (Slovakia, Central Europe) during the last decades. (b) Monthly averages of ^{137}Cs activity concentrations for different years during the last decade (notice higher ^{137}Cs levels during winter months). (Adapted from Povinec et al. (2012))

6.1.2 Chernobyl Accident

The Chernobyl accident occurred on 26 April 1986 at around 1:30 with melting of the nuclear reactor unit No. 4 of the Chernobyl Nuclear Power Plant (NPP) (formerly Soviet Union and presently Ukraine) (Fig. 6.10). The NPP consisted of four units housing

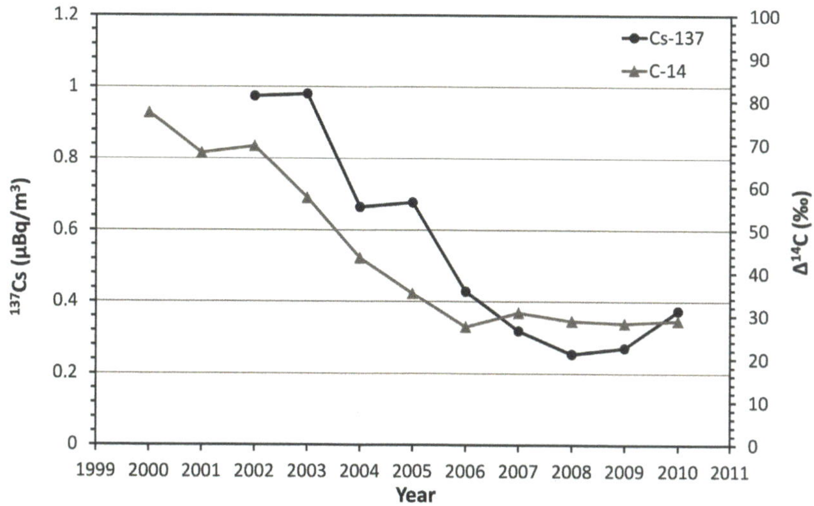

Figure 6.9 Yearly averages of ^{14}C and ^{137}Cs concentrations in the Bratislava air (notice almost constant values during the recent years documenting equilibrium between the deposition and release of radionuclides). (Adapted from Povinec et al. (2012))

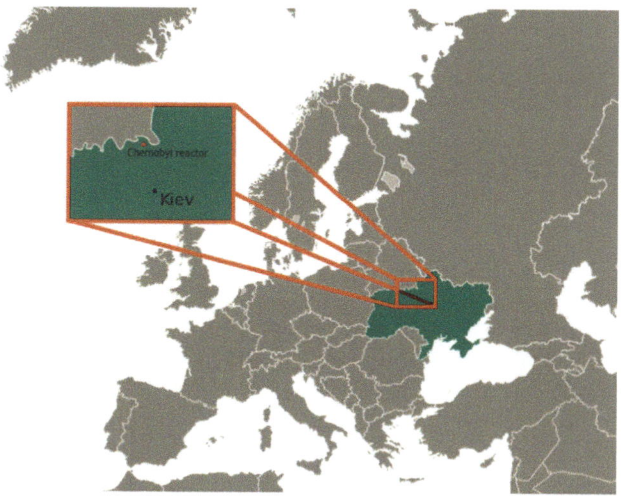

Figure 6.10 Location of the Chernobyl NPP in Ukraine.

RBMK-1000 reactors of 1000 MWe. The reactor No. 4 has been heavily damaged during the accident (Fig. 6.11).

Because of the specific meteorological conditions, the first radioactive clouds went in northwest direction affecting Ukraine, Belarus, Lithuania, Sweden,

Figure 6.11 Aerial view of the damaged reactor No. 4 of the Chernobyl NPP. (Source: Reprinted from Wikipedia)

Finland, and so on. Later, after 29 April, several waves of air masses containing fission and activation products contaminated the Central and West Europe, as well as the global atmosphere (IAEA, 2003, 2006; UNSCEAR, 2008). The most contaminated areas were mostly within the (later declared) 30 km zone around the Chernobyl NPP, but also areas in the north Belarus and Russia. The European countries such as Sweden and Finland also got high ^{137}Cs deposition rates. Other countries (e.g. Austria, but even Scotland) due to rains, which occurred during the passage of radioactive clouds, got higher surface ^{137}Cs contamination than the rest of Europe (Lujanienė et al., 2012b; Fig. 6.12 and Table 6.2).

The radionuclide releases from the damaged nuclear reactor were controlled by their boiling points, and the majority of the radioactivity present

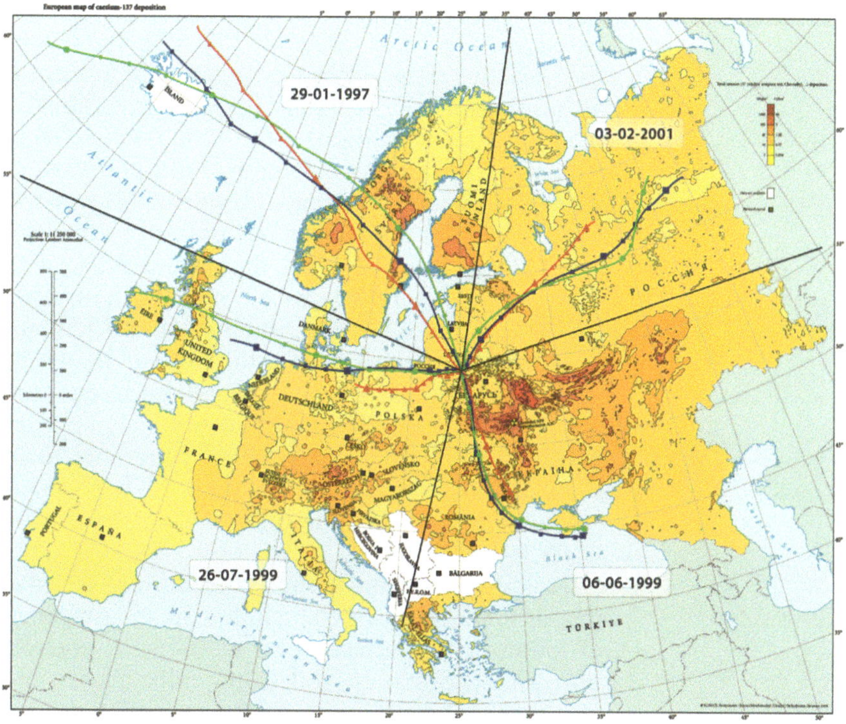

Figure 6.12 Chernobyl and global fallout ^{137}Cs deposition over the Europe (scale: 1–1500 kBq/m^2). (Modified after Lujanienė et al. (2012b))

Table 6.2. The Most Contaminated Areas (in square kilometers) by ^{137}Cs Fallout after the Chernobyl Accident (after IAEA (2006))

Country	37–185 kBq/m^2	185–555 kBq/m^2	555–1480 kBq/m^2	>1480 kBq/m^2
Russia	49,800	5700	2100	300
Belarus	29,900	10,200	4200	2200
Ukraine	37,200	3200	900	600
Sweden	12,000			
Finland	11,500			

in the core was retained in the reactor. All the noble gases, such as ^{85}Kr, ^{133}Xe and ^{135}Xe, contained within the reactor were released immediately into the atmosphere by the first steam explosion. Around 55% of the radioactive iodine (about 1760 PBq of ^{131}I) was released, as a mixture of vapor, solid particles, and organic iodine compounds. About 85 PBq of ^{137}Cs and ^{132}Te were released in aerosol form (Tables 6.1 and 6.2). It was estimated that about 3.5% of fuel material was released to the environment, representing about 6 t of fragmented fuel. The total atmospheric release was estimated to be around 5200 PBq. Most of the particles had the aerodynamic diameter from 0.3 to 1.5 μm, however, large particles of 10 μm were also found, which contained mainly nonvolatile elements (such as ^{95}Zr, ^{95}Nb, ^{140}La, and ^{144}Ce, but also transuranics) (IAEA, 2003, 2006; UNSCEAR, 2008).

An extensive monitoring program was carried out in many countries which included the measurements of radionuclides attached on aerosols, in precipitation, grass, vegetation, soil, water, milk, cheese, meat and other food products. As the radionuclide levels were high, mostly nondestructive gamma-ray spectrometry was used for radionuclide analyses.

Several radionuclides were detected on aerosol filters, e.g. ^{95}Zr, ^{95}Nb, ^{99}Mo, ^{99m}Tc, ^{103}Ru, ^{103}Rh, ^{106}Ru, ^{106}Rh, ^{125}Sb, ^{131}I, ^{132}I, ^{132}Te, ^{134}Cs, ^{137}Cs, ^{140}Ba, ^{140}La, ^{141}Ce, and ^{144}Ce (e.g. Povinec et al., 1988). The dominant radionuclides for example in the Bratislava surface air (Central Europe) were ^{132}Te + ^{132}I, ^{131}I, ^{103}Ru + ^{103}Rh and ^{137}Cs with the maximum measured concentrations on 30 April in the afternoon and being 98.2, 14.3 and 6.2 Bq/m^3, respectively (Povinec et al., 1988).

Three dominant peaks were observed in the radionuclide records, indicating that radioactive clouds reached the Central Europe territory three times. A comparison of pre-Chernobyl and post-Chernobyl ^{137}Cs activity concentrations in the Bratislava air (Povinec et al., 2012; Fig. 6.8) indicates that during the Chernobyl time, its concentration increased by about 200,000 times. The peak level was about factor of 10 higher than the global fallout peak observed in 1963, due to fallout after large-scale atmospheric nuclear weapons tests carried out in 1961–1962 at Novaya Zemlya Island in the Kara Sea. The low ^{137}Cs concentrations were observed again a few years after the Chernobyl accident, however, without typical summer maxima observed in early 1980s, which were due to a transport of global fallout ^{137}Cs from the lower stratosphere to the upper troposphere. The ^{137}Cs concentrations observed in the 1990s and later in the ground-level air were due to a resuspension of the global fallout and Chernobyl cesium from the soil (Fig. 6.13) (Povinec et al., 2012; Pham et al., 2011, 2013; Sýkora et al., 2012).

A long record of $^{239,240}Pu$ monthly activity concentrations is available for Vilnius (Lujanienė et al., 2012b), which ranged from 0.9 to 300 nBq/m^3 (Fig. 6.14). The highest $^{239,240}Pu$ activity concentration within the 1995–2011 record with the $^{240}Pu/^{239}Pu$ atom ratio close to the Chernobyl value was found in 1995, and it was attributed to the transport of "hot" particles from the areas contaminated after the Chernobyl accident.

Figure 6.15 shows a detailed $^{239,240}Pu$ and ^{241}Am monthly record for 2005–2006. The $^{239,240}Pu$ activity concentrations ranged during this period from 2.2 to 49 nBq/m^3, ^{241}Am activities varied from 0.3 to 25 nBq/m^3. The $^{241}Am/^{239,240}Pu$ ratio ranged from 0.19 ± 0.02

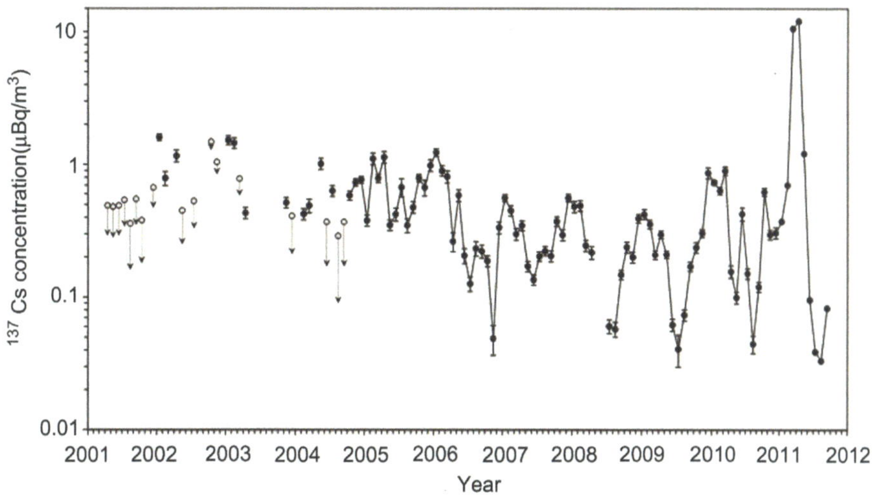

Figure 6.13 Pre-Fukushima measurements of ^{137}Cs activity on aerosols in the Bratislava air (Slovakia, Central Europe). (Adapted from Povinec et al. (2012b))

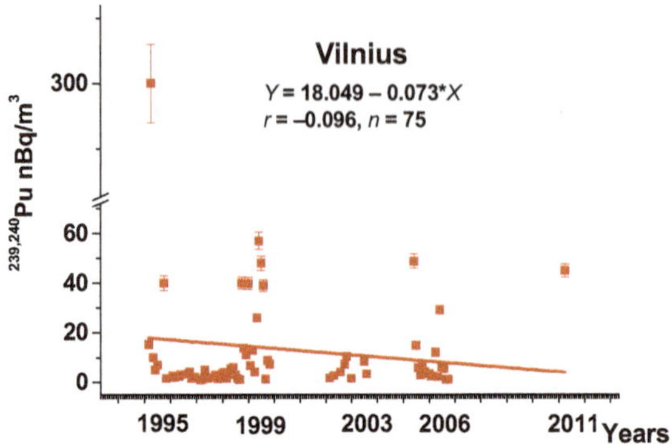

Figure 6.14 Pu activity record on aerosols collected in Vilnius, Lithuania. (Adapted from Lujanienė et al. (2012a))

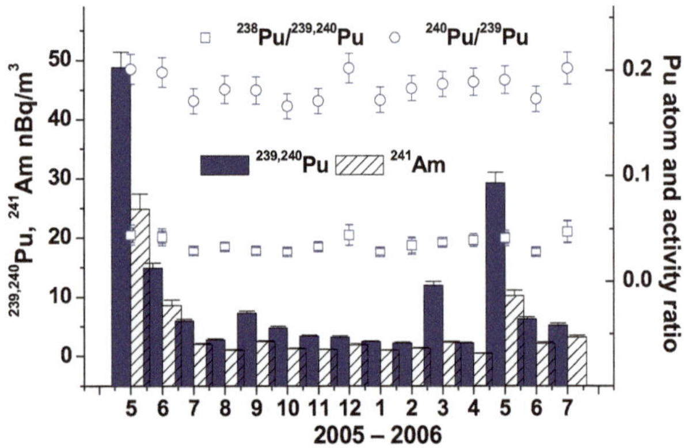

Figure 6.15 $^{239,240}Pu$ and ^{241}Am activities of aerosols collected in Vilnius, Lithuania. (Adapted from Lujanienė et al. (2012a))

to 0.65 ± 0.06 with the mean value of 0.40. The $^{238}Pu/^{239,240}Pu$ activity ratio ranged from 0.028 ± 0.02 to 0.042 ± 0.03 (a mean 0.033 ± 0.02, a typical value for global fallout plutonium), and the $^{240}Pu/^{239}Pu$ atom ratio varied from 0.164 ± 0.015 to 0.207 ± 0.006 (a mean value of 0.185 ± 0.021, also typical for global fallout, 0.180 ± 0.007). These data are in good agreement with results obtained in Prague (Hölgye, 2008).

The absence of correlations of the $^{239,240}Pu$ activity concentration total suspended particle (TSP) concentration ($r = 0.24$ and $n = 49$) in Vilnius in 1995–1999 could indicate that Pu did not originate from a local resuspension of soil particles. However, the results obtained in 2005–2006 showed a different tendency as compared to the 1995–1999 period. There is a good correlation between the $^{239,240}Pu$ activity concentration and total suspended particle (TSP) ($r = 0.83$, $n = 15$, and $p > 0.99$), as well as mineral dust (MD) concentration ($r = 0.66$, $n = 15$, and $p > 0.99$) (Lujanienė et al., 2012b).

The following hypothesis could explain the origin of Pu isotopes detected at the Vilnius sampling station during these two periods. It seems that Pu isotopes in samples collected in Vilnius in 2005–2006 derived from global fallout, while high activities found in aerosol filters in 1995 were due to transport of "hot" particles from the Chernobyl region. Similar conclusions have been obtained by analyzing Pu data in aerosol samples collected in Finland, Germany and the Czech Republic (Hölgye, 2008; Pöllänen et al., 1997; Wershofen et al., 2001). Recent studies have confirmed that particles raised from the ground can be transported over long distances (Choi et al., 2006; Hirose et al., 2010). However, chemical composition of particles carrying Pu isotopes in such a case should be different from that derived from the global fallout. A peculiar contamination of the Chernobyl restricted zone by Pu isotopes was discussed in a recent review (Matsunaga and Nagao, 2009). It was pointed out that the environmental behavior of the Chernobyl-originated Pu was related to the dissolution of the released fuel particles and its distribution over the surface soil layer with low infiltration capacity. In addition, the observed relative immobility of plutonium was attributed to its affinity for soil organics, most probably humic substances. Although it was concluded that Pu solid speciation needed further characterization, the transport of Pu, associated with both nuclear fuel and natural soil organics, could be expected from the Chernobyl zone. For this reason, it is not surprising that the correlation of the Pu activity with the components of the earth crust was not observed in the post-Chernobyl period (Lujanienė et al., 2012b).

The frequency distribution of the $^{240}Pu/^{239}Pu$ atom ratios ($n = 74$) in aerosol samples collected in Vilnius during the period of 1995–2006 presented in Fig. 6.16 indicates a log-normal distribution. The more precise analyses showed the bimodal distribution of the $^{240}Pu/^{239}Pu$ atom ratio. The first mode with

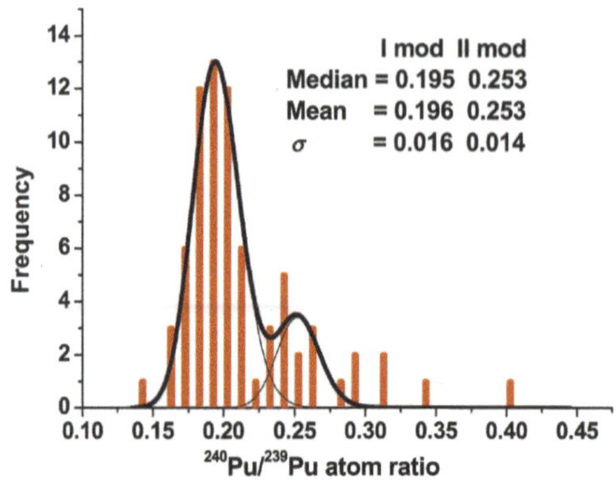

Figure 6.16 Frequency distribution of $^{240}Pu/^{239}Pu$ atom ratios in aerosol samples collected in Vilnius (Lithuania) during the period 1995–2006. (Adapted from Lujanienė et al. (2012a))

the median value of 0.195 and variations from 0.155 to 0.215 corresponds well to the global fallout ratio of 0.180 ± 0.007, characteristic of the Northern Hemisphere (Kelley et al., 1999). The second mode showed the median value of 0.253 with the range from 0.225 to 0.285, which could be most probably attributed to the nuclear power reactor plutonium with characteristic $^{240}Pu/^{239}Pu$ atom ratio of 0.23–0.67 (Warneke et al., 2002). It should be noted that the $^{240}Pu/^{239}Pu$ atom ratio of the first mode is mainly derived from global fallout, and can be attributed to the atmospheric nuclear weapons tests period. However, it is also evident that the Pu atom ratio of the second mode did not originate from a single source, but probably from several sources such as the Chernobyl accident, and nuclear reprocessing facilities, including resuspension of Pu from soil. It can be seen, however, that the contribution of the latter period to the airborne Pu observed in Europe is rather small. Apart from the two modes, there are a few values which cannot fit any mode. The presence of these values is indicating a transport of Pu isotopes from the Chernobyl-contaminated zone (e.g. the $^{240}Pu/^{239}Pu$ atom ratio of 0.40), while the low $^{240}Pu/^{239}Pu$ atom ratio of 0.141 ± 0.010 detected in June, 1997 in Vilnius can be explained by the contribution of the weapon-grade plutonium with the typical $^{240}Pu/^{239}Pu$ atom ratio of 0.01–0.07 (Lujanienė et al., 2012b).

The Chernobyl accident differed from expectations based on severe accident analyses done for reactors in that the release was prolonged, it varied in rate and radionuclide composition over time, and the meteorological conditions were complex. The release of radionuclides from the plant did not occur in a single massive event, but it occurred as a protracted process over a few days. The initial release contained fission products, enriched in volatile radionuclides such as tellurium, iodine and cesium, together with noble gases. Later releases consisted of all types of fission and activation products. The radioactive material was carried away in the form of gases and aerosols. The complex meteorological conditions with varying characteristics of radioactive releases from the reactor led to a complex pattern of atmospheric radionuclide transport. It was confirmed that the radioactive material released in the reactor accident may be transported on long distances from the reactor.

To prevent a further release of radioactive materials from the damaged nuclear reactor, a sarcophagus was built around the reactor in 1986 (Fig. 6.17). Owing

Figure 6.17 A view of the sarcophagus covering the damaged nuclear reactor No.4 of the Chernobyl NPP, which was constructed in November 1986 (P.P. Povinec between two Chernobyl colleagues visiting Chernobyl in 1987; Credit: P.P. Povinec).

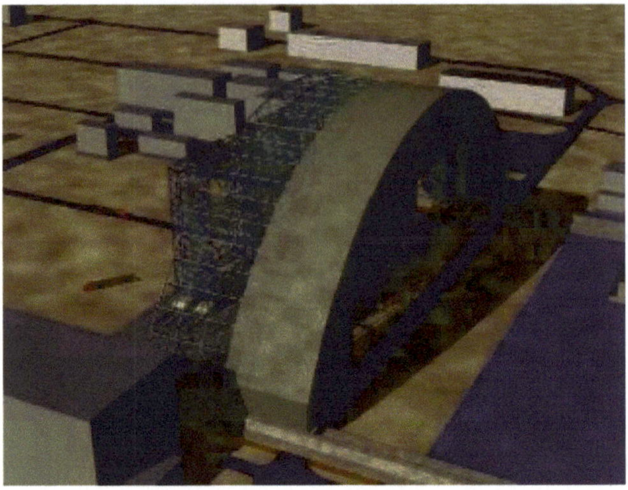

Figure 6.18 New Safe Confinement of the nuclear reactor No. 4 of the Chernobyl NPP. (A computer image, Source: Reprinted from Wikipedia)

to its aging, a new shelter of the sarcophagus—a New Safe Confinement—is presently under construction (Fig. 6.18), which is expected to be completed in 2015.

6.2 World Ocean

6.2.1 Anthropogenic Radionuclides in the World Oceans

The anthropogenic radionuclides in surface seawater in the world oceans (Pacific, Atlantic, Indian Oceans, Mediterranean Sea and marginal seas) have originated mainly from the global fallout in the 1950s and 1960s. As another radioactivity input into the ocean, the discharge from the nuclear reprocessing plants (Sellafield and La Hague) was a dominant source to the North Atlantic Ocean and its marginal seas in the 1970s and 1980s. Concentrations of anthropogenic radionuclides (^{137}Cs, ^{90}Sr, plutonium isotopes, ^{3}H and others) in seawater have been measured since the mid-1950s. Marine radioactivity observation projects such as Geochemical Ocean Sections Study (Bowen et al., 1980) and Worldwide Marine Radioactivity Studies (Povinec et al., 2003,

2005) revealed the vertical and spatial distributions of anthropogenic radionuclides in the world oceans.

Global distribution of ^{137}Cs in surface water followed global fallout pattern until the early 1970s, when mid-latitude maxima in both Northern and Southern Hemispheres were observed (Miyake et al., 1988). Thereafter, meridional gradient of ^{137}Cs activity in surface water tends to be smaller because the global circulation of surface water homogenized ^{137}Cs activity. In 2000s, a few Becquerel per cubic meter of ^{137}Cs in surface water were observed worldwide except in the North Atlantic region where radioactive liquid discharges from nuclear reprocessing plants have been increasing concentrations of ^{137}Cs in surface water. The impact of the Chernobyl accident on the marine radioactivity was strong in the Baltic and Black Seas, and negligible (but measurable) in the north-western Pacific Ocean.

Although a dominant source of plutonium was global fallout due to the atmospheric nuclear testing as did ^{137}Cs, oceanic behaviors of plutonium differ from that of ^{137}Cs. During 1970–2010, plutonium concentrations in the surface waters of the open ocean have decreased with sea area-dependent residence time, which is generally shorter than that of the corresponding surface ^{137}Cs (Hirose et al., 1992; Hirose and Aoyama, 2003a). Vertical profiles of plutonium in the water column show surface minimum, mid-depth maximum (500–800 m depth) and decrease with increasing depth, which differ from those of ^{137}Cs (Livingston et al., 2001). Plutonium, which is a typical particle-reactive radionuclide in contrast to ^{137}Cs, moves vertically with sinking biogenic particles and is regenerated in deep waters as a result of microbial decomposition of sinking particles (Hirose, 2009). Biogeochemical models have reproduced the vertical profile of plutonium with surface minimum, mid-depth maximum, and decrease with depth (Hirose, 1997; Tsumune et al., 2003a). However, plutonium behavior in oceanic water is more complicated because of such physical processes as advection and upwelling. For example, the observation that the plutonium maximum layer in the mid-latitude region of the North Pacific deepened with time (Livingston and Povinec, 2002) has been explained by a simple one-dimensional biogeochemical model (Hirose, 1997); however, biogeochemical processes alone cannot be used to explain a decrease of inventory in the corresponding site, where advection may play a significant role. The North Pacific deep waters (more than 2000 m depth) contain a significant amount of plutonium (Bowen et al., 1980), the input processes of which are still unknown.

6.2.2 Temporal and Horizontal Distributions of ^{137}Cs in Surface Water of the World Ocean

Since the 1950s, ^{137}Cs concentrations in the surface waters of the global ocean have been extensively measured by numerous studies. This chapter summarizes the results of spatial and temporal variations in ^{137}Cs concentrations in the surface waters in the global ocean for the period from 1957 to just before the Fukushima accident based on the analysis of the ^{137}Cs concentrations in the Historical Artificial Radionuclides in Marine Environment, HAM database and update (Aoyama et al., 2004). The global oceans have been divided into 37 latitudinal boxes on the basis of known ocean current systems, latitudinal and longitudinal distributions of ^{137}Cs concentrations, the distribution of global fallout, location of nuclear reprocessing plants sites, fallout from the Chernobyl accident, and availability of recent data in order to study the distribution of ^{137}Cs concentrations in surface seawater (IAEA, 2005; Inomata et al., 2009; Povinec et al., 2004, 2005). Figures 6.19 and 6.20 and Table 6.3 provide the graphic presentation of 37 boxes and a list of names of the boxes used in this chapter.

Cesium-137, with a half-life of 30.17 years, is the most abundant anthropogenic radionuclide in the marine environment. The main sources of ^{137}Cs are global and local fallouts from the atmospheric nuclear weapons tests, discharges from the nuclear reprocessing plants, and fallout from the Chernobyl accident (UNSCEAR, 2008). Sixty-nine percent of ^{137}Cs is derived from global fallout, 21% from local fallout, 7% from reprocessing plant discharge, and

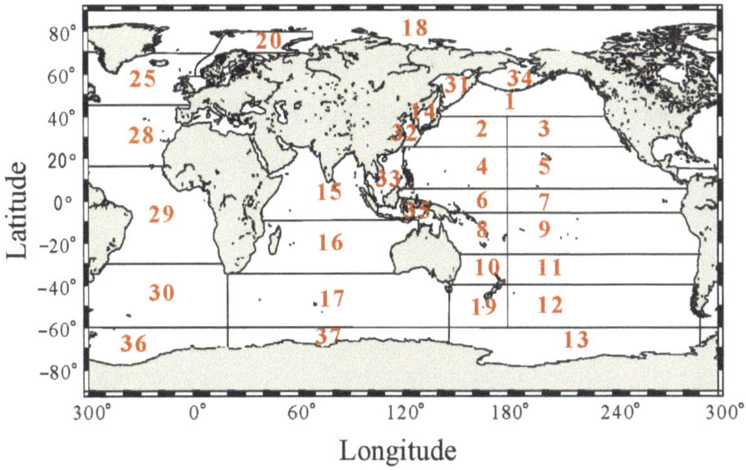

Figure 6.19 Latitudinal boxes for the Pacific, Indian, and Atlantic oceans.

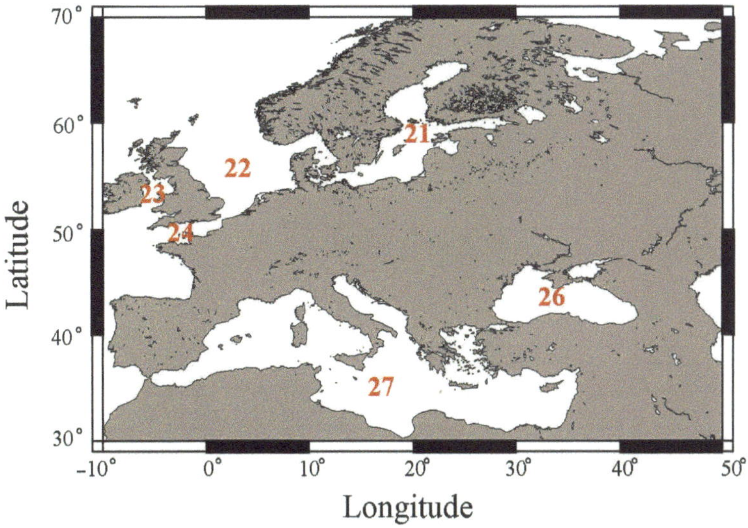

Figure 6.20 Latitudinal boxes in the northern European Seas.

3% from the Chernobyl accident (Livingston and Povinec, 2000). Other sources, such as sea dumping of nuclear waste, routine discharges from NPPs, nuclear reactors, satellites failures, lost nuclear weapons, and the use of radioisotopes in industry and medicine, are minor contributors to the radioactive contamination of the global ocean (Livingston and Povinec, 2000; Povinec et al., 2004).

Table 6.3. Box Numbers, Geographic Regions, and Designations (Abbreviation) in HAM-Global Database

Box	Area	Code
1	Subarctic North Pacific Ocean	Subarctic NPO
2	Western North Pacific Ocean	Western NPO
3	Eastern North Pacific Ocean	Eastern NPO
4	Western subtropical North Pacific Ocean	Subtropical western NPO
5	Eastern subtropical North Pacific Ocean	Subtropical eastern NPO
6	Western equatorial Pacific Ocean	Equatorial western PO
7	Eastern equatorial Pacific Ocean	Equatorial eastern PO
8	Western subtropical South Pacific Ocean	Subtropical western SPO
9	Eastern subtropical South Pacific Ocean	Subtropical eastern SPO
10	Western South Pacific Ocean	Western SPO
11	Eastern South Pacific Ocean	Eastern SPO
12	Eastern Southern Ocean	Eastern SO
13	Antarctic Ocean	ANT
14	Sea of Japan	SOJ
15	Arabian Sea	ARB
16	Indian Ocean	IO
17	Southern Ocean	SO
18	Arctic Ocean	AO
19	Middle Southern Ocean	MSO
20	Barents Sea	BARE
21	Baltic Sea	BALT
22	North Sea	NORS
23	Irish Sea	IRIS
24	English Channel	ENGC
25	Northern North Atlantic Ocean	NNA
26	Black Sea	BLAS
27	Mediterranean Sea	MEDS
28	North Atlantic Ocean	NAO
29	Central Atlantic Ocean	CAO
30	South Atlantic Ocean	SAO
31	Sea of Okhotsk	SOO
32	East China Sea	ECS
33	South China Sea	SCS
34	Bering Sea	BERS
35	Indonesian Archipelago	IARC
36	Atlantic sector of Antarctic	AOANT
37	Indian sector of Antarctic	IOANT

Inomata (2010) described the general trend of surface ^{137}Cs activity in the world ocean until 2005. We will follow her work in this chapter, however, update about recent trend, just before the Fukushima accident on 11 March 2011. The temporal variations of ^{137}Cs concentrations in the surface seawater are summarized in Fig. 6.21. The temporal variations of ^{137}Cs concentrations in the surface seawater in the North Pacific Ocean are shown in Fig. 6.21(a) and those in its marginal seas are shown in Fig. 6.21(b), respectively. General trend of ^{137}Cs concentration in

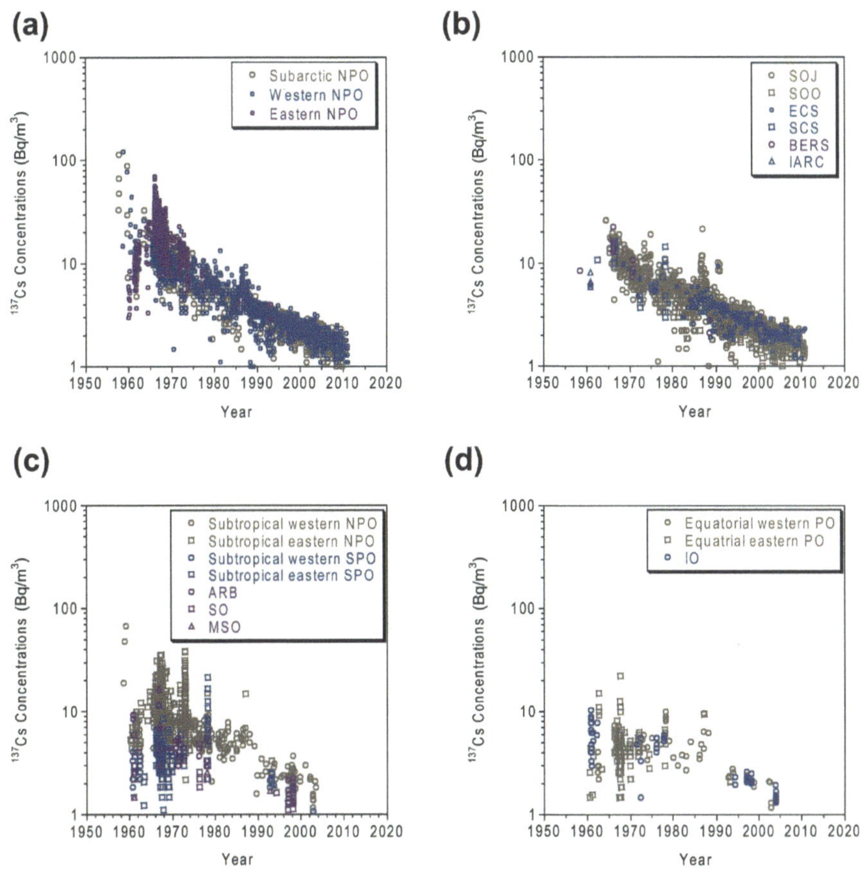

Figure 6.21 The ^{137}Cs concentrations in surface seawater in each box. (a) Boxes 1–3 (subarctic NPO, western NPO, and eastern NPO); (b) boxes 14 (SOJ) and 31–35 (SOO, ECS, SCS, BERS, IARC); (c) boxes 4, 5, 8, 9, 15, 17, and 19 (subtropical western NPO, subtropical eastern NPO, subtropical western SPO, subtropical eastern SPO, Arabian Sea, and Southern Ocean); (d) boxes 6, 7, and 16 (Equatorial western PO, Equatorial eastern PO, and IO);

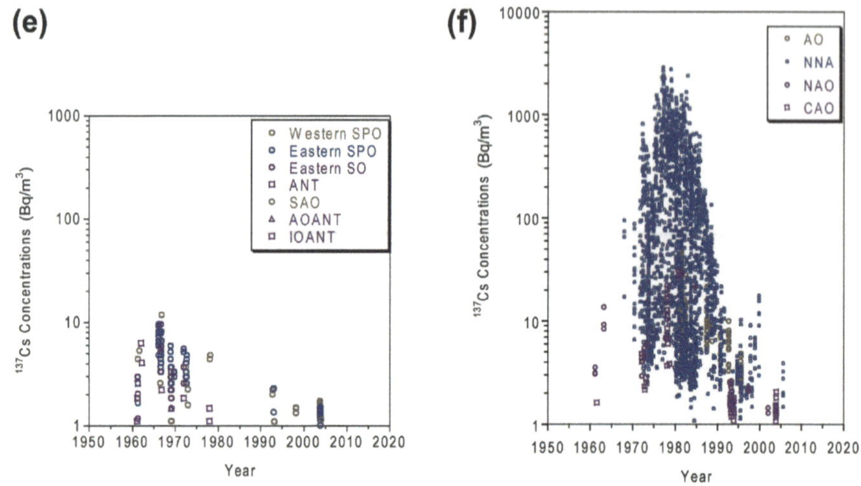

Figure 6.21 Continued (e) boxes 10–13, 30, 36, and 37 (Western SPO, eastern SPO, ANT, SAO, AOANT, and IOANT); and (f) boxes 18 (AO), 25 (NNA), and 28 and 29 (NAO and CAO).

these regions is similar. Before 1970, during the period of greater radioactive deposition, the decrease rate of ^{137}Cs concentrations in surface seawater was faster. The delay between the time of maximum concentration in the eastern North Pacific Ocean compared with the western North Pacific Ocean suggests that ^{137}Cs deposited on the sea surface in the western North Pacific Ocean was transported eastward and remained in the shallow surface mixed layer in the eastern North Pacific Ocean (Inomata, 2010). These processes have been recognized through the modeling of surface water ^{137}Cs concentration histories with the Ocean Global Circulation Model (Tsumune et al., 2003a, 2003b). After 1970, the ^{137}Cs concentrations in surface seawater decreased exponentially. The peak of the Chernobyl accident was also observed in the North Pacific Ocean and its marginal seas (Aoyama and Hirose, 1995). In the 2000s, concentrations of ^{137}Cs were almost constant in the North Pacific Ocean and its marginal seas (Hirose and Aoyama, 2003b). ^{137}Cs concentrations after 2000 ranged from 1 Bq/m^3 to 2 Bq/m^3 in the western North Pacific Ocean.

Concentrations of ^{137}Cs in each sea segment in the subtropical North Pacific Ocean and subtropical

South Pacific Ocean were almost at constant level before the 1990s (Fig. 6.21(c)). In these sea segments, there was a large latitudinal gradient with high values in the subtropical western and eastern North Pacific Ocean compared with those in areas to the equatorial western and eastern Pacific Ocean (Fig. 6.21(d)), subtropical western and eastern South Pacific Oceans. These suggest, after the radioactive decay of ^{137}Cs has been corrected for, that concentrations in the equatorial western and eastern Pacific Ocean have gradually increased through southward transport of ^{137}Cs-enriched water from the mid-latitudes of the North Pacific Ocean by the processes of global ocean circulation (Aoyama and Hirose, 1995). As shown in Fig. 6.21(d), concentrations of ^{137}Cs in the equatorial Pacific Ocean and Indian Ocean were almost at constant level. Considering that global and local fallouts of ^{137}Cs were small in this area, the almost constant levels can be explained by the influx of Pacific Ocean waters through the Indonesian Seas and the system of equatorial currents that recirculate surface water masses in the region (Povinec et al., 2005, 2011).

Although the observed data were scant in the South Pacific, the Antarctic, and the South Atlantic Oceans, it is expected that surface ^{137}Cs levels decreased after 1971 because the major input due to global fallout occurred in the mid 1960s and the major 22 French nuclear weapons tests were conducted until 1971 (Fig. 6.21(e)). The temporal variations of ^{137}Cs concentrations in these regions can be primarily explained by a small number of nuclear explosions in the eastern Southern Hemisphere, limited interhemispheric transport of the atmospheric nuclear weapons test-derived ^{137}Cs from the Northern Hemisphere to the Southern Hemisphere, vertical mixing of the surface water masses, and the influence of less contaminated Antarctic waters (Aarkrog, 1994; Bourlat et al., 1995; Yamada et al., 2007). The Antarctic Ocean is the area in the global ocean least contaminated by ^{137}Cs (Aarkrog, 1994).

Figure 6.21(f) shows the temporal variation of ^{137}Cs concentration in the surface seawater in the Arctic and the North Atlantic Oceans. The higher concentrations in the northern North Atlantic and

North Atlantic Oceans in the 1960s were due to the global fallout. After the 1970s, concentrations of ^{137}Cs were influenced by releases from the nuclear fuel reprocessing plants in Sellafield and La Hague. The ^{137}Cs concentrations in the Arctic Ocean, northern North Atlantic Ocean, and North Atlantic Ocean (north of 20°N) did not exponentially decrease over time. This indicates that the ^{137}Cs released from the nuclear fuel reprocessing plants was transported by prevailing currents from the Irish Sea and English Channel through the North Channel to the North Sea, northern North Atlantic Ocean, Barents Sea, and Arctic Ocean (Povinec et al., 2003). In the 2000s, compared with the concentrations in the Pacific and the Indian Oceans, the ^{137}Cs concentrations in the North Atlantic Ocean were still high.

Horizontal distribution of ^{137}Cs concentrations in surface seawater from 1957 to 2011 is shown in Figs 6.22–6.27 for the periods 1957–1964, 1960–1969, 1970–1979, 1980–1989, 1990–1999, and 2000 to just before Fukushima accident, respectively (in this case, these data were not radioactive decay corrected). Because the data density was not high enough for drawing contour lines at wide area in the world ocean, the ^{137}Cs data were plotted as dots with corresponding color scale.

The effect of global fallout was large in 1957–1965 as shown in Fig. 6.22. The highest concentrations

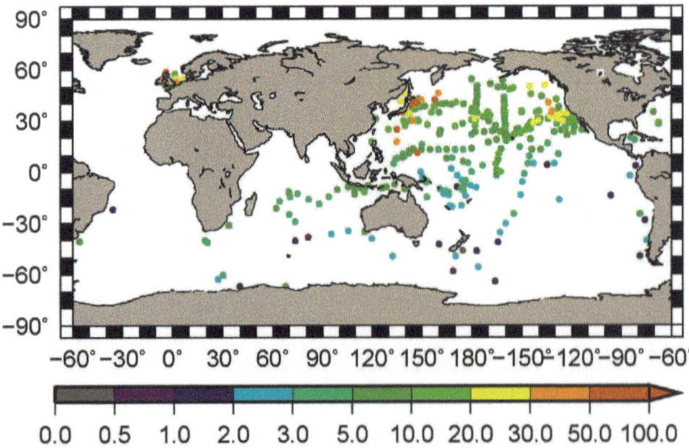

Figure 6.22 The ^{137}Cs in surface waters of the global ocean in the period from 1957 to 1964 (unit: Bq/m^3).

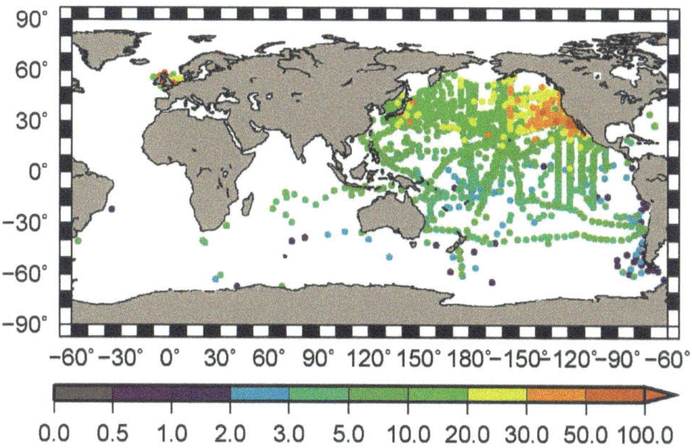

Figure 6.23 The same as Fig. 6.4 but from 1960 to 1969 (unit: Bq/m^3).

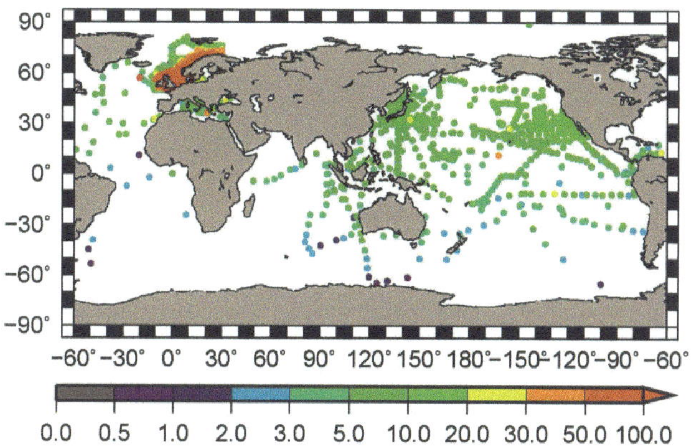

Figure 6.24 The same as Fig. 6.4 but from 1970 to 1979 (unit: Bq/m^3).

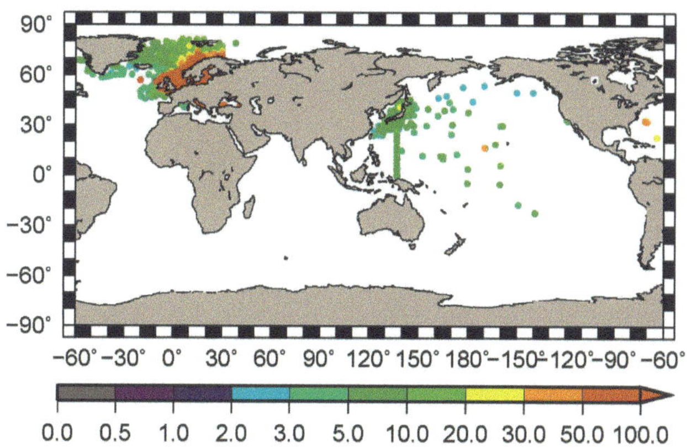

Figure 6.25 The same as Fig. 6.4 but from 1980 to 1989 (unit: Bq/m^3).

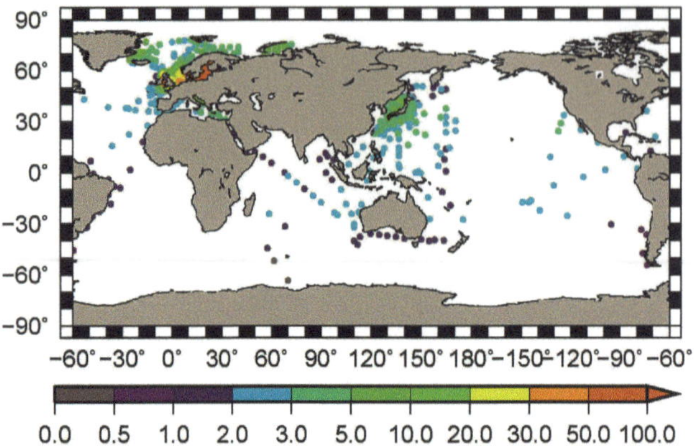

Figure 6.26 The same as Fig. 6.4 but from 1990 to 1999 (unit: Bq/m^3).

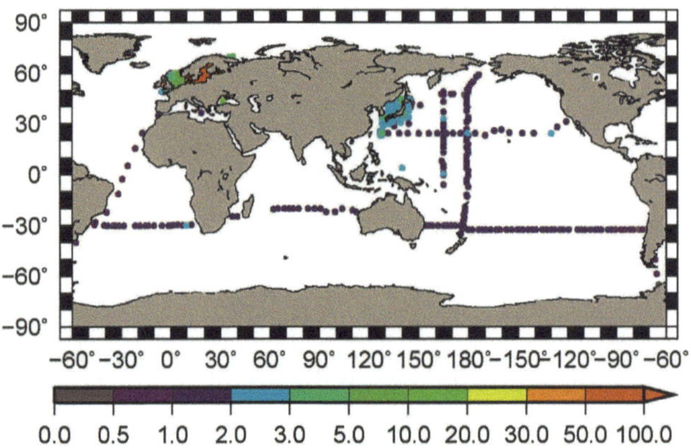

Figure 6.27 The same as Fig. 6.4 but from 1999 to 2011 (unit: Bq/m^3).

were found in the western North Pacific Ocean (30–45°N and 135–155°E) and the North Atlantic Ocean (30–50°N). These areas correspond to cross-overs of areas where larger precipitation amounts and higher stratosphere–troposphere exchange were expected (Aoyama and Hirose, 2003; Aoyama et al., 2006). The effect of local fallout originating from nuclear weapon tests carried out at the Bikini Atoll can clearly be seen in the western part of the equatorial Pacific Ocean (Miyake et al., 1963). The higher concentrations moved to the eastern North Pacific Ocean in the 1960s (Fig. 6.23). As shown in Fig. 6.24, the differences in ^{137}Cs concentrations in surface seawater between mid-latitudes and the equatorial region in the 1970s became smaller (Aoyama et al., 2001). In the North Atlantic Ocean and its northeast marginal seas, liquid discharges from the nuclear fuel reprocessing plants had significant impacts after the 1970s. In the 1980s, the high concentrations in the Baltic, Black, and Mediterranean Seas were due to the fallout from the accident at Chernobyl in April 1986 as shown in Fig. 6.25. The concentrations in the North Atlantic Ocean and its northeast marginal seas were also high due to the release of liquid discharges from the nuclear fuel reprocessing plants. In the 1990s (Fig. 6.26), releases from the nuclear fuel reprocessing plants still influenced the northeast marginal seas of the North Atlantic Ocean.

In the Pacific and Indian Oceans, latitudinal difference in ^{137}Cs concentrations in surface seawater became small. During the period from 2000 to 2011 (Fig. 6.27), latitudinal difference in ^{137}Cs concentrations in surface seawater in the global ocean became small, and concentrations of ^{137}Cs tended to be almost constant in the surface seawater of the global ocean; however, a weak mid-latitude maximum of ^{137}Cs concentration can be still observed. Details of this issue will be discussed later.

6.2.2.1 Meridional Distribution

Meridional distribution of ^{137}Cs concentration in surface seawater from 1957 to 2011 is shown in Figs 6.28–6.33 for the periods 1957–1964, 1960–1969, 1970–1979, 1980–1989, 1990–1999, and 2000 to just before Fukushima accident (as well as horizontal distributions which were already discussed in previous paragraph). These data were not corrected for radioactive decay. The regions with box numbers 18, 20, 21, 22, 23, 24 and 25 were affected with liquid discharges from the fuel reprocessing plants (shown in light blue).

In 1957–1964 at a high global fallout period, meridional distribution of ^{137}Cs in surface water (Fig. 6.28) was high at mid-latitudes in the Northern Hemisphere, and low in the Southern Hemisphere, as a results of spatial distribution of global fallout (Aoyama et al. 2006). The observed higher ^{137}Cs activities were due to liquid discharges from the nuclear

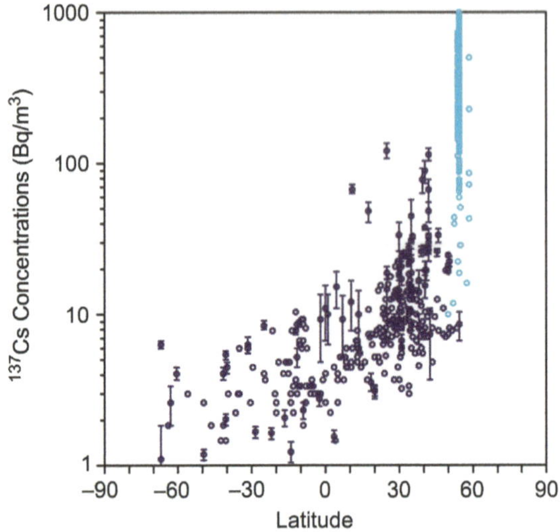

Figure 6.28 Meridional distribution of ^{137}Cs in surface waters of the global ocean in the period from 1957 to 1964. Light blue displays the area, of which box numbers are 18, 20, 21, 22, 23, 24 and 25, where liquid discharge affected ^{137}Cs activity in surface water Dark blue displays the area of which box numbers are 1-37 except 18, 20, 21, 23, 24 and 25 (unit: Bq/m^3).

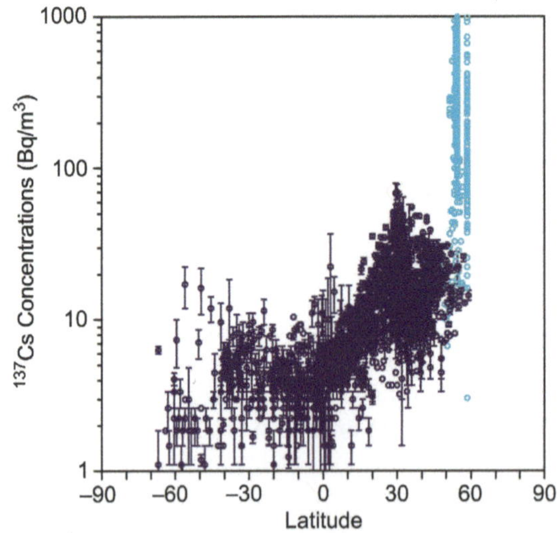

Figure 6.29 The same as Fig. 6.10 but from 1960 to 1969 (unit: Bq/m^3).

fuel reprocessing plant at Sellafield, UK as shown in Fig. 6.28 (Guéguéniat et al., 1996).

In 1960s, mid-latitude maxima were observed at both the Northern Hemisphere and the Southern

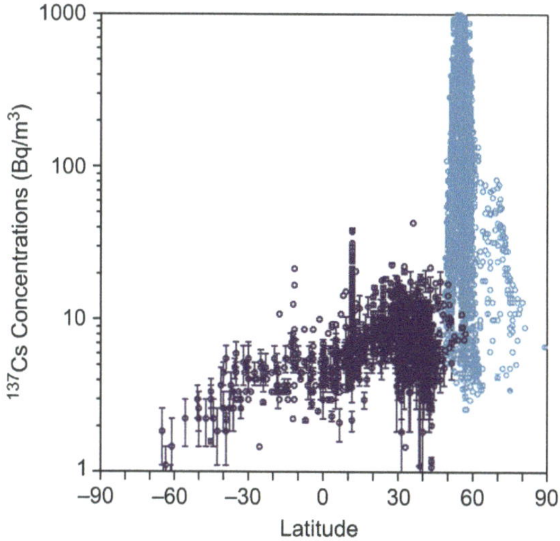

Figure 6.30 The same as Fig. 6.10 but from 1970 to 1979 (unit: Bq/m^3).

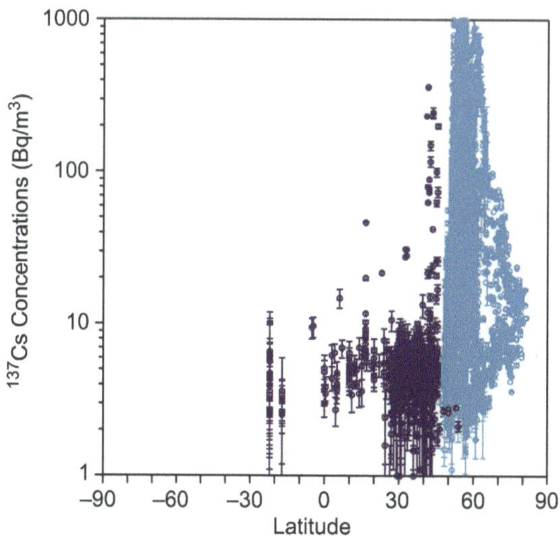

Figure 6.31 The same as Fig. 6.10 but from 1980 to 1989 (unit: Bq/m^3).

Hemisphere as shown in Fig. 6.29. Higher ^{137}Cs activities were also observed, which were due to liquid discharges from the nuclear fuel reprocessing plant at Sellafield, UK (similarly as in 1957–1964).

In 1970s, higher ^{137}Cs activity was also observed which originated from nuclear fuel reprocessing plants at Sellafield, UK and la Hague, France. The

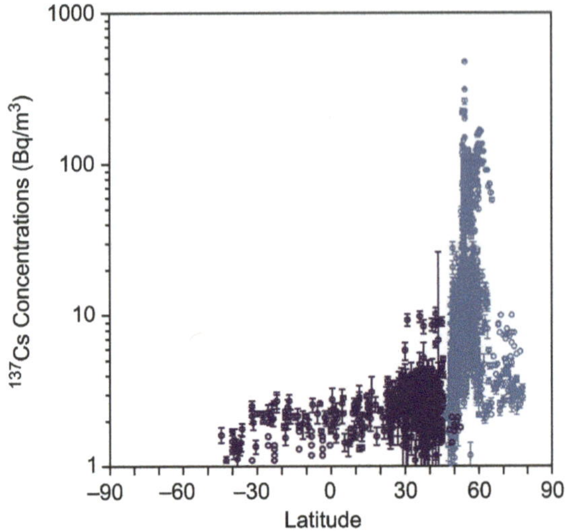

Figure 6.32 The same as Fig. 6.10 but from 1990 to 1999 (unit: Bq/m^3).

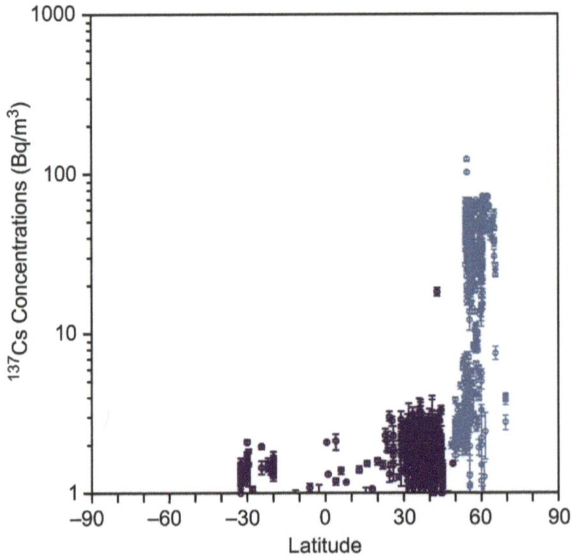

Figure 6.33 The same as Fig. 6.10 but from 1999 to 2011 (unit: Bq/m^3).

observations for higher ^{137}Cs from nuclear reprocessing plant in the North Atlantic Ocean were enhanced and northward extent of them was clearly observed as shown in Fig. 6.30.

In 1980s and 1990s as shown in Figs 6.31 and 6.32, differences between ^{137}Cs concentration in surface water at the Northern Hemisphere and Southern Hemisphere became small as discussed previously. The enhanced observations for higher ^{137}Cs from nuclear reprocessing plant in the North Atlantic Ocean were continued in 1980s and 1990s and northward extent of them was still clearly observed (Figs 6.31 and 6.32).

In the period from 2000 to 2011, differences between ^{137}Cs concentration in surface water at the Northern Hemisphere and Southern Hemisphere became smaller as shown in Fig. 6.33. The ^{137}Cs concentration in surface water ranged from 1 to 3 Bq/m^3 except in the North Atlantic where due to liquid discharges from nuclear reprocessing plants ^{137}Cs activities up to 100 Bq/m^3 were observed.

6.2.3 Impact of the Chernobyl Accident on the World Ocean

Approximately two-thirds of the 85 PBq ^{137}Cs released by the Chernobyl accident in 1986 were deposited outside the former Soviet Union (IAEA, 2003, 2006). Although most of this activity fell over land, a significant part went to the sea. In particular, the European marginal seas received a substantial amount of the Chernobyl debris. The primary route of contamination of these seas was atmospheric fallout, with smaller inputs from riverine transport occurring over the years following the accident. The Baltic Sea was the most contaminated. The total inventory from the accident was calculated to be 4.5 PBq ^{137}Cs (IAEA, 2005, 2006). Since Chernobyl, the Baltic Sea has been a main source of fresh inflow of ^{137}Cs to the north-eastern Atlantic Ocean.

The Black Sea received about 3.2 PBq of ^{137}Cs from the Chernobyl accident (IAEA, 2005, 2006), thus doubling the existing inventory of ^{137}Cs from global fallout. The amount of ^{90}Sr increased by 19% in comparison with the pre-Chernobyl period and was estimated to be about 1.8 PBq. The outflow from the Black Sea is the main source of the Chernobyl-derived ^{137}Cs in the Mediterranean Sea. However, the inflow is less than the decay of the present ^{137}Cs

Chernobyl Mediterranean Sea inventory of 3–5 PBq. The Mediterranean Sea also received direct fallout from the accident. From German and British measurements, it has been estimated (IAEA, 2005) that the North Sea received 1.2 PBq ^{137}Cs from Chernobyl. The Chernobyl inventory in the north-eastern Atlantic was estimated at about 6 PBq ^{137}Cs. In the summer of 1987, surface seawater samples collected in the Greenland, Norwegian and Barents Seas and off the west coast of Norway and the Faroe Islands all contained Chernobyl-derived ^{137}Cs. Impact of the Chernobyl accident on the concentrations of ^{137}Cs in surface waters of the Baltic Sea is documented in Fig. 6.34 (Povinec et al., 2003).

The total Chernobyl ^{137}Cs input to the world ocean is estimated at about 16 PBq. It has been assumed that the input of ^{90}Sr to the sea was negligible. The inputs of ^{137}Cs were estimated at 10 PBq in the 30–60°N latitude belt and at 6 PBq in the 60–90°N belt. Nearly all ^{137}Cs of Chernobyl origin went to the North Atlantic and Arctic Oceans. The relative input to the ocean thus became significantly less than that from nuclear weapons fallout, because the Chernobyl accident was a tropospheric event, first of all contaminating the surrounding European land areas.

As can be seen from Fig. 6.35, the Chernobyl accident impacted ^{137}Cs concentrations in surface waters of the north-western Pacific Ocean, as there is a clear peak in the ^{137}Cs time series.

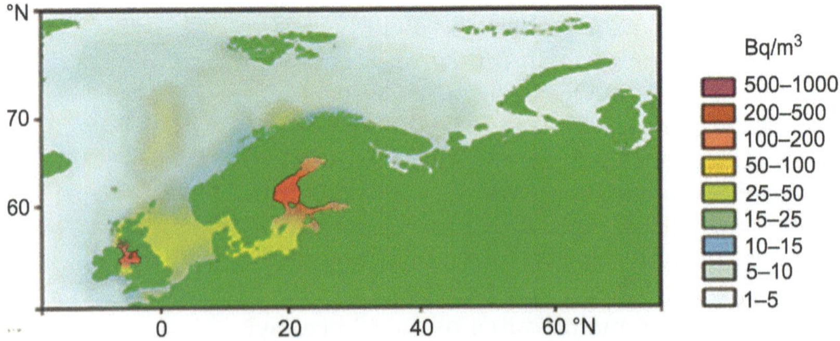

Figure 6.34 Distribution of ^{137}Cs in surface waters of the North European Seas after the Chernobyl accident (1986–1990). (Modified from Povinec et al. (2003))

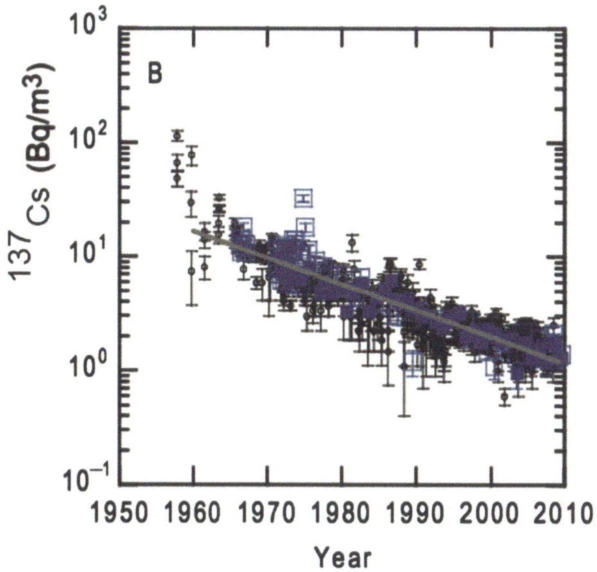

Figure 6.35 ^{137}Cs surface water time series in the north-western Pacific Ocean. (Data from the HAM database)

6.2.4 Radionuclide Residence Times in the Pacific Ocean

The latitudinal distribution of the apparent half-residence times of ^{137}Cs in surface water is shown in Fig. 6.36 (Inomata et al., 2009). The apparent half-residence times in each box ranged from 4.5 to 36.8 years (Table 6.4). The most prominent feature is that the apparent half-residence times in the western and eastern Pacific Ocean show a marked latitudinal distribution with longer in the equatorial area and shorter in the mid-high latitude area, although the times are short in the subtropical western Pacific Ocean. The large uncertainty in the equatorial and southern hemisphere (between 20°N and 20°S; Box 5, 7, 9) was due to the small data as well as large sea area. The latitudinal distribution of the apparent half-residence times in the Indian Ocean is similar to those in the Pacific Ocean. The present estimation is similar to those in the previous study (Povinec et al., 2005). The latitudinal distribution of physical oceanographic half-residence times of ^{137}Cs in

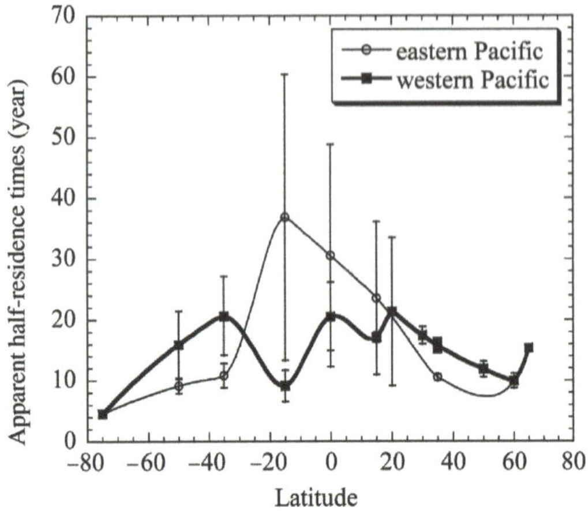

Figure 6.36 Latitudinal distribution of apparent half-residence times of ^{137}Cs in surface waters of different ocean areas: eastern Pacific Ocean (boxes 1, 3, 5, 7, 9, 11–13, and 34) and western Pacific Ocean (boxes 1, 2, 4, 6, 8, 10, 13, 14, 17, and 31–34). The error bars mean 95% confidence level.

surface seawater in each box is similar to those in the apparent half-residence times. The Pacific Ocean half-residence times ranged from 4.3 to 106 years, except for the sea areas between 10°N and 20°S in the eastern Pacific Ocean (the equatorial eastern Pacific Ocean and subtropical eastern Subtropical Pacific Ocean) where we got negative values.

The latitudinal variation of the apparent and half-residence times in the western and eastern Pacific Ocean may reflect the pathway of basin-scale surface flow. The ^{137}Cs deposited in the subarctic North Pacific Ocean and western North Pacific Ocean was transported to eastward (eastern North Pacific Ocean) with penetrating under the surface layer. Following then, water with high ^{137}Cs moved southward according to the North Pacific subtropical gyre and subducted in the subtropical central and east North Pacific Ocean. Furthermore, the ^{137}Cs was westward transported with upwelling, which resulted in increasing of ^{137}Cs concentrations in the equatorial Pacific Ocean. The longer apparent half-residence times and negative values of Pacific Ocean half-residence times in the equatorial Pacific Ocean suggest the supply of ^{137}Cs from high ^{137}Cs-contaminated sea area.

Table 6.4. The Apparent and Physical Oceanographic Half-Residence Times for Each Box

Box	Latitude	Start (Year)	End (Year)	Apparent Half-Residence Times	Physical Oceanographic Half-Residence Times
1	65	1970	2005	15.3 ± 0.5	31.2 ± 1.1
2	35	1970	2005	15.1 ± 0.4	30.2 ± 0.8
3	35	1970	1992	10.5 ± 0.5	16.1 ± 0.8
4	15	1970	2003	17.1 ± 0.8	39.4 ± 1.9
5	15	1970	1992	23.5 ± 12.6	106 ± 57
6	0	1972	1998	20.5 ± 5.6	63.9 ± 17.4
7	0	1971	1993	30.5 ± 18.3	
8	−15	1976	1993	9.1 ± 2.6	13.0 ± 3.8
9	−15	1971	1993	36.8 ± 23.5	
10	−35	1972	1998	20.6 ± 6.5	64.7 ± 20.4
11	−35	1971	1992	10.8 ± 2.0	16.8 ± 3.0
12	−50	1971	1992	9.1 ± 1.2	13.0 ± 1.7
13	−75	1972	1994	4.5 ± 0.6	5.4 ± 0.8
14	35	1970	2005	16.3 ± 0.5	35.6 ± 1.0
31	50	1978	2004	11.8 ± 1.3	19.3 ± 2.1
32	30	1971	2005	17.4 ± 1.4	41.1 ± 3.4
33	20	1978	1997	21.3 ± 12.2	31.0 ± 17.7
34	60	1970	1988	9.9 ± 1.2	14.8 ± 1.8

References

Aarkrog, A., Lippert, J., 1959. Environmental Radioactivity at Riso, 1 April 1958–31 March 1959, Riso-R-9 Report, Riso National Laboratory, Roskilde, Denmark.

Aarkrog, A., 1994. Radioactivity in polar regions—Main sources. J. Environ. Radioact. 25, 21–35. http://dx.doi.org/10.1016/0265-931X(94)90005-1.

Aoyama, M., Hirose, K., 1995. The temporal and spatial variation of ^{137}Cs Concentration in the western North Pacific and its marginal seas during the period from 1979 to 1988. J. Environ. Radioact. 29, 57–74. http://dx.doi.org/10.1016/0265-931X(94)00050-7.

Aoyama, M., Hirose, K., Miyao, T., Igarashi, Y., Povinec, P.P., 2001. ^{137}Cs activity in surface water in the western North Pacific. J. Radioanal. Nucl. Chem. 248, 789–793. http://dx.doi.org/10.1023/A:1010669501814.

Aoyama, M., Hirose, K., 2003. Temporal variation of ^{137}Cs water column inventory in the North Pacific since the 1960s. J. Environ. Radioact. 69, 107–117. http://dx.doi.org/10.1016/S0265-931X(03)00089-4.

Aoyama, M., Hirose, K., 2004. Artificial Radionuclides database in the Pacific Ocean: Ham database. The Scientific World JOURNAL 4, 200–215.

Aoyama, M., Hirose, K., Igarashi, Y., 2006. Re-construction and updating our understanding on the global weapons tests ^{137}Cs fallout. J. Environ. Monit. 8, 431–438.

Arimoto, R., Webb, J.L., Conley, M., 2005. Radioactive contamination of atmospheric dust over southeastern New Mexico. Atmos. Environ. 39, 4745–4754.

Bourlat, Y., Millies-Lacroix, J.C., Nazard, R., 1995. Determination of plutonium radioactivity in Mururoa lagoon water. J. Radioanal. Nucl. Chem. 197, 387–408. http://dx.doi.org/10.1007/BF02036013.

Bowen, V.T., Noshkin, V.E., Livingston, H.D., Volchok, H.L., 1980. Fallout radionuclides in the Pacific Ocean: vertical and horizontal distributions, largely from GEOSECS stations. Earth Planetary Sci. Lett. 49, 411–434.

Choi, M.S., Lee, D.S., Choi, J.C., Cha, H.J., Yi, H.I., 2006. $^{239,240}Pu$ concentration and isotope ratio ($^{240}Pu/^{239}Pu$) in aerosols during high dust (yellow sand) period. Korean Sci. Total Environ. 370, 262–270.

Crooks, R.N., Osmond, R.D.G., Owers, M.J., Fisher, E.M.R., 1959. The Deposition of Fission Products from Distant Nuclear Test Explosions Results to Mid-1959. AERE-R 3094. HMSO, London.

Ehhalt, D.H., 1973. Turnover times of ^{137}Cs and HTO in the troposphere and removal rates of natural particles and water vapor. J. Geophys. Res. 78, 7076–7086.

Fujiwara, H., Fukuyama, T., Shirato, Y., Ohkuro, T., Taniyama, I., Zhang, T.-H., 2007. Deposition of atmospheric ^{137}Cs in Japan associated with the Asian dust event of March 2002. Sci. Total Environ. 384, 306–315.

Guéguéniat, P., Hermann, J., Kershaw, P., Bailly du Bois, P., Baron, Y., 1996. Artificial radioactivity in the English channel and the North Sea. In: Guéguéniat, P., Germain, P., Métiver, H. (Eds.), Radionuclides in the Oceans: Inputs and Inventories. Les Edition de Physique, Les Ulis, pp. 121–154.
HASL, 1976. Health and Safety Laboratory Environmental Quarterly, HASL-298, 1954–1975. Energy Research & Development Administration. Health and Safety Laboratory, New York.
Hirose, K., 1997. Complexation-scavenging of plutonium in the ocean. Radioprotection-colloq. 32, C2-225–230.
Hirose, K., 2009. Plutonium in the ocean environment: its distribution and behavior. J. Nucl. Radiochemical Sci. 10, R7–R16.
Hirose, K., Aoyama, M., Katsuragi, Y., Sugimura, Y., 1987. Annual deposition of Sr-90, Cs-137 and Pu-239,240 from the 1961–1980 nuclear explosions: a simple model. J. Meteor. Soc. Japan 65, 259–277.
Hirose, K., Aoyama, M., 2003a. Analysis of ^{137}Cs and ^{239,240}Pu concentrations in surface waters of the Pacific Ocean. Deep-Sea Res. II 50, 2675–2700.
Hirose, K., Aoyama, M., 2003b. Present background levels of surface ^{137}Cs and ^{239,240}Pu concentrations in the Pacific. J. Environ. Radioact. 69, 53–60. http://dx.doi.org/10.1016/S0265-931X(03)00086-9.
Hirose, K., Sugimura, Y., Aoyama, M., 1992. Plutonium and ^{137}Cs in the western North Pacific: estimation of residence time of plutonium in surface waters. Appl. Radiat. Isot. 43, 349–359.
Hirose, K., Igarashi, Y., Aoyama, M., Kim, C.K., Kim, C.S., Chang, B.W., 2003. Recent trends of plutonium fallout observed in Japan: plutonium as a proxy for desertification. J. Environ. Monit. 5, 1–7.
Hirose, K., Kim, C.K., Kim, C.S., Chang, B.W., Igarashi, Y., Aoyama, M., 2004. Plutonium deposition observed in Daejeon, Korea: wet and dry depositions of plutonium. Sci. Total Environ. 332, 243–252.
Hirose, K., Igarashi, Y., Aoyama, M., 2007. Recent trends of plutonium fallout observed in Japan: comparison with natural lithogenic radionuclides, thorium isotopes. J. Radioanal. Nucl. Chem. 273, 115–118.
Hirose, K., Igarashi, Y., Aoyama, M., 2008. Analysis of 50 years records of atmospheric deposition of long-lived radionuclides in Japan. Appl. Radiat. Isot. 66, 1675–1678.
Hirose, K., Igarashi, Y., Aoyama, M., Inomata, Y., 2010. Depositional behaviors of plutonium and thorium at Tsukuba and Mt. Haruna in Japan indicate the sources of atmospheric dust. J. Environ. Radioact. 101, 106–112.
Hölgye, Z., 2008. Plutonium isotopes in surface air of Prague in 1986–2006. J. Environ. Radioact. 10, 1653–1655.
IAEA, International Atomic Energy Agency, 2003. Chernobyl's Legacy: Health, Environmental and Socio-Economic Impacts. IAEA, Vienna.

IAEA, International Atomic Energy Agency, 2005. Worldwide Marine Radioactivity Studies (WOMARS). Radionuclide Levels in Oceans and Sea. IAEA-TECDOC-1429. IAEA, Vienna.

IAEA, International Atomic Energy Agency, 2006. Environmental Consequences of the Chernobyl Accident and Their Remediation: Twenty Years of Experience. IAEA, Vienna.

Igarashi, Y., Aoyama, M., Hirose, K., Miyao, T., Yabuki, S., 2001. Is it possible to use ^{90}Sr and ^{137}Cs as tracers for the aeolian transport? Water Air Soil Pollut. 130, 349–354.

Igarashi, Y., Aoyama, M., Hirose, K., Miyao, T., Nemoto, K., Tomita, M., Fujikawa, T., 2003. Resuspension: decadal monitoring time series of the anthropogenic radioactivity deposition in Japan. J. Radiat. Res. 44, 319–328.

Igarashi, Y., Aoyama, M., Hirose, K., Povinec, P., Yabuki, S., 2005. What anthropogenic radionuclides (^{90}Sr and ^{137}Cs) in atmospheric deposition, surface soils and aeolian dusts suggest for dust transport over Japan. Water Air Soil Pollut. 5, 51–69.

Igarashi, Y., Inomata, Y., Aoyama, M., Hirose, K., Takahashi, H., Shinoda, Y., Sugimoto, N., Shimizu, A., Chiba, M., 2009. Possible change in Asian dust source suggested by atmospheric anthropogenic radionuclides during the 2000s. Atmos. Environ. 43, 2971–2980.

Inomata, Y., Aoyama, M., Hirose, K., 2009. Analysis of 50-y record of surface ^{137}Cs concentrations in the global ocean using the HAM-global database. J. Environ. Monit. 11, 116–125.

Inomata, Y., 2010. Global trends in cesium distribution. In: Radionuclides in the Environment. John Wiley & Sons, Ltd, Singapore.

Junge, C.E., Chagnon, C.W., Manson, J.E., 1961. Stratospheric aerosols. J. Atmos. Sci. 16, 81–108.

Junge, C.E., Manson, J.E., 1961. Stratospheric aerosol studies. J. Geophys. Res. 66, 2163–2182.

Karlsson, L., Hernandez, F., Rodríguez, S., López-Pérez, M., Hernandez-Armas, J., Alonso-Pérez, S., Cuevas, E., 2008. Using ^{137}Cs and ^{40}K to identify natural Saharan dust contributions to PM_{10} concentrations and air quality impairment in the Canary Islands. Atmos. Environ. 42 7032–7042.

Katsuragi, Y., 1983. A study of ^{90}Sr fallout in Japan. Pap. Meteor. Geophys. 33, 277–291.

Kelley, J.M., Bond, L.A., Beasley, T.M., 1999. Global distribution of Pu isotopes and Np-237. Sci. Total Environ. 238, 483–500.

Krey, P.W., Krajewksi, B.T., 1970. Comparison of atmospheric transport model calculations with observations of radioactive debris. J. Geophys. Res. 75, 2901–2908.

Kuroda, P.K., Miyake, Y., Nemoto, J., 1965. Strontium isotopes: global circulation after the Chinese nuclear explosion of 14 May 1965. Science 150, 1289–1290.

Lee, S.H., Pham, M.K., Povinec, P.P., 2002. Radionuclide variations in the air over Monaco. J. Radioanal. Nucl. Chem. 254, 445–453.

Livingston, H.D., Povinec, P.P., 2000. Anthropogenic marine radioactivity. Ocean Coastal Manage. 43, 689–712,. http://dx.doi.org/10.1016/S0964-5691(00)00054-5.

Livingston, H.D., Povinec, P.P., Ito, T., Togawa, O., 2001. The behaviour of plutonium in the Pacific Ocean. In: Pu in the Environment. Elsevier, Amsterdam, pp. 267–292.
Livingston, H.D., Povinec, P.P., 2002. Millennium perspective on the contribution of global fallout radionuclides to ocean science. Health Phys. 82, 656–668.
Lujanienė, G., Byčenkienė, S., Povinec, P.P., Gera, M., 2012a. Radionuclides from the Fukushima accident in the air over Lithuania: measurement and modeling approaches. J. Environ. Radioact. 114, 71–80.
Lujanienė, G., Valiulis, D., Byčenkienė, S., Šakalys, J., Povinec, P.P., 2012b. Plutonium isotopes and ^{241}Am in the atmosphere of Lithuania: a comparison of different source terms. Atmos. Environ. 61, 419–427.
Machta, L., List, R.J., Huber, L.F., 1956. World-wide travel of atomic debris. Science 124, 474.
Martell, E.A., 1966. The size distribution and interaction radioactive and natural aerosols in the stratosphere. Tellus 18, 486–498.
Masson, O., Piga, D., Gurriaran, R., D'Amico, D., 2010. Impact of an exceptional Saharan dust outbreak in France: PM_{10} and artificial radionuclides concentrations in air and in dust deposit. Atmos. Environ. 44, 2478–2486.
Matsunaga, T., Nagao, S., 2009. Environmental behavior of plutonium isotopes studied in the area affected by the Chernobyl accident. Humic Subst. Res. 5/6, 19–33.
Miyake, Y., Saruhashi, K., Katsuragi, Y., Kanazawa, T., Tsunogai, S., 1963. Deposition of Cs-137 and Sr-90 in Tokyo through the end of July 1963. Pap. Meteor. Geophys. 14, 58–65.
Miyake, Y., Saruhashi, K., Sugimura, Y., Kanazawa, T., Hirose, K., 1988. Contents of ^{137}Cs, plutonium and americium isotopes in the Southern Ocean waters. Pap. Meteor. Geophys. 39, 95–113.
Pham, M.K., LaRosa, J.J., Lee, S.-H., Oregioni, B., Povinec, P.P., 2005. Deposition of Saharan dust in Monaco rain 2001–2002: radionuclides and elemental composition. Physica Scripta 71 (118), 14–17.
Pham, M.K., Betti, M., Nies, H., Povinec, P.P., 2011. Temporal changes of ^{7}Be, ^{137}Cs and ^{210}Pb activity concentrations in surface air at Monaco and their correlation with meteorological parameters. J. Environ. Radioact. 102, 1045–1054.
Pham, M.K., Povinec, P.P., Nies, H., Betti, M., 2013. Dry and wet deposition of ^{7}Be, ^{210}Pb and ^{137}Cs in Monaco air during 1997–2010: seasonal variations of deposition fluxes. J. Environ. Radioact. 120, 45–57.
Povinec, P., Chudý, M., Sýkora, I., Szarka, J., Pikna, M., Holý, K., 1988. Aerosol radioactivity monitoring in Bratislava following the Chernobyl accident. J. Radioanal. Nucl. Chem. Lett. 126, 467–478.
Povinec, P.P., Bailly Du Bois, P., Kershaw, P.J., Nies, H., Scotto, P., 2003. Temporal and spatial trends in the distribution of ^{137}Cs in surface waters of Northern European Seas – a record

of 40 years of investigations. Deep-Sea Res. Part II Topical Stud. Oceanogr. 50, 2785–2801.

Povinec, P.P., Hirose, K., Honda, T., Ito, T., Scott, E.M., Togawa, O., 2004. Spatial distribution of ^{3}H, ^{90}Sr, ^{137}Cs and $^{239,240}Pu$ in surface waters of the Pacific and Indian Oceans – GLOMARD database. J. Environ. Radioact. 76, 113–137.

Povinec, P.P., Aarkrog, A., Buesseler, K.O., Delfanti, R., Hirose, K., Hong, G.H., Ito, T., Livingston, H.D., Nies, H., Noshkin, V.E., Shima, S., Togawa, O., 2005. ^{90}Sr, ^{137}Cs and $^{239,240}Pu$ concentration surface water time series in the Pacific and Indian Oceans – WOMARS results. J. Environ. Radioact. 81, 63–87.

Povinec, P.P., Breier, R., Coppola, L., Groening, M., Jeandel, C., Jull, A.J.T., Kieser, W.E., Top, Z., 2011. Tracing of water masses using a multi-isotope approach in the southern Indian Ocean. Earth Planetary Sci. Lett. 302, 14–26.

Povinec, P.P., Holý, K., Chudý, M., Šivo, A., Sýkora, I., Ješkovský, M., Richtáriková, M., 2012. Long-term variations of ^{14}C and ^{137}Cs in the Bratislava air – implications of different atmospheric transport processes. J. Environ. Radioact. 108, 33–40.

Pöllänen, R., Valkama, I., Toivonen, H., 1997. Transport of radioactive particles from the Chernobyl accident. Atmos. Environ. 31, 3575–3590.

Reiter, E.R., Bauer, E., 1975. Residence Times of Atmospheric Pollutant. CIAP Monogr.1. US Department of Transportation, Washington, DC.

Rosner, G., Winkler, R., 2001. Long-term variation (1986–1998) of post-Chernobyl ^{90}Sr, ^{137}Cs, ^{238}Pu and $^{239,240}Pu$ concentrations in air, depositions to ground, resuspension factors and resuspension rates in south Germany. Sci. Total Environ. 273, 11–25.

Sýkora, I., Povinec, P.P., Brest'áková, L., Florek, M., Holý, K., Masarik, J., 2012. Resuspension processes control variations of ^{137}Cs activity concentrations in the ground-level air. J. Radioanal. Nucl. Chem. 293, 595–599.

Tsumune, D., Aoyama, M., Hirose, K., 2003a. Numerical simulation of ^{137}Cs and $^{239,240}Pu$ concentrations by an ocean general circulation model. J. Environ. Radioact. 69, 61–84. http://dx.doi.org/10.1016/S0265-931X(03)00087-0.

Tsumune, D., Aoyama, M., Hirose, K., 2003b. Behavior of ^{137}Cs concentrations in the North Pacific in an ocean general circulation model. J. Geophys. Res. Oceans 108. http://dx.doi.org/10.1029/2002JC001434.

UNSCEAR, 1982. United Nations Scientific Committee on the Effects of Atomic Radiation. Ionizing Radiation: Sources and Biological Effects. Report to the General Assembly, United Nations Scientific Committee on the Effects of Atomic Radiation, p. 773.

UNSCEAR, 1993. United Nations Scientific Committee on the Effects of Atomic Radiation. Sources and Effects of Ionizing Radiation. Report to the General Assembly. United Nations, New York, p. 922.

UNSCEAR, 2000. United Nations Scientific Committee on the Effects of Atomic Radiation. Sources and Effects of Ionizing Radiation, New York.

UNSCEAR, 2002. United Nations Scientific Committee on the Effects of Atomic Radiation. Sources and Effects of Ionizing Radiation, New York.

UNSCEAR, 2008. United Nations Scientific Committee on the Effects of Atomic Radiation, 2008. Sources and Effects of Ionizing Radiation, New York.

Warneke, T., Croudace, I.W., Warwick, P.E., Taylor, R.N., 2002. A new ground-level fallout record of uranium and plutonium isotopes for northern temperate latitudes. Earth Planetary Sci. Lett. 203, 1047–1057.

Wershofen, H., Arnold, D., Steinkopff, Th, 2001. Measurement of plutonium isotopes in ground-level air in Northern-Germany – history and recent results. Nukleonika 46, 155–159.

Yamada, M., Zheng, J., Wang, Z.-L., 2007. $^{240}Pu/^{239}Pu$ atom ratios in seawater from Sagami Bay, western Northwest Pacific Ocean: sources and scavenging. J. Environ. Radioact. 98, 274–284. http://dx.doi.org/10.1016/j.jenvrad.2007.05.005.

7

ESTIMATION OF RADIATION DOSES

CHAPTER OUTLINE

7.1 Introduction

Releases of radionuclides during the Fukushima accident were controlled by the physical and chemical properties of radioactive elements in the cores of nuclear reactors, mainly by their boiling

Fukushima Accident. http://dx.doi.org/10.1016/B978-0-12-408132-1.00007-3

characteristics. Therefore, the majority of the radionuclides present in the reactor core were retained there, as we already discussed in previous chapters. In the intense heat and pressure of the melting reactors, a reaction between the nuclear fuel metal cladding and the remaining water surrounding them produced explosive hydrogen gas, which resulted in several hydrogen-air explosions (MEXT, 2011a,b; Tanabe, 2012a,b).

There were therefore two main forms of radionuclide releases to the environment:

- atmospheric releases as a result of venting of radioactive gases stored in the reactor containments, and later due to hydrogen-air explosions and
- direct releases of contaminated fresh water and later also seawater to coastal waters offshore Fukushima Dai-ichi Nuclear Power Plant (NPP).

While the atmospheric releases due to dry and wet depositions had an impact on the both terrestrial and marine environment, the liquid releases to the sea affected only the marine environment (Povinec et al., 2012a).

The public often perceives plutonium as a particularly dangerous nuclear fuel material, however, its radiological effects in the atmosphere, terrestrial and marine environments are much smaller than those from fission products. Particularly dangerous are highly radioactive compounds that accumulate in the food chain, such as isotopes of iodine, cesium and strontium. On the other hand, the release rates of plutonium isotopes (as well as strontium) during reactor accidents are much smaller than that of iodine and cesium isotopes.

This has also been the case during the Fukushima accident when much smaller concentrations of strontium isotopes (and especially of plutonium isotopes) were measured in the environment (Povinec et al., 2012a; Zheng et al., 2012; Lujaniené et al., 2012). The situation in the marine environment was quite different as during the first week after the accident the releases of strontium isotopes were comparable with that of cesium isotopes; however, later their levels were much lower than those of cesium isotopes (Povinec et al., 2012a).

Therefore in this chapter, we shall focus only on these radionuclides present in the environment after the Fukushima accident, which could deliver the highest doses to the public:

- ^{131}I and ^{134,137}Cs—as the most important radionuclides for the contamination of the atmosphere;
- ^{134,137}Cs—as the most important radionuclides for the terrestrial contamination;
- ^{134,137}Cs—as the most important radionuclides for the contamination of the marine environment.

The human body absorbs iodine and cesium readily, which are after inhalation or ingestion transported by the blood to the human organs. The ^{131}I is rapidly absorbed by the thyroid, and leaves only after its radioactive decay with a half-life of 8 days. Cesium is absorbed by muscles, and due to its long physical half-life (30 years for ^{137}Cs and 2.1 years for ^{134}Cs), it remains in the body until it is excreted (10–100 days biological half-life).

While in the body the beta- and gamma-radiation from ^{131}I and Cs isotopes can do significant damage, mainly to DNA. Children who ingest ^{131}I can develop thyroid cancer 10 or more years later; adults seem relatively more resistant. However, it is still not yet clear how much damage could be caused during exposure to low-level radionuclide concentrations (NAS, 2006; ICRP, 2005, 2007).

There have been three main pathways for contamination of the public during and after the Fukushima accident from terrestrial sources:

1. inhalation of the contaminated air,
2. external irradiation from radioactive clouds, and from dry and wet deposition of radionuclides, and from a contact with contaminated materials and objects, and
3. ingestion of the contaminated terrestrial food.

The main contamination pathway from marine sources has been an ingestion of the contaminated seafood.

7.2 Radiation Doses to the Japanese Population

7.2.1 Radiation Doses to the Public from Terrestrial Sources

7.2.1.1 Radiation Doses to the Public from Inhalation

Concerns about possibility of repeated hydrogen-air explosions, and atmospheric venting of radioactive gasses from the reactor containments, led the Japanese government following the International Atomic Energy Agency's (IAEA, 2002, 2009, 2011a,b) recommendations to organize a 20 km radius evacuation area around the plant (presently the restricted area), which has been later increased at some places to a larger evacuation zone (NSIA, 2011; NAIIC, 2012).

Due to specific meteorological conditions in Japan—winds follow usually the direction from west to east—most of the contaminated air blown over the Pacific Ocean, America to Europe (Masson et al., 2011) with smaller impact on the Japanese population. There were, however, cases when the wind changed the direction from the Fukushima Dai-ichi NPP to the northwest, which had the main contamination impact on the terrestrial environment (Lyons and Colton, 2012) (as we already discussed in more detail in Chapters 3 and 5).

Atmospheric emissions of large amounts (around 150 PBq) of short-lived radionuclides (such as ^{131}I, ^{132}I, ^{132}Te, and others), as well as emissions (around 15 PBq) of medium-lived ^{134}Cs and long-lived ^{137}Cs were of great importance for possible delivery of radiation doses to inhabitants (Chino et al., 2011; Kawamura et al., 2011). A major concern was the internal radiation exposure following incorporation and uptake of radioiodine in the thyroid, which occurred through inhalation of contaminated air and ingestion of contaminated food and drinks (ICRP, 2009).

Therefore a screening of evacuated residents on possible radioactive contamination was the first action carried out by the local governments and institutions. Following the IAEA's (2009, 2010) recommended screening level of 1 μSv/h (dose rate at a distance of 10 cm) as the operational intervention level for decontamination of body surface for residents, decontaminations were carried out on about 100 people (PMJC, 2011). Internal exposure doses were also estimated using the whole- body counting, as well as the bioassay method using urine as the contaminated medium. The internal exposure dose due to ^{134}Cs and ^{137}Cs was assessed to be below 1 mSv for all the checked residents. A specific attention was given to children as they have higher risk of thyroid cancer. From individual monitoring of about 1000 children up to 15 years of age for a thyroid exposure, no child was found exceeding the screening level of 0.2 μSv/h.

The whole-body counting showed that the internal exposure levels of Japanese residents were much lower than previously estimated (Hayano et al., 2013). Between 12 and 20 months after the Fukushima accident, the ^{137}Cs detection frequency was 1.0% (0.09% among children). In the town of Miharu, where they measured 95% of the children (ages 6–15) enrolled in town-operated schools, the ^{137}Cs body burdens of all children were below the detection limit of 300 Bq/body in the fall of 2012. This has been the first sampling-bias-free assessment of the internal exposure of children in Fukushima. This does not mean, however, that Fukushima residents are free of internal exposure risks, as evidenced by a small number of senior citizens whose body burden exceeded 100 Bq/kg. This study indicates that well-organized food screening with subsequent whole-body counting has been very efficient in minimizing the low-level internal exposure of the Japanese citizens.

Priest (2012) estimated radiation doses to the public from inhalation of radionuclides in the air. He used in his estimation the main atmospheric releases of radionuclides from the damaged Fukushima Daiichi NPP (Chapter 5), measured by TEPCO (2011a,b, 2012) as gamma-dose rates as a function of time (UTC, Coordinated Universal Time):

- 14 March, ~2 am: hydrogen explosion at Unit 3.
- 14 March, 9:10 pm–15 March, ~2 am: hydrogen explosion at Unit 2 and fire/explosion at Unit 4.
- 15 March, 1:30 pm: unidentified cause.

- 16 March, 1:00 am: fire in Unit 4 and steam releases from Unit 3.
- After 18 March: smaller unidentified releases.

Radionuclides used in the dose estimation were ^{131}I (vapor and particulate phase), ^{132}I (particulate phase), ^{132}Te, ^{134}Cs, ^{136}Cs and ^{137}Cs. The dose rates were estimated for Takasaki (the Comprehensive Nuclear-Test-Ban Treaty Organization (CTBTO) station), Chiba (Japan Chemical Analysis Center (JCAC) station) and Tokyo, as well as for other prefectures except Fukushima Prefecture, where monitoring data were not available. The highest committed effective doses from inhalation were obtained for Mito (Ibaraki Prefecture, 0.32 mSv), Miyagi (Senday, 0.18 mSv), Tochigi (Utsunomiya, 0.18 mSv) and Gunma (Takasaki/Maebashi, 0.12 mSv). For other prefectures, the inhalation doses were below 0.1 mSv (PMJC, 2011a,b).

The total dose rates (including inhalation, gamma-doses from radioactive clouds and from deposited radionuclides) were by about 20% higher than the doses from inhalation. The total doses for the largest towns were 0.1 mSv for Tokyo and 0.08 mSv for Chiba. The thyroid doses were from 5.9 mSv at Mito down to 2.3 mSv at Gunma (1.4 mSv for Tokyo and 1.1 mSv for Chiba).

The Japan Atomic Energy Agency (JAEA, 2011) came with estimations of doses (except of Fukushima Prefecture) between 0.01 and 1 mSv. It is expected that the total radiation doses during 1 year exposition after the Fukushima accident will not exceed 1 mSv. They will be therefore below the existing dose limit of 1 mSv accepted worldwide (ICRP, 1991; IAEA, 2006, 2011b) for the public, and they will not pose any health hazards to the Japanese population (WHO, 2012, 2013).

7.2.1.2 Total Radiation Doses to the Public Estimated by the WHO

As recently the World Health Organization (WHO, 2013) reviewed in very comprehensive way an impact of the Fukushima accident on the health of the Japanese population, we shall not assess in detail the radiation doses from terrestrial sources, but summarize only shortly the results presented in the WHO report. The readers looking for more detail information on the terrestrial impact of the Fukushima accident are referred, therefore, to the WHO (2013) report, as well as to the previous report WHO (2012), where preliminary estimations of radiation doses to the public were made. Doses in the following areas were considered in the WHO (2013) report:

- Locations within Fukushima Prefecture (outside of the 20 km evacuation zone) where doses were likely to be among the highest of those received by the general population.
- The rest of Fukushima Prefecture.
- The prefectures in Japan nearest Fukushima.
- The rest of Japan.
- Countries neighboring Japan.
- The rest of the world.

Doses within a 20 km radius around Fukushima Dai-ichi NPP (the restricted zone) were not assessed in the WHO preliminary dose estimation as most people in that area were rapidly evacuated. However, a certain dose may have been received prior to evacuation by these people as well.

The estimated effective doses in the first year after the Fukushima accident are described in the WHO (2013) report as follows:

- In the most affected areas of Fukushima prefecture, the estimated effective doses are within 10–50 mSv.
- In the rest of Fukushima prefecture, the estimated effective doses are within 1–10 mSv.
- In prefectures neighboring Fukushima, the estimated effective doses are within 0.1–10 mSv.
- In all other Japanese prefectures, the effective doses are estimated to be within 0.1–1 mSv.
- In the rest of the world, estimated effective doses are usually far below 0.01 mSv.

The exposure pathways which contributed most to effective doses varied with distance from Fukushima. In the most affected regions by the radioactive plume, the external dose from ground shine was important, but with increasing distance from the Fukushima site, the ingestion of food became the main contributor.

The estimated thyroid doses in the first year estimated in WHO (2013) report are as follows:

- In the most affected area of Fukushima Prefecture, the estimated thyroid doses are within

10–100 mSv (with the exception of one example location where the upper bound of the estimated thyroid doses to infants is 200 mSv).

- In the rest of Fukushima Prefecture, the estimated thyroid doses are within 1–10 mSv for adults and 10–100 mSv for children and infants.
- In other Japanese prefectures, the estimated thyroid doses are within 1–10 mSv for all age groups.
- In the rest of the world, estimated thyroid doses are usually far below 0.01 mSv.

The exposure pathways that contribute most to the thyroid dose vary with distance from Fukushima. In the more affected regions, inhalation, external irradiation from the radioactive cloud and from ground shine are important as well, but with increasing distance from the Fukushima site (i.e. when overall exposure is very low), the ingestion of food becomes again the main contributor.

In the two most affected hot spots in the Fukushima Prefecture (Iitate village and Namie town), the preliminary estimated radiation effective doses for the first year ranged from 12 to 25 mSv. They could raise cancer by 7% in males, and breast cancer by 6% in females exposed as infants. Lifetime risks for all solid cancers are expected to rise by 4% in females exposed as infants. The assessment does not expect an increase in miscarriages, stillbirths or physical and mental disorders in babies born after the accident.

The estimated lifetime risk for thyroid cancer increases by up to around 70% over baseline rates in females exposed as infants. These percentages represent estimated relative increases over the baseline rates, not the absolute risks for developing such cancers. Due to the low baseline rates of thyroid cancer, even a large relative increase represents a small absolute increase in risks.

Radiation doses in the other locations of the Fukushima Prefecture were estimated to be between 3 and 5 mSv, by about a factor of two above the natural background. In those locations, the increased risk of cancer over a lifetime could be at half of the level predicted for the most radioactive zones. In other locations, no observable increases in cancer incidence are expected.

However, these estimations are so low that statistically they will not be detectable in epidemiologic studies. The report is still only a partial health assessment as collective cancer risks have not been covered yet, especially in a time span of several tens of years (Beyea et al., 2013).

7.2.2 Radiation Doses to the Public from Marine Sources

We shall discuss in greater detail the radiation doses to the public from marine sources. The assessment of radiation doses to the public may come from ingestion of seafood collected either in coastal waters and/or in the open ocean. The radiation doses can be considerably higher for public living close to the Fukushima Daichi NPP, where inhalation and surface contamination should be taken into account, as well as a consumption of contaminated food from agricultural production.

Following the IAEA's Marine Radioactivity Dose Assessment (MARDOS) (IAEA, 1995; Aarkrog et al., 1997) project, the radiation doses from consumption of marine food were calculated by two different methods:

1. Using the estimated activity concentrations of ^{137}Cs in seawater and by applying recommended concentration factors.
2. Using the estimated radionuclide concentrations in the marine products, e.g. in fish and shellfish.

As usually available data on radioactivity of marine biota are very sparse, the first method is generally preferred because it may better integrate radiation doses to the public from seafood collected both in coastal waters, as well as from the open ocean. As we see later, however, the Fukushima case was very different due to an influence of the strong Kuroshio Current which transported contaminated coastal waters to the open ocean.

The effective dose commitment (S) from consumption of marine products collected in a sea region (method 1) was calculated (following Nakano and Povinec, 2012) using the formula

$$S = \sum_{j,k} = (\mathrm{DC})_j(\mathrm{IN})_k(\mathrm{CF})_{j,k}(C_W)_j,$$

where

$(DC)_j$ represents the dose coefficient for a radionuclide j (Sv/Bq);

$(IN)_k$ represents the averaged intake rate of a marine product k (kg/year);

$(CF)_{j,k}$ represents the concentration factor for a radionuclide j and a product k;

$(C_W)_j$ represents the concentration of a radionuclide j in seawater (Bq/kg).

The dose coefficients, $(DC)_j$, for specific nuclides j were obtained from the International Commission of Radiological Protection (ICRP) report (ICRP, 1995). In the case of ^{137}Cs and ^{134}Cs, the values of 1.3×10^{-8} and 1.9×10^{-8} Sv/Bq were used, respectively. The averaged intake rate, $(IN)_k$, of a marine product k by the Japanese public was estimated from the statistical records of the Ministry of Health, Labor and Welfare (MHLW), which have been available for 2008 (MHLW, 2010). Table 7.1 also lists the average intake rates $(IN)_k$ used in the calculations. Generally, it is known that people in the world consume smaller amounts of marine products than the Japanese population. The concentration factor, $(CF)_{j,k}$, is defined as a massic activity of a marine product (Bq/kg) per radionuclide concentration in seawater (Bq/kg) at the equilibrium conditions. Table 7.1 also lists IAEA recommended $(CF)_{j,k}$ values (IAEA, 2004) used in the presented calculations.

The effective dose commitment from consumption of marine products collected in a sea region (method 2) was calculated using the approach

Table 7.1. Diet Habits and Concentration Factors

Seafood	Food intake (g/day)	Concentration factors ^{134}Cs	Concentration factors ^{137}Cs
Fish	64	100	100
Crustaceans	5.4	50	50
Shellfish	3.5	60	60
Seaweed	10	50	50

described by Aarkrog et al. (1997) when directly measured radionuclide concentrations in seafood were used. In the case of individual dose rates, calculations were done using the formula

$$S = \sum_{j,k} (\mathrm{CF})_j (\mathrm{IN})_k (C_f)_j$$

where

$(\mathrm{CF})_j$ is the dose conversion factor for radionuclide j (Sv/Bq);

$(\mathrm{IN})_k$ is the averaged intake rate of a marine product k (kg/year);

$(C_f)_j$ is the concentration of a radionuclide j in seafood (Bq/kg).

Our approach will be very conservative, as we shall expect that the marine products will be consumed as complete samples. This is correct, e.g. in the case of shellfish and seaweed, however, in the case of fish only about 50% is really consumed. Also we do not take into account loses due to cooking, which can represent about 70% of the total marine product.

The estimated doses should be compared with pre-Fukushima doses calculated using radionuclide data sets stored in Global Marine Radioactivity Database/Marine Information System (GLOMARD/MARIS) databases developed by the IAEA's Marine Environment Laboratories in Monaco (Povinec et al., 2004, 2005, 2006).

7.2.2.1 Radiation Doses from Consumption of Seafood Collected in Coastal Waters

The most sensitive part of the dose calculation using the method 1 will be a correct assessment of radionuclide concentrations in seawater. As we discussed in Chapter 5, the radionuclide concentrations differed by several orders of magnitude due to different release rate scenarios (from about 1 kBq/m^3 to about 90 MBq/m^3). As the radionuclide concentrations varied with time, as well as with the distance from the shore, it would be rather difficult to calculate, in this case, effective dose commitments. If we take a rather conservative approach expecting that the average ^{137}Cs (and similarly also ^{134}Cs) activity concentration off Fukushima during most of the year 2011 was about 10 kBq/m^3, then the effective dose commitment calculated using the method 1 would be about 0.3 mSv/year from ingestion of ^{137}Cs in fish (and similar for ingestion of ^{134}Cs in fish). The total effective dose commitment from ingestion of both radionuclides in fish, shellfish and seaweed will be about 0.7 mSv/year.

We mentioned in Chapter 5 that there has been a rapid seawater transport due to the Kuroshio Current from the coast to the open Pacific Ocean (Nakano and Povinec, 2012). Therefore in such situation, and for some fish species, which migrate in the ocean for large distances, it will be difficult to use for calculation of radiation doses the method 1—radionuclide concentrations in seawater. It would be probably better to integrate the data on radionuclide content in fish and other seafood, and for the dose calculation to use the method 2—radionuclide concentrations in marine biota.

However, there are also problems with application of the method 2 due to rapid changes of radionuclide contents in seawater and corresponding fish. Fish can migrate several tens of kilometers, and some types of fish even more (e.g. Yellow tuna or Bluefin tuna can migrate from Japan to California; Madigan et al., 2013). Therefore, the radionuclide concentrations in seawater, which affected the fish under consideration, could change over the migration distance by several orders of magnitude.

This of course should not be a problem for other types of seafood, such as seaweed and shellfish, which are either fixed on the seafloor (e.g. seaweed), or the travelling distances are small (e.g. crustaceans). It is therefore almost impossible to get reasonable average radionuclide concentrations in fish, and also the data set would be too small for getting required statistical significance of the dose assessment.

The ^{137}Cs concentrations varied in fish within several orders of magnitude, from about 2 Bq/kg ww (wet weight) to about 5 kBq/kg ww, but the majority of ^{137}Cs concentrations were in the interval 10–1000 Bq/kg ww (see Chapter 5, Fig. 5.30, for more details). It seems therefore that the most suitable way to calculate the radiation doses from seafood in highly contaminated areas would be to use the maximum permissible radionuclide concentrations in a given type of the seafood. The resulting dose rates can be then adjusted for possible deviations, both in the radionuclide levels in the seafood, as well as in the consumption rate of seafood by a hypothetical group of people, which could be under a radiation stress. When more data on radionuclide levels in seafood will be available, then more precise dose assessment based on the real radionuclide concentrations could be carried out.

The Japanese Government has applied very strict regulations for radionuclide content in seafood. As can be seen from Table 7.2, the Japanese limits (both for ^{134}Cs and ^{137}Cs) due to the Fukushima accident were decreased from 500 to 100 Bq/kg ww (MHLW, 2012), so they became by about a factor of 4–10 lower than for other Asia countries. The Codex value of 1000 Bq/kg ww, which has been accepted by most of the world countries, is assuring the maximum effective dose limit to population of 1 mSv/year. The Japanese approach, which has been thus very conservative, has been based on large consumption amounts of seafood, as well on the fact to make provisional regulation limits to be safer from the point

Table 7.2. Regulations for Radionuclide Content in Seafood Adapted by Japan and Some Other Countries

Country	$^{134}Cs + ^{137}Cs$ (Bq/kg ww)
Codex	1000
Japan	500 ↦ 100
EU	1250 ↦ 100
USA	1200
Hong Kong	1000
Malaysia	1000
Philippines	1000
China	800
Thailand	500
Singapore	500
South Korea	370

of view of total dose commitments from other contamination sources, such as inhalation, external irradiation and the ingestion of terrestrial food.

National and regional institutions have been carrying extensive monitoring programs to exclude those seafood items, which were over the radionuclide concentration limit. This has also been done outside of Japan. For example, the monitoring program of the European Union did not find any seafood, which would be over the radionuclide limit (EC, 2012).

The approach based on the regulation limits is, of course, having a weak point as there could be a hypothetical group of people (e.g. a fisherman family), which will not be covered by a radionuclide screening, and the delivered radiation doses could be thus higher. However, due to a heavy propaganda from both the central and local governments, there is only a small chance that some of the inhabitants would not be informed about precautions when eating seafood from highly contaminated coastal waters (PMJC, 2011a,b).

The estimated effective dose commitments from consumption of marine products collected in Japan coastal waters in the Pacific Ocean (calculated using the methods 1 and 2) are presented in Table 7.3. If we

Table 7.3. Estimated Effective Dose Commitments (EDC) from Consumption of Seafood Collected in Japan Coastal Waters of the Pacific Ocean

Seafood	EDC—method 1 @ $^{134,137}Cs$ activity in seawater of 1 Bq/kg (mSv/year)	EDC—method 2 @ ^{137}Cs activity in fish of 1000 Bq/kg (mSv/year)
Fish	0.6	0.6
Shellfish	0.04	~0.05
Seaweed	0.08	~0.05
Total	0.72	0.7

use in the method 2 as the average ^{137}Cs content in fish of 100 Bq/kg ww (the limit adopted by the Japanese government), the resulting effective dose commitment will be about 0.03 mSv/year. If we use for the ^{137}Cs limit the Codex value of 1000 Bq/kg ww (adopted in most of the countries), then the effective dose commitment will be about 0.3 mSv/year. We see from Table 7.3 that both the methods 1 and 2 gave similar results. If we do summation of doses for ^{137}Cs and ^{134}Cs, and also for other pathways (shellfish and seaweed), we get for the total dose a value 0.7 mSv/year.

We also calculated effective dose commitment for a critical group tentatively consuming fish with ^{137}Cs content of 1000 Bq/kg ww, and the seafood amounts by a factor of four higher as the Japanese average per year (i.e. the total consumption of seafood of 100 kg/year). The annual dose from ^{137}Cs will be then about 1.3 mSv, or the total dose including ^{134}Cs and other pathways will be about 2.9 mSv/year, slightly higher than the world average dose from natural sources (2.4 mSv/year).

It is expected that uncertainties of the calculated doses will be within a range of ±50%, similarly to other dose assessment exercises (Aarkrog et al., 1997; IAEA, 1995), however, a change in the radionuclide source term may have a greater effect.

7.2.2.2 Radiation Doses from Consumption of Seafood from the Open Pacific Ocean

In the open Pacific Ocean, as we discussed in Chapter 5, it is enough if we take for the estimation of radiation doses only ^{134}Cs and ^{137}Cs. The highest ^{134}Cs and ^{137}Cs concentrations $(C_W)_j$ in the open northwest Pacific Ocean (around 25 Bq/m^3), predicted by the model simulations were used for the dose estimation (Nakano and Povinec, 2012). Then, the concentrations in marine biota were calculated using the method 1 by multiplying radionuclide concentrations in water with the concentration factors $(CF)_{j,k}$. No loss by market dilution and cooking was considered.

The effective dose commitment for ingestion of ^{134}Cs and ^{137}Cs in fish caught in 2012 at the open north-western Pacific Ocean was estimated to be 1.8 μSv/year (1.0 and 0.8 μSv/year, respectively), and about 2 μSv/year for all pathways (Table 7.4). Most of

Table 7.4. Estimated Effective Dose Commitments (EDC) from Consumption of Seafood Collected in the Open Pacific Ocean after the Fukushima Accident (^{134}Cs and ^{137}Cs Content in Seawater of 25 Bq/kg), and during the Pre-Fukushima Time (^{137}Cs Content in Seawater of 1 mBq/kg)

Seafood	EDC—method 1 @ $^{134,137}Cs$ activity in seawater of 25 mBq/kg (μSv/year)	EDC—method 1 @ ^{137}Cs activity in seawater of 1 mBq/kg (μSv/year)
Fish	1.8	0.03
Shellfish	0.06	0.001
Seaweed	0.12	0.002
Total	2	0.033

the dose will originate from fish consumption. The estimated dose is much lower than the dose limit of 1 mSv/year recommended by ICRP and IAEA. Because the dose is proportional to the consumption amount of contaminated seafood, most public in the world will get lower doses than the Japanese population.

There have been many radionuclides released from the Fukushima Dai-ichi NPP. Using the atmospheric release rates for 30 radionuclides published by NISA (2011a), the internal dose from these radionuclides was also estimated. Assuming that the environmental behavior of these radionuclides is similar to Cs, the concentration (C_W) in seawater for a nuclide j can be obtained from the following equation (Nakano and Povinec, 2012):

$$(C_W)_j = \frac{Q_j}{Q_{Cs137}} \cdot \frac{e^{-\lambda_j t}}{e^{-\lambda_{Cs137} t}} \cdot (C_W)_{Cs137},$$

where Q is the amount of atmospheric release of radionuclide, λ is the decay constant, and t (= 1 year)

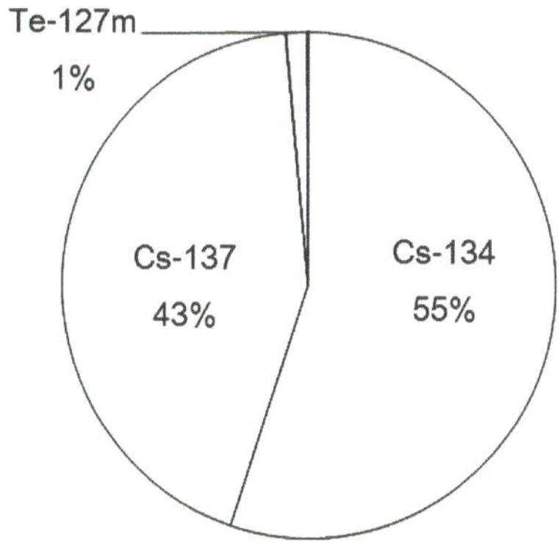

Figure 7.1 Comparison of internal dose contributions from several radionuclides released into the marine environment after the Fukushima accident. (Source: After Nakano and Povinec (2012))

is the elapsed time after the release of radionuclides from the Fukushima Dai-ichi NPP.

A comparison of internal dose contributions from several radionuclides released into the marine environment is presented in Fig. 7.1. As the contribution from ^{134}Cs and ^{137}Cs to the total dose delivered by 30 radionuclides represents 98%, it has been confirmed that the Cs isotopes are of the main interest for the dose assessment. The monitoring of radionuclide levels and assessment of doses in the future should be therefore focusing on ^{134}Cs and ^{137}Cs in the coastal marine environment, and probably also on ^{90}Sr, as there are no data available to assess contribution of this radionuclide to the radiation doses of the Japanese population (Povinec et al., 2012a).

7.2.2.3 Comparison of Radiation Doses from Consumption of Seafood– Natural vs Anthropogenic Radionuclides

The most comprehensive comparison of radiation doses from the consumption of seafood collected in the world ocean and the adjacent seas was carried out in the framework of the MARDOS project carried out by the IAEA's Marine Environment Laboratories in Monaco (Aarkrog et al., 1997; IAEA, 1995). Two most significant radionuclides for marine dose assessment were studied:

- ^{137}Cs (a gamma-emitter)—representing the anthro pogenic radioactivity sources (mainly global fallout, discharges from nuclear reprocessing facilities, and the impact of the Chernobyl accident on the marine environment).

- ^{210}Po (natural alpha-emitter in the ^{238}U decay chain)—representing the natural radioactivity sources.

Data concerning ^{137}Cs in the oceans seem to be quite reliable. The behavior of this radionuclide in the marine environment is well known, its concentration factors are well established, and predictions can easily be done about its fate in the case of accidental releases (IAEA, 1995; UNSCEAR, 2008a,b,c,d).

Data referring to ^{210}Po are much less abundant and concentration factors from the literature do not seem to be supported enough by field studies, at least in some cases. A wide range of ^{210}Po concentrations in the marine environment can be observed and the reason for such ranges of values is not yet sufficiently understood (IAEA, 1995).

The collective effective dose commitments estimated for the consumption of seafood collected in Food and Agriculture Organization (FAO) of United Nations fishing areas (FAO, 1988) of the world ocean in 2000 are presented in Fig. 7.2. It can be seen that the dominant contribution is due to ^{210}Po, even in such areas as the European seas (3300 man Sv for ^{210}Po vs 56 man Sv for ^{137}Cs), which were affected by radioactive discharges from the nuclear reprocessing facilities in Sellafield and La Hague. In the north-western Pacific fishing area, the ^{210}Po dominates again over the ^{137}Cs (16,300 man Sv for ^{210}Po vs 18 man Sv for ^{137}Cs), confirming that the consumption of seafood in this part of the world is much higher than in other fishing areas, as the ^{210}Po activity concentration in seawater (around 1 Bq/m^3) is uniform over the world ocean (Aarkrog et al., 1997; IAEA, 1995).

The collective effective dose commitment from fish and shellfish caught in 2000 for global population was estimated to be

- for ^{137}Cs—100 man Sv for fish and 7 man Sv for shellfish;
- for ^{210}Po—10,000 man Sv for fish, and 20,000 man Sv for shellfish.

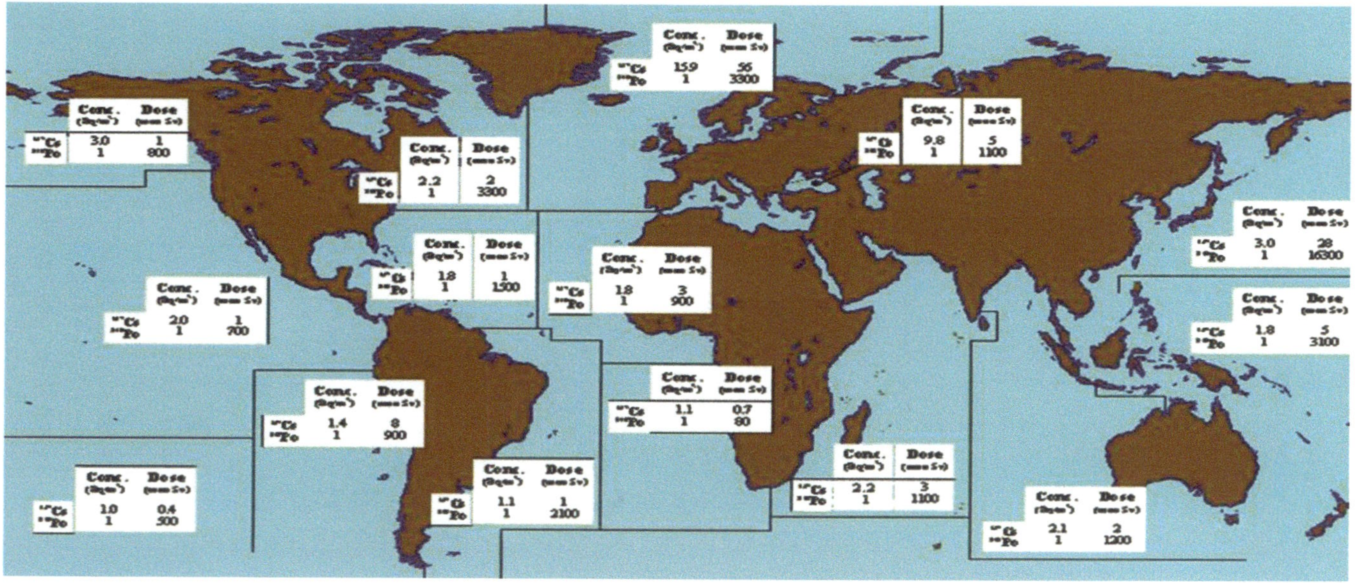

Figure 7.2 Average ^{137}Cs and ^{210}Po concentrations in surface waters of the FAO fishing areas of the world ocean, and the collective effective dose commitments from ingestion of ^{137}Cs and ^{210}Po in fish and shellfish (adjusted to the year 2000). *C*—activity concentration (in Bq/m^3) and *S*—collective effective dose commitment (in man Sv). (Modified from Aarkrog et al. 1997)

The contribution of ^{137}Cs to the collective effective dose commitment from fish and shellfish consumption is negligible, below 1% of that for ^{210}Po.

The integral dose to the population from 2000 onwards (assuming that there will be no changes in predicted input of these nuclides to the oceans, and that the effective residence time for radiocesium is 25 years) was calculated according to the formula,

$$D(\propto) = \int_0^\infty D_i e^{\frac{\ln 2}{25} t} dt,$$

where

$D(\propto)$ is the integrated dose from 2000 to infinity;

D_i is the collective effective dose commitment from consumption during 2000.

The dose due to ^{137}Cs becomes 3500 man Sv from consumption of fish and 300 man Sv from consumption of shellfish. These figures can be compared with the doses of 10,000 man Sv received in 1 year by consumption of ^{210}Po in fish and 20,000 man Sv by consumption of shellfish.

The results obtained in the framework of the MARDOS projects provided the most complete data available on radionuclide levels in the marine environment and on doses to world population from marine radioactivity through ingestion of marine foods. The results have been used as the international reference source on the average radionuclide levels in the marine environment and the corresponding collective committed effective doses from fish and shellfish to the public.

A comparison of radiation doses from consumption of seafood contaminated by the Fukushima accident with pre-Fukushima doses evaluated in the MARDOS project suggests these conclusions:

- Individual dose commitment from consumption of ^{134}Cs and ^{137}Cs in seafood collected in Japan coastal waters of the Pacific Ocean in 2011 may reach about 0.7 mSv/year, but more probably only about 0.07 mSv/year as the Japanese government implemented a strict regulation on ^{137}Cs content in seafood (limit 100 Bq/kg ww). Although these doses are by about four orders of magnitude higher than the pre-Fukushima dose (0.03 μSv/year calculated for the ^{137}Cs content in seawater of 1 Bq/m^3), they are below the maximum permissible annual dose to the public from external sources (1 mSv/year), or the world average dose from natural sources (2.4 mSv/year). The estimated doses are comparable with the annual dose due to ^{210}Po ingestion in seafood (0.7 mSv/year) or in fish (0.05 mSv/year).

- Individual dose commitment from consumption of ^{137}Cs in fish caught in the open Northwestern Pacific Ocean in 2012 is 0.8 μSv/year (together with ^{134}Cs and other pathways it is about 2 μSv/year), what is about 30 times above the global fallout background. This dose is, however, about 60 times lower than the dose due to the consumption of natural ^{210}Po in fish (0.05 mSv/year), and 350 times lower than the dose due to the consumption of natural ^{210}Po in fish and shellfish (0.7 mSv/year).

7.2.2.4 Other Radionuclides and Pathways for Dose Deliveries from Marine Sources

There are other radionuclides, which could be considered radiologically important, e.g. radionuclides from nuclear weapons testing in the atmosphere such as ^{90}Sr, ^{239,240}Pu, ^{241}Am, ^{3}H, and ^{14}C are still present in measurable quantities in the world ocean (as discussed in Chapter 6). In surface ocean waters contaminated only by global fallout, the concentrations relative to ^{137}Cs are as follows: $^{90}Sr/^{137}Cs \approx 0.66$, $^{239,240}Pu/^{137}Cs \approx 2{-}3 \times 10^{-3}$ and $^{241}Am/^{137}Cs \approx 1 \times 10^{-3}$(Livingston and Povinec, 2000, 2002). Their contribution to doses from ingestion of seafood is, however, only in the order of 25% of the dose from ^{137}Cs. The ^{137}Cs represents therefore the most important environmental radionuclide of anthropogenic origin as there is not a natural analogue of this fission product in the environment. The ^{137}Cs concentrations in the atmosphere due to atmospheric weapons testing were by about factor of 100 higher than its levels observed during the Chernobyl accident. Average global dose from nuclear bomb tests is estimated to be around 10 mSv, what is similar as the intervention dose for governmental action proposed by IAEA. This dose is also similar to the dose expected for inhabitants living in the 100 km zone around the Chernobyl (UNSCEAR, 1991). The total collective dose commitment from ^{137}Cs in marine foods due to all nuclear weapons tests in the atmosphere can be estimated to be 9000 man Sv. According to UNSCEAR (1993), this dose is about 1.3% of the total collective dose commitment from global fallout ^{137}Cs in human diet.

Although other radionuclides, e.g. tritium increased in 1963 by three orders of magnitude above its cosmogenic background (i.e. produced in the atmosphere by interactions of cosmic rays), its dose impact on the population was very low due to its soft beta-rays (18.6 keV) emitted during its decay (UNSCEAR, 2008a,b,c,d).

Smaller doses but lasting for about 10,000 years will be delivered by the long-lived ^{14}C (half-life 5730 years), which was produced during nuclear weapons test in the atmosphere at concentrations by about a factor of two higher than its cosmogenic levels. This radionuclide was actually a trigger for large-scale protests against atmospheric nuclear weapons tests, which led to the signing of the antitest treaty in 1963.

Other marine pathways that discussed in this chapter till now may also be of interest, particularly radionuclides in edible seaweed and transuranics in shellfish. On a global scale, although the consumption of seaweed is generally low, this pathway is of some regional importance, e.g. especially, in Japan. In a global context, the contribution from pathways other than that for fish/shellfish ingestion is believed to be $<10\%$ of the collective dose from ^{137}Cs and ^{210}Po in the marine environment.

The radiation doses from marine pathways by naturally occurring radionuclides other than ^{210}Po are about one-third of the dose from ^{210}Po, the main contributors to this dose being ^{210}Pb, ^{40}K, ^{87}Rb, $^{226,228}Ra$, $^{235,238}U$ and ^{14}C. Table 7.5 clearly demonstrates that natural radionuclides are the dominant species in the marine environment.

The doses to man from anthropogenic radionuclides in the marine environment are generally 1–2 orders of magnitude less than the doses from such radionuclides in the terrestrial environment.

Table 7.5. Comparison of Natural (^{40}K) and Anthropogenic (^{137}Cs) Levels in the World Ocean and the Adjacent Seas during Pre- and Post-Fukushima Time

Sea	^{40}K (Bq/kg)	^{137}Cs (@ 2000) (Bq/kg)	^{137}Cs (@ 2012) (Bq/kg)
World Ocean	12	0.002[1]	
Dead Sea	180	0.002[1]	
Baltic Sea	11.5	0.05[2]	
Irish Sea	11.5	0.05[3]	
North Sea	11.5	0.02[3]	
Open Pacific Ocean	12	0.002[1]	0.025
Offshore Fukushima	11.5	0.002[1]	10

[1]Global fallout.
[2]Chernobyl impact.
[3]Impact of the Sellafield and La Hague reprocessing facilities.

Compared with doses from natural radionuclides, the doses from anthropogenic radionuclides in the marine environment are therefore insignificant (Aarkrog et al., 1997; IAEA, 1995).

7.2.3 Comparison of Fukushima-Derived Radiation Doses from Terrestrial and Marine Environments

The dominant radiation doses to the public are usually from inhalation and from external irradiation from radioactive clouds and radionuclides deposited on the ground. With increasing distance from the Fukushima NPP, these doses will be decreasing and the doses from ingestion of contaminated food will dominate, however, they will be usually much lower than from inhalation (Hamada and Ogino, 2012), as we already discussed. The doses received from ^{137}Cs via marine foods are in general lower than those received from terrestrial foods. If the terrestrial and the marine environments received the same deposition of ^{137}Cs per unit area, the dose commitment received by man from the seafood will typically be 2 orders of magnitude less than that received from the terrestrial food chain (IAEA, 1995; Aarkrog et al., 1997).

This is also supported with the data obtained for the Fukushima case when in the two most affected hot spots in the Fukushima Prefecture (Iitate village and Namie town), the estimated radiation effective doses for the first year ranged from 12 to 25 mSv. On the other hand, the doses from ingestion of seafood were estimated to be below 0.7 mSv/year. More information on the terrestrial doses may be found in the recently published WHO (2013) report.

7.3 Comparison of Radiation Doses to the Public from Various Sources

For having better information on the radiation doses to the public from the Fukushima accident, we shall also compare the radiation doses to the public from other two sources of anthropogenic radionuclides in the environment:

1. Chernobyl accident
2. Applications of radiation and radionuclides in medicine.

7.3.1 Radiation Doses due to the Chernobyl Accident

Although both the Chernobyl and the Fukushima accidents were classified equally as no. 7 on the International Nuclear Event Scale (INES), (IAEA, 2009), the impact of the Fukushima accident on the total environment (citizens, fauna and flora) has been much smaller. The Chernobyl accident affected mainly the people working in the Chernobyl area during emergency actions, plus evacuees from the contaminated zone (about 135,000 people, including 50,000 from the Pripyat town), and residents of control zone and other contaminated zones in Ukraine, Belarus and Russia (IAEA, 2006; UNSCEAR, 2008a,b,c,d, 2011):

About 600,000 people who took part in the liquidation of the accident (called liquidators) got during

1986–1988 doses around 100 mSv (UNSCEAR, 1988, 2008a,b,c,d; WHO, 1988, 2006).

Residents of control zones (mostly living in the area without permission) got during 1986–2005 average accumulated doses over 50 mSv. Evacuees from the contaminated zone got an average dose in 1986 of 30 mSv. Residents of other contaminated areas got during 1986–2005 an average accumulated dose of 10–20 mSv.

Unfortunately, there were many deaths due to an overexposition to radiation:

- 139 workers died with acute irradiation syndrome soon after the accident;
- 31 workers died several months after the accident;
- 10 people (including nine children) died several years after the accident.

There are still about 4000 people, who are in danger because of a potential developing of a disease due to radiation as the latency period for solid cancers caused by excess radiation exposure is more than 10 years (IAEA, 2006; UNSCEAR, 2008a,b,c,d).

The individual doses to the European population outside the former Soviet Union were generally below the limit of 1 mSv/year. For example, for the Central Europe (Slovakia), which may represent a typical European country, the total dose (inhalation, ingestion and external irradiation) was below 0.2 mSv (Fig. 7.3), what represented <10% of the dose from all radiation sources. About 60% of the dose is

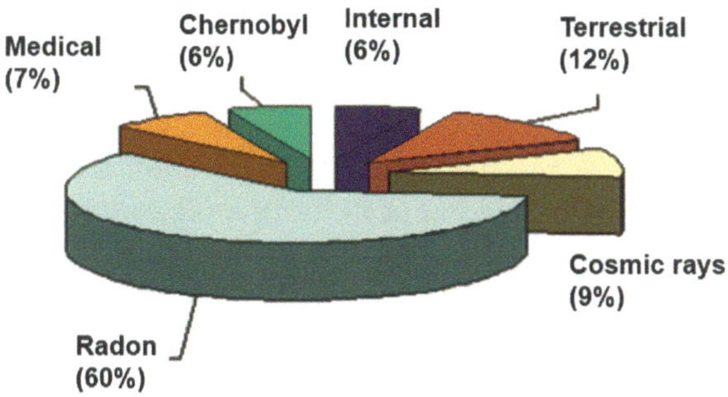

Figure 7.3 Comparison of radiation doses from Chernobyl and natural sources of ionizing radiation (including medical applications) to citizens of Slovakia (Central Europe). (Modified from Cabánekova and Vladár (1988))

delivered by radon and its daughter products in dwellings (Cabáneková and Vladár, 1998).

The global collective dose commitments from ^{137}Cs in marine foods contaminated by the Chernobyl accident has been estimated to be 2000 man Sv. On the other hand, liquid radioactive discharges from the nuclear reprocessing facilities in Sellafield (UK) and La Hague (France) contributed until 1984 approximately 3000 man Sv (CEC, 1989). The total collective dose commitment from marine-derived ^{137}Cs from global fallout, liquid radioactive discharges in Europe, and the Chernobyl accident is 1.4×10^4 man Sv, which corresponds to half of the dose received in 1 year from ^{210}Po in marine foods (IAEA, 1995).

7.3.2 Radiation Doses from Other Sources

General radiation doses for the public from various sources are listed below:

- 0.1 μSv—one banana equivalent dose
- 0.25 μSv—US limit on effective dose from a single airport security screening
- 5–10 μSv—a dental X-ray test
- 0.01–0.1 mSv—a long-range oversea flight
- 0.1 mSv—a simple X-ray test
- 0.4–0.6 mSv—a mammogram
- 1 mSv/year—the maximum effective dose commitment for the public
- 2.4 mSv/year—the worldwide average of the effective dose commitment for the public from natural sources of radioactivity in the environment
- 10–30 mSv—a CT scan or similar nuclear medicine diagnostics
- 10 mSv/year—the intervention level for a governmental action, e.g. evacuation, food control, etc.
- 50 mSv/year—the lowest effective dose commitment for any statistical risk of cancer
- 50–100 mSv—maximum exposure for professionals (in some cases increased up to 200 mSv, as was also done during the Fukushima accident)
- 0.5–1 Sv—mild radiation sickness
- >1 Sv—serious health problems (over 2–3 Sv—death).

The health effects due to the radiation overdoses are divided into two groups (ICRP, 2007):

- Deterministic—starting with a threshold, death damage of cells—e.g. at doses above 500 mSv, we could expect serious health problems.
- Stochastic—no threshold apply, information damage of cells may occur; multistress factors could enlarge the impact.

The radiation doses in prefectures around the Fukushima Dai-ichi NPP were well below the deterministic levels, and therefore health effects are not expected to occur in the general population (WHO, 2012, 2013). The impact of the Fukushima accident was also kept well below the 50 mSv/year limit for the statistical risk of cancer (ICRP, 1995, 2005, 2012). In some areas around the Fukushima NPP, however, the intervention level of 10 mSv was reached which required governmental action on the evacuation and food control, as it has been done.

A well-designed long-term epidemiological study should be developed which could help a better estimation of the radiation impact of the Fukushima accident on the Japanese population. This was not the case after the Chernobyl accident when residents from contaminated areas spread over large territory of the former Soviet Union. However, chances to detect any irregularities in such an epidemiological study will be very low as the expected effects will be within statistical uncertainties of existing cancer incidence from sources other than the Fukushima radiation impact.

7.4 Radiation Doses to Biota

In recent years, there has been a clear shift in radioecological studies from the egocentric (also called anthropocentric) approach to the ecocentric approach, supported both by the ICRP and the IAEA. Why in the first case a man was in the center of the interest, the second approach is treating the man as a part of the total environment, i.e. humans, fauna and flora form one complex system. Our aim is therefore to better understand the effects of radiation on the environmental components, from individual organisms to populations of species and ecosystems

(ICRP, 2003, 2008; UNSCEAR, 2011; Bréchignac, 2012; Bréchignac and Doi, 2009). The studies have included preventing or minimizing effects leading to life shortening or reduction in reproductive potential of flora and fauna species, preservation of species, maintaining and ensuring appropriate quality of habitat (Kryshev and Ryazantsev, 2010; Kryshev et al., 2012).

Studying the Fukushima impact on the terrestrial and marine biota may represent therefore such an environmental laboratory to investigate the radioactivity impact on the total environment. Unfortunately, up to now, we do not have enough radionuclide data to test this new approach. An interesting paper was published recently in the terrestrial radioactivity sector on the impact of the Fukushima accident on the physiological and genetic damage to the pale grass blue *Zizeeriamaha*, a common lycaenid butterfly in Japan (Hiyama et al., 2012). The first-voltine adults in the Fukushima area which were collected in May 2011 showed relatively mild abnormalities, however, adult butterflies which were collected in September 2011 showed much more abnormalities. Similar abnormalities were experimentally reproduced in individuals from a noncontaminated area by external and internal low-dose exposures. The authors concluded that anthropogenic radionuclides released from the damaged Fukushima Dai-ichi NPP caused physiological and genetic damage to this species. The heritable germ-line genetic damage caused by low-dose exposure due to radioactive contamination in a species of butterfly may have invaluable implications for the possible future effects of radiation on animals.

In the marine sector, Kryshev et al. (2012) recently studied radiation doses to biota (pelagic fish, shellfish and algae) due to high concentrations of ^{131}I and ^{134,137}Cs in seawater off Fukushima Dai-ichi NPP. Using a concentration factor approach and the dynamic radioecological model (ECOMOD; Sazykina, 2000) they got comparable results indicating the following:

- The main pathway was internal irradiation due to ^{134}Cs and ^{137}Cs giving dose rates of about 20 μGy/h for each of them in fish and shellfish, and 10 and

20 μGy/h for ^{134}Cs and ^{137}Cs in algae. The contribution due to ^{131}I was about 10 times lower, except for algae where it was about 700 μGy/h.

- The external irradiation from water was about 10 times lower, except for ^{131}I on algae where it was lower by about a factor of 500.
- The external irradiation from all three radionuclides in sediments was by about a factor of five lower than from internal irradiation from seawater, except for algae where it was lower by about a factor of 500.

The total dose rates to fish and shellfish in coastal waters during the largest radionuclide releases were around 1 mGy/day (based on the ECOMOD model, Kryshev et al., 2012). For the later period (May 2011), the dose rates decreased to about 0.5 mGy/day. As the highest dose rates did not exceed the baseline safe level of 10 mGy/day postulated for the marine organisms, harmful effects are unexpected for the marine ecosystem (IAEA, 2007). However, some radiation effects may appear in individual marine organisms (living close to the discharge canals, NISA, 2011b) under conditions of chronic exposure. Also some marine organisms living in the bottom sediments could get radiation doses close to postulated limits. Radiation doses for fish living about 30 km far from the coast were around 0.003 mGy/day, i.e. by about two orders of magnitude lower than in coastal waters in May 2011.

When comparing the Fukushima impact on the aquatic environment with the Chernobyl accident (a cooling pond of the Chernobyl NPP in April–May 1986), we see that the radiation doses to local fish were by about a factor of five lower than in the case of Fukushima (IAEA, 2006). We may conclude on the basis of marine radioactivity studies that the Fukushima accident did not result in any significant damage of the marine ecosystem (Kryshev et al., 2012).

References

Aarkrog, A., Baxter, M.S., Bettencourt, A.O., Bojanowski, R., Bologa, A., Charmasson, S., Cunha, I., Delfanti, R., Duran, E., Holm, E., Jeffree, R., Livingston, H.D., Mahapanyawong, S., Nies, H., Osvath, I., Pingyu, Li, Povinec, P.P., Sanchez, A., Smith, J.N., Swift, D.A., 1997. Comparison of doses from ^{137}Cs

and ^{210}Po in marine food: a major international study. J. Environ. Radioact. 34, 69–90.
Beyea, J., Lyman, E., von Hippel, F.N., 2013. Energy Environ. Sci.. http://dx.doi.org/10.1039/c2ee24183h.
Bréchignac, F., 2012. Environment protection: the current challenge in radioecology. In: Plastino, W., Povinec, P.P. (Eds.), Environmental Radioactivity – New Frontiers and Developments. Societa Italiana di Fisica, Bologna, pp. 3–18.
Bréchignac, F., Doi, M., 2009. Challenging the current strategy of radioecological protection of the environment: arguments for ecosystem approach. J. Environ. Radioact. 10, 1125–1134.
Cabáneková, H., Vladár, M., 1998. The monitoring of the air contamination in the territory of Slovak Republic. J. Radioecol. 6, 3–8.
CEC, Commission of the European Communities, 1989. The Radiological Exposure of the Population of the European Community from Radioactivity in North European Marine Waters – Project MARINA. Report EUR 12483, Brussels.
Chino, M., Nakayama, H., Nagai, H., Terada, H., Katata, G., Yamazawa, H., 2011. Preliminary estimation of released amounts of ^{131}I and ^{137}Cs accidentally discharged from the Fukushima Daiichi nuclear power plant into the atmosphere. J. Nucl. Sci. Technol. 48, 1129–1134.
EC, European Commission, 2012. https://webgate.ec.europa.eu/maritimeforum.
FAO, Food and Agriculture Organization of UN, 1988. Fishery Statistics, vols. 70, 71. FAO, Roma.
Hamada, N., Ogino, H., 2012. Food safety regulations: what we learned from the Fukushima nuclear accident. J. Environ. Radioact. 111, 83–99 (Proceedings of the Japan AcSeries).
Hayano, R.S., Tsubokura, M., Miyazaki, M., Satou, H., Sato, K., Masaki, S., Sakuma, Y., 2013. Internal radiocesium contamination of adults and children in Fukushima 7 to 20 months after the Fukushima NPP accident as measured by extensive whole-body-counter surveys. Proc. Jpn. Acad. B 89, 157–163.
Hiyama, A., Nohara, C., Kinjo, S., Taira, W., Gima, S., Tanahara, A., Otaki, J.M., 2012. The biological impacts of the Fukushima nuclear accident on the pale grass blue butterfly. Sci. Rep. 2, 570. http://dx.doi.org/10.1038/srep00570.
IAEA, International Atomic Energy Agency, 1995. Sources of Radioactivity in the Marine Environment and Their Relative Contributions to Overall Dose Assessment from Marine Radioactivity. IAEA TECDOC Series No. 838, Vienna.
IAEA, International Atomic Energy Agency, 2002. Requirements GS–R2 Preparedness and Response for a Nuclear or Radiological Emergency. Jointly Sponsored by FAO, IAEA, ILO, OECD/NEA, PAHO, OCHA, WHO. IAEA, Vienna.
IAEA, International Atomic Energy Agency, 2004. Sediment Distribution Coefficients and Concentration Factors for Radionuclides for Biota in the Marine Environment. Technical Report Series No. 422. IAEA, Vienna.

IAEA, International Atomic Energy Agency, 2005. Environmental and Source Monitoring for Purposes of Radiation Protection. IAEA Safety Standards for Protecting People and the Environment, Safety Guide No. RS-G-1.8. IAEA, Vienna.
IAEA, International Atomic Energy Agency, 2006. Environmental Consequences of the Chernobyl Accident and Their Remediation: Twenty Years of Experience. Report of the Chernobyl Forum Expert Group. IAEA, Vienna. http://www.pub.iaea.org/mtcd/publications/pdf/pub1239_web.pdf.
IAEA, International Atomic Energy Agency, 2007. Arrangements for Preparedness for a Nuclear or Radiological emergency. IAEA Safety Standards for Protecting People and the Environment, Safety Guide No. GS-G-2.1. IAEA, Vienna.
IAEA, International Atomic Energy Agency, 2009. INES The International Nuclear and Radiological Event Scale User's Manual. IAEA, Vienna.
IAEA, International Atomic Energy Agency, 2010. Joint Radiation Emergency Management Plan of the International Organizations. Emergency Preparedness and Response. IAEA, Vienna. http://www.pub.iaea.org/books/IAEABooks/8453/Joint–Radiation–Emergency–Management–Plan–of–the–International–Organizations–EPR–JPLAN–2010–Emergency–Preparedness–and–Response.
IAEA, International Atomic Energy Agency, 2011a. Radiation Protection and Safety of Radiation Sources: International Basic Safety Standards (BSS), Interim ed. IAEA safety standards series, Vienna. http://www.pub.iaea.org/MTCD/publications/PDF/p.1531interim_web.pdf (accessed 16.11.12.).
IAEA, International Atomic Energy Agency, 2011b. Final Report of the International Mission to Support the Remediation of Large Contaminated Areas Off-site of the Fukushima Daiichi Nuclear Power Plant (NPP). Document NE/NEFW/2011. IAEA, Vienna. http://www.iaea.org/news- center/focus/fukushima/final_report151111.pdf (accessed 16.11.12.).
ICRP, International Commission on Radiological Protection, 1991. 1990 Recommendations of the International Commission on Radiological Protection. ICRP Publication 60. ICRP, Paris.
ICRP, International Commission on Radiological Protection, 1995. Age-Dependent Doses to the Members of the Public from Intake of Radionuclides – Part 5: Compilation of Ingestion and Inhalation Coefficients. Publication 72. ICRP, Paris.
ICRP, International Commission on Radiological Protection, 2003. Protection of Non-Human Species from Ionizing Radiation. Proposal for a Framework for the Assessment of Ionizing Radiation in the Environment. Annals of the ICRP, No. 91. Pergamon Press, Oxford.
ICRP, International Commission on Radiological Protection, 2005. Low-Dose Extrapolation of Radiation-Related Cancer Risk. Annals of the ICRP, 99 (35), 4. ICRP, Ottawa.
ICRP, International Commission on Radiological Protection, 2007. The 2007 Recommendations of the International Commission

on Radiological Protection. Ottawa, International Commission on Radiological Protection, Annals of the ICRP, 103 (37), 2–4. ICRP, Oxford.

ICRP, International Commission on Radiological Protection, 2008. ICRP Environmental Protection: The Concept and Use of Reference Animals and Plants. Annals of ICRP, No. 108. Elsevier, Amsterdam.

ICRP, International Commission on Radiological Protection, 2009. Application of the commission's recommendations to the protection of people living in long–term contaminated areas after a nuclear accident or a radiation emergency. Annals of the ICRP, 111 (39), 3. ICRP, Ottawa.

ICRP, International Commission on Radiological Protection, 2012. Statement on Tissue Reactions/Early and Late Effects of Radiation in Normal Tissues and Organs – Threshold Doses for Tissue Reactions in a Radiation Protection Context. Annals of the ICRP, 118 (41), 1–2. ICRP, Ottawa.

JAEA, Japan Atomic Energy Agency, 2011. Results of the environmental radiation monitoring following the accident at the Fukushima Daiichi Nuclear Power Plant. JAEA-Review 2011-035, 2011/08. JAEA, Tokyo.

Kawamura, H., Kobayashi, T., Furuno, A., In, T., Ishikawa, Y., Nakayama, T., Shima, S., Awaji, T., 2011. Preliminary numerical experiments on oceanic dispersion of ^{131}I, and ^{137}Cs discharged into the ocean because of the Fukushima Dai-ichi nuclear, power plant disaster. J. Nucl. Sci. Technol. 48, 1349–1356.

Kryshev, I.I., Kryshev, A.I., Sazykina, T.G., 2012. Dynamics of radiation exposure to marine biota in the area of the Fukushima NPP in March–May 2011. J. Environ. Radioact. 114, 157–161.

Kryshev, I.I., Ryazantsev, E.P., 2010. Ecological Safety of the Nuclear Energy Sector of Russia. IzdAt, Moscow (in Russian).

Livingston, H.D., Povinec, P.P., 2000. Anthropogenic marine radioactivity. Ocean Coastal Management 43, 689–712.

Livingston, H.D., Povinec, P.P., 2002. A millennium perspective on the contribution of global fallout radionuclides to ocean science. Health Phys. 82, 656–668.

Lujanienė, G., Valiulis, D., Byčenkienė, S., Šakalys, J., Povinec, P.P., 2012. Plutonium isotopes and ^{241}Am in the atmosphere of Lithuania: a comparison of different source terms. Atm. Environ. 61, 419–427.

Lyons, C., Colton, D., 2012. Aerial measuring system in Japan. Health Phys. 102, 509–515.

Madigan, D.J., Baumann, Z., Snodgrass, O.E., Ergül, H.A., Dewar, H., Fisher, N.S., 2013. Radiocesium in Pacific bluefin tuna Thunnus orientalis in 2012 validates new tracer technique. Environ. Sci. Technol. 47, 2287–2294.

Masson, O., Baeza, A., Bieringer, J., Brudecki, K., Bucci, S., Cappai, M., Carvalho, F.P., Connan, O., Cosma, C., Dalheimer, A., Didier, D., Depuydt, G., De Geer, L.E., De Vismes, A., Gini, L., Groppi, F., Gudnason, K., Gurriaran, R., Hainz, D., Halldórsson, Ó, Hammond, D., Hanley, O., Holý, K., Homoki, Z., Ioannidou, A., Isajenko, K., Jankovic, M., Katzlberger, C., Kettunen, M., Kierepko, R., Kontro, R.,

Kwakman, P.J., Lecomte, M., Leon Vintro, L., Leppänen, A.P., Lind, B., Lujaniene, G., McGinnity, P., Mc Mahon, C., Malá, H., Manenti, S., Manolopoulou, M., Mattila, A., Mauring, A., Mietelski, J.W., Møller, B., Nielsen, S.P., Nikolic, J., Overwater, R.M., Pálsson, S.E., Papastefanou, C., Penev, I., Pham, M.K., Povinec, P.P., Ramebäck, H., Reis, M.C., Ringer, W., Rodriguez, A., Rulík, P., Saey, P.R., Samsonov, V., Schlosser, C., Sgorbati, G., Silobritiene, B.V., Söderström, C., Sogni, R., Solier, L., Sonck, M., Steinhauser, G., Steinkopff, T., Steinmann, P., Stoulos, S., Sýkora, I., Todorovic, D., Tooloutalaie, N., Tositti, L., Tschiersch, J., Ugron, A., Vagena, E., Vargas, A., Wershofen, H., Zhukova, O., 2011. Tracking of airborne radionuclides from the damaged Fukushima Dai-ichi nuclear reactors by European networks. Environ. Sci. Technol. 45, 7670–7677.

MEXT, Ministry of Education, Culture, Sports, Science and Technology, 2011a. Results of airborne monitoring in restricted areas and deliberate evacuation areas. http://radioactivity.mext.go.jp/ja/contents/4000/3708/24/1305819_0616_1.pdf; http://radioactivity.mext.go.jp/ja/contents/4000/3709/24/1305819_0616_2.pdf; http://radioactivity.mext.go.jp/en/list/203/list-1.html.

MEXT, Ministry of Education, Culture, Sports, Science and Technology, 2011b. Enlargement of national environmental radioactivity monitoring at emergency of the Fukushima Daiichi and Daini nuclear power plants. http://www.mext.go.jp/b_menu/houdou/23/03/1303843.htm.

MHLW, Ministry of Health, Labour and Welfare, 2010. National health and nutrition examination survey in 2008. http://www.mhlw.go.jp/bunya/.

MHLW, Ministry of Health, Labour and Welfare, 2012. New standard limits for radionuclides in foods. http://www.mhlw.go.jp/english/topics/2011eq/dl/new_standard.pdf.

Nakano, M., Povinec, P.P., 2012. Long-term simulations of the ^{137}Cs dispersion from the Fukushima accident in the world ocean. J. Environ. Radioact. 111, 109–115.

NAIIC, Nuclear Accident Independent Investigation Commission, 2012. The Official Report of the Fukushima Nuclear Accident Independent Investigation Commission. Tokyo.

NAS, 2006. Health risks from exposure to low levels of ionizing radiation: BEIR VII phase 2. Washington, DC, National Academy of Science – National Research Council Committee to assess health risks from exposure to low levels of ionizing radiation. http://www.nap.edu/openbook.php?isbn=030909156X (accessed 25.11.12.).

NISA, Nuclear and Industrial Safety Agency, 2011a. Regarding the Evaluation of the Conditions on Reactor Cores of Units 1, 2 and 3 Related to the Accident at Fukushima Daiichi Nuclear Power Plant. Tokyo, Nuclear and Industrial Safety Agency, 2011 (www.nisa.meti.go.jp/english/press/2011/06/en20110615_5.pdf, (accessed 21.11.12.).

NISA, Nuclear and Industrial Safety Agency, 2011b. Regarding the discharge of the waste water, of which the concentration of radioactive materials exceeds the concentration limit by the notification, to the sea. http://www.nisa.meti.go.jp/english/files/en20110416-10.pdf (accessed on 20 June 2011).

PMJC, Prime Minister of Japan and His Cabinet, 2011. Report of Japanese Government to the IAEA Ministerial Conference on Nuclear Safety-The accident at TEPCO's Fukushima Nuclear Power Stations. http://www.kantei.go.jp/jp/Topics/2011/iaea_houkokusho.html.

PMJC, Prime Minister of Japan and His Cabinet, 2011a. Report of Japanese government to the IAEA Ministerial Conference on nuclear safety – the accident at TEPCO's Fukushima nuclear power stations. http://www.kantei.go.jp/jp/Topics/2011/iaea_houkokusho.html.

PMJC, Prime Minister of Japan and His Cabinet, 2011b. Government actions to ensure the safety of beef and other food.http://www.kantei.go.jp/foreign/kan/topics/201108/measures_beef.pdf.

Povinec, P.P., Hirose, K., Aoyama, M., 2012a. Radiostronium in the western North Pacific: Characteristics, behavior, and the Fukushima impact. Environ. Sci. Technol. 46, 10356–10363.

Povinec, P.P., Hirose, K., Honda, T., Ito, T., Scott, M.E., Togawa, O., 2004. Spatial distribution of ^{3}H, ^{90}Sr, ^{137}Cs and ^{239,240}Pu in surface waters of the Pacific and Indian Oceans – GLOMARD database. J. Environ. Radioact. 76, 113–137.

Povinec, P.P., Aarkrog, A., Buesseler, K.O., Delfanti, R., Hirose, K., Hong, G.-H., Ito, T., Livingston, H.D., Nies, H., Noshkin, V.E., Shima, S., Togawa, O., 2005. ^{90}Sr, ^{137}Cs and ^{239,240}Pu concentration surface water time series in the Pacific and Indian Oceans – WOMARS results. J. Environ. Radioact. 81, 63–87.

Povinec, P.P., Scotto, Ph., Osvath, I., Ramadan, H., 2006. The Marine Information System (MARIS). Isotopes in Environmental Studies. Aquatic Forum 2004. IAEA, Vienna. 68–69.

Priest, N.D., 2012. Radiation doses received by adult Japanese populations living outside Fukushima Prefecture during March 2011, following the Fukushima 1 nuclear power plant failures. J. Environ. Radioact. 114, 162–170.

Sazykina, T.G., 2000. ECOMOD – an ecological approach to radioecological modeling. J. Environ. Radioact. 50, 207–220.

Tanabe, F., 2012a. A scenario of large amount of radioactive materials discharge to the air from the unit 2 reactor in the Fukushima Daiichi NPP accident. J. Nucl. Sci. Technol. 49, 360–365.

Tanabe, F., 2012b. Analysis of core melt and re-melt in the Fukushima Daiichi nuclear reactors. J. Nucl. Sci. Technol. 49, 18–36.

TEPCO, Tokyo Electric Power Company, 2011a. Fukushima Daiichi nuclear power, station unit 2: countermeasures to stop the outflow of contaminated water, and the water amount flowed out into the sea. http://www.tepco.co.jp/en/, press/corp-com/release/11042103-e.html (accessed 20.06.11.).

TEPCO, Tokyo Electric Power Company, 2011b. Countermeasures to outflow of, radioactive water off the site near water intake of unit 3 at Fukushima Daiichi, nuclear power station. http://www.tepco.co.jp/en/press/corp-com/release/, 11052102-e.html (accessed 20.06.11.).

TEPCO, Tokyo Electric Power Company, 2012. Final Report of Investigation Committee on the Accident at Fukushima Nuclear Power Plants of Tokyo Electric Power Company. Tokyo Electric Power Company, Tokyo. http://icanps.go.jp/eng/final-report.html.

UNSCEAR, United Nations Scientific Committee on the Effects of Atomic Radiation, 1988. Exposures from the Chernobyl accident. Annex D. In: Sources, Effects and Risks of Ionizing Radiation, vol. II. UNSCEAR, Vienna. http://www.unscear.org/docs/reports/1988/1988i_unscear.pdf.

UNSCEAR, United Nations Scientific Committee on the Effects of Atomic Radiation, 1991. Report to the General Assembly. Sources and effects of ionizing radiation. Scientific Annex J. In: Exposures and Effects of the Chernobyl Accident, vol. II. UNSCEAR, New York.

UNSCEAR, United Nations Scientific Committee on the Effects of Atomic Radiation, 1993. Ionizing Radiation: Sources and Biological Effects. New York, United Nations.

UNSCEAR, United Nations Scientific Committee on the Effects of Atomic Radiation, 2008a. Non-targeted and delayed effects of exposure to ionizing radiation. Annex C. In: Effects of Ionizing Radiation, vol. II. UNSCEAR, Vienna (UNSCEAR 2006 Report to the General Assembly). http://www.unscear.org/docs/reports/2006/09_81160_Report_Annex_C_2006_Web.pdf.

UNSCEAR, United Nations Scientific Committee on the Effects of Atomic Radiation, 2008b. Epidemiological studies of radiation and cancer. Annex A. In: Effects of Ionizing Radiation, vol. I. UNSCEAR, Vienna. http://www.unscear.org/docs/reports/2006/07_82087_Report_Annex_A_2006_Web_corr.pdf.

UNSCEAR, United Nations Scientific Committee on the Effects of Atomic Radiation, 2008c. Exposures from the Chernobyl accident. Annex D. In: Sources, Effects and Risks of Ionizing Radiation, vol. II. UNSCEAR, Vienna. http://www.unscear.org/docs/reports/1988/1988i_unscear.pdf.

UNSCEAR, United Nations Scientific Committee on the Effects of Atomic Radiation, 2008d. Exposures of the public and workers from various sources of radiation. Annex B In: Sources, Effects and Risks of Ionizing Radiation, vol. II. UNSCEAR, Vienna. http://www.unscear.org/docs/reports/1998/1988i_unscear.pdf.

UNSCEAR, United Nations Scientific Committee on the Effects of Atomic Radiation, 2011. Health effects due to radiation from the Chernobyl accident. Annex D. In: Sources and Effects of Ionizing Radiation. UNSCEAR 2008 Report to the General Assembly, vol. II. UNSCEAR, Vienna. http://www.unscear.org/docs/reports/2008/11_80076_Report_2008_Annex_D.pdf (Effects of Ionizing Radiation. vol. I. UNSCEAR, Vienna). http://www.unscear.org/docs/reports/2008/09-86753_Report_2008_Annex_B.pdf.

WHO, World Health Organization, 1988. Derived Intervention Levels for Radionuclides in Food: Guidelines for Application after Widespread Radioactive Contamination Resulting from a Major Radiation Accident. WHO, Geneva. http://apps.who.int/iris/bitstream/10665/40421/1/9241542330_eng. pdf.

WHO, World Health Organization, 2006. Health Effects of the Chernobyl Accident and Special Health Care Programmes. Report of the United Nations Chernobyl Forum Expert Group "Health". WHO, Geneva. http://www.who.int/ionizing_radiation/chernobyl/assessment_mitigation/en/index.html.

WHO, World Health Organization, 2012. Preliminary Dose Estimation from the Nuclear Accident after the 2011 Great East Japan Earthquake and Tsunami. WHO, Geneva. http://www.who.int/ionizing_radiation/pub_meet/fukushima_dose_assessment/en/index.html.

WHO, World Health Organization, 2013. Health Risk Assessment from the Nuclear Accident after the 2011 Great East Japan Earthquake and Tsunami. WHO, Geneva.

Zheng, J.Z., et al., 2012. Isotopic evidence of plutonium release into the environment from the Fukushima DNPP accident. Sci. Rep. 2, 304–309.

8

SUMMARY

The main observations about the impact of the Fukushima accident on the terrestrial and marine environment and the Japanese and the world population may be summarized as follows:

- After the Great East Japan earthquake and subsequent Tohoku tsunami with unexpected high waves on 11 March 2011, the Fukushima Daiichi nuclear power plant (NPP) by force majeure had undergone a total damage of the electrical network and the emergency diesel generators, resulting in the loss of ability to cool three nuclear reactors which were in operation, as well as the fuel storage pools, what resulted in numerous explosions and a damage of the Fukushima Daiichi NPP.
- The source term estimated in this book for releases of radionuclides during the Fukushima accident consisted mainly of atmospheric releases (^{131}I of about 150 PBq and ^{134}Cs and ^{137}Cs, each of about 15 PBq) and of the direct releases to coastal waters of about 4 PBq (for ^{134}Cs and ^{137}Cs equally). The deposition from the atmosphere to the north-western Pacific Ocean was estimated to be about 5 PBq for ^{134}Cs and ^{137}Cs equally. The second important long-lived radionuclide after ^{137}Cs, which was released directly to the ocean, was ^{90}Sr with an estimated total activity of about 1 PBq.
- The radionuclide levels observed in the atmosphere at the Fukushima varied for ^{131}I (particulate form) between 1 and 10,000 Bq/m^3, and from 0.1 to 200 Bq/m^3 for ^{137}Cs. The dose rates at the surface air varied at the Fukushima site between 0.04 μSv/h (at the beginning) to maximum values at 10 μSv/h (in the middle of March 2011), and then down to 0.3 μSv/h at the end of April 2011.

Fukushima Accident. http://dx.doi.org/10.1016/B978-0-12-408132-1.00008-5

- The radionuclide levels in the atmosphere of Europe varied for ^{131}I (in the particulate form) between 0.01 and 6 mBq/m^3 (up to 11 mBq/m^3 for the gaseous form), and for ^{134}Cs and ^{137}Cs (equally) between 0.001 mBq/m^3 (at the end of May 2011) to about 1.5 mBq/m^3 during peak time in the middle of March 2011.
- The radionuclide levels in seawater offshore Fukushima varied (equally for ^{137}Cs and ^{134}Cs) between about 1 kBq/m^3 to about 90 MBq/m^3. The average ^{137}Cs (and similarly also ^{134}Cs) activity concentration off Fukushima during most of the year 2011 was about 10 kBq/m^3.
- Global ocean circulation modeling has shown that the radioactive plume in the Pacific Ocean has been due to the Kuroshio Current moving fast from the Japanese coast to the east, reaching maximum ^{137}Cs activity concentrations in surface north-western Pacific waters of about 25 Bq/m^3. After 3–4 years, the plume will reach the Californian coast with ^{137}Cs concentrations of about 3 Bq/m^3, what is by a factor of three higher that the present background from testing of nuclear weapons in the atmosphere (global fallout).
- The ^{137}Cs concentrations in fish caught offshore Fukushima varied within several orders of magnitude, from about 2 Bq/kg ww (wet weight) to about 5 kBq/kg ww, but the majority of ^{137}Cs concentrations were in the interval 10–1000 Bq/kg ww. The Japanese Government applied very strict regulations for radionuclide content in seafood. The Japanese limit for ^{134}Cs and ^{137}Cs in seafood after the Fukushima accident was decreased from 500 to 100 Bq/kg ww, so they became by about a factor of 4–10 lower than for other Asia countries. The Codex value of 1000 Bq/kg ww, which has been adopted by most of the world countries, is assuring that the maximum effective dose limit to population will be well below the International Committee on Radiological Protection and the International Atomic Energy Agency recommended maximum value for the citizens of 1 mSv/year.
- Radiation doses to the Japanese population from inhalation of contaminated air, from external

irradiation, terrestrial and marine food contamination were estimated and compared with other sources of anthropogenic (global fallout, Chernobyl accident) and natural radionuclides (e.g. radionuclides in food, cosmic radiation) and medical applications (X-ray tests, CT-tests, etc.) of ionizing radiation. The doses from inhalation, the external exposure from radioactive clouds and deposited radionuclides, and ingestion of terrestrial food were, as estimated by the World Health Organization (WHO, 2012, 2013), generally below 25 mSv; therefore, they should not cause health damage to the Japanese population.

- If we take a rather conservative approach expecting that the average ^{137}Cs (and similarly also ^{134}Cs) activity concentration in surface seawater off Fukushima during most of the year 2011 was about 10 kBq/m^3, then the effective dose commitment would be 0.3 mSv/year from ingestion of ^{137}Cs in fish (and similar for ingestion of ^{134}Cs in fish). The total effective dose commitment from ingestion of both radionuclides in fish, shellfish and seaweed will then be 0.7 mSv/year. If we take a similar conservative approach and take for the average ^{137}Cs concentration in fish a value of 1000 Bq/kg (i.e. 10 times higher than the screening level adopted by the Japanese government), then the estimated effective dose commitment will be 0.3 mSv/year. If we do summation of doses for ^{137}Cs and ^{134}Cs, and also for other pathways (shellfish and seaweed), we get for the total dose a value of 0.7 mSv/year. Although this dose is by about four orders of magnitude higher than the pre-Fukushima dose (0.03 μSv/year calculated for the ^{137}Cs content in seawater of 1 Bq/m^3), it is below the maximum permissible annual dose to the public from external sources (1 mSv/year), or the world average dose from natural sources (2.4 mSv/year). This dose is, however, comparable with annual dose due to natural ^{210}Po ingestion in seafood (0.7 mSv/year).
- Individual effective dose commitment from consumption of ^{137}Cs in fish caught in the open north-western Pacific Ocean in 2012 was estimated to be 0.8 μSv/year (together with ^{134}Cs

and other pathways it is about 2 μSv/year), what is about 30 times above the global fallout background. This dose is, however, 60 times lower than the dose due to the consumption of natural ^{210}Po in fish (0.05 mSv/year), and 350 times lower than the dose due to the consumption of natural ^{210}Po in fish and shellfish (0.7 mSv/year).

- The radiation doses to the world population from inhalation and ingestion of terrestrial and marine foods were by several orders of magnitude lower than the recommended maximum limit of 1 mSv/year.
- The estimated total radiation doses to fish and shellfish in coastal waters during the largest radionuclide releases were by a factor of 10 lower than the baseline safe level postulated for the marine organisms, therefore no harmful effects are expected for the marine ecosystem as well.
- A well-designed long-term epidemiological study should be developed which could help a better estimation of the radiation impact of the Fukushima accident on the Japanese population. This was not the case after the Chernobyl accident when residents from contaminated areas spread over large territory of the former Soviet Union. However, chances to detect any irregularities in such an epidemiological study in Japan will still be very low as the expected effects will be probably within statistical uncertainties of existing cancer incidence from sources other than the Fukushima radiation impact.
- The Fukushima accident has slowed down the renaissance in constructing nuclear power plants, which started about 25 years after the Chernobyl accident. However, due to further needs in developments of new energy sources and strong limitations from climate change and air pollution impacts, these needs cannot be realized without new nuclear technologies. The new, fourth generation of nuclear reactors, together with developments of Th-based reactors, offer much safer and economically more competitive technologies, which will meet growing demands on new energy sources.

References

WHO, World Health Organization, 2012. Preliminary Dose Estimation from the Nuclear Accident after the 2011 Great East Japan Earthquake and Tsunami. WHO, Geneva. http://www.who.int/ionizing_radiation/pub_meet/fukushima_dose_assessment/en/index.html.

WHO, World Health Organization, 2013. Health Risk Assessment from the Nuclear Accident after the 2011 Great East Japan Earthquake and Tsunami. WHO, Geneva.

Epilogue

Unfortunately, environmental disasters and accompanying accidents are part of our life and we must therefore find ways to minimize their impacts on the environment and the society. The Fukushima accident was a typical combination of environmental disaster—in this case the strong Tohoku earthquake, followed by an unexpectedly high tsunami—plus a technological accident. It has been very often in the past that people were doing mistakes, and surely the same situation will also be in the future. What is important, however, that we should learn from these mistakes, and try not to repeat them again in future.

So what are the lessons we should learn from the Fukushima accident?

(i) Environmental:

- Earthquakes—we need clearly to be prepared in the active seismic regions of the Earth that earthquakes with magnitude of nine should be expected. The designers of new nuclear power plants should therefore take this into account. Financial aspects of such newly constructed plants should not dominate over the plant safety. The seismic aspects should be, however, taken into account more rigorously also in regions with a lower seismic activity, where long-term seismic records (over several centuries) should be taken into account.
- Tsunami—the main problem in the Fukushima accident was in the low height of the protection barriers against tsunami. It is believed that a barrier of 15 m high would protect the Fukushima plant against the tsunami. If a longer tsunami record (not only for the last 100 years) would be taken into account, when designing and constructing the Fukushima plant, the accident in the Fukushima NPP would not occur. Now it is evident that 15 m tsunami, similar to those observed in 2011, was observed already in 869. Therefore, even 1000–years-old records of the tsunami history should be studied

when constructing nuclear power plants in coastal regions.
- Optimum design—coastal vs inland construction of the nuclear power plants should be carefully evaluated and an optimum decision should be taken.

(ii) Technological:
- There has been a general trend in operation of nuclear power plants to prolong their active life, mainly because of financial aspects. Surely, this has not been a good tendency as a new generation of plants is having higher safety standards than the older one. This has also been the case of the Fukushima Dai-ichi nuclear plant, whose prototype was designed in the late 1950s. The unit 1 (460 MWel) was constructed in 1967, and it commenced commercial electrical production in 1971. We may compare here a technological progress done, e.g. in the construction of family cars. Everybody would agree that 50-years-old cars are in many aspects, including a safety, incomparable with presently constructed cars.
- Too many nuclear power plants at one site—here again financial aspects should not take over the safety aspects, or, the safety first! Management of nuclear accidents and a danger of multiaccidents in the case of high concentration of plants should pose strict limits on the number of plants at one site.

(iii) Future developments:
- The Chernobyl accident had very negative impact on the future development of the nuclear industry. However, after more than 20 years after this accident, there was a general understanding that a nuclear "renaissance" is there as many new nuclear power plants were under consideration. These activities stop after the Fukushima accident, and only slowly will develop again.
- Partly due to the Chernobyl accident, but also due to the existence of a nuclear phobia in the public, politicians stopped in many

cases developments of new generation of nuclear reactors. Unfortunately, this is a general trend that politicians think only in 4–5 years intervals (done by elections). Society should, however, more follow scientific advice, not a populist behavior of politicians.

- Even in such difficult conditions, there have been some developments and new, more efficient and safer nuclear reactors have been developed, including the fourth generation of nuclear reactors. There are, however, under development (unfortunately more theoretically than experimentally) new revolutionary types of reactors, e.g. thorium salt reactors, in which operational safety would be comparable with non-nuclear installations, and problems with radioactive wastes would be minimized. Also the efficiency of new generation of nuclear reactors would guarantee that there will be enough nuclear fuel for several 100 years. Therefore, even slow progress in fusion reactors (presently there is under construction an ITER—International Thermonuclear Reactor in Cadarache, France) would not cause a short of energy in the future.
- Renewable energy sources vs nuclear energy—this is nowadays a very frequently discussed topic, however, serious estimates show that this is not a real source of energy for the future fast development of our civilization. It may be good for rich countries which may spend much more money on energy supplies, but surely not for the new rapidly developing world such as China and India. And if these countries really should decrease their contributions to carbon dioxide releases into the atmosphere, there is only the nuclear way how to increase the energy sources and simultaneously to decrease carbon dioxide content in the atmosphere. Therefore, it looks highly probable that in the future a new generation of fission, and even a first commercial fusion reactor, will be operating in energy-thirsty China or India.

Surely, the Fukushima accident was a terrible disaster, which complicated life of tens of thousands of inhabitants of the Fukushima and the neighbor districts as about 150,000 people has to be evacuated from the impacted zone. The nuclear part of the disaster need not happen if more rigorous anti-tsunami precautions would be applied in the past when constructing the Fukushima Dai-ichi nuclear power plant. On the other hand, we must also say that from 15,873 Japanese citizens who died, 6114 who were injured, and 2744 who are still missing because of the earthquake and resulting tsunami disaster, nobody died because of radiological effects of the nuclear accident. Thanks to regulations applied by the Japanese government (a large-scale evacuation of inhabitants, and radioactivity control of terrestrial and marine food), it is expected that there will not be any casualty due to inhalation of the contaminated air with radionuclides, direct exposition to radiation from the air and soil, or the consumption of the radioactive food, as predicted by the World Health Organization in the report, published in February 2013.

INDEX

Note: Page numbers with "*f*" denote figures; "*t*" tables.

A

B

C

D

J

K

L

M

N

O

P

FSC
www.fsc.org
MIX
Papier | Fördert gute Waldnutzung
FSC® C083411

Kolibri 360 GmbH
Ferdinand-Jühlke-Straße 7
99095 Erfurt, Deutschland
produktsicherheit@kolibri360.de